Cell and Tissue Culture in Forestry

FORESTRY SCIENCES

Baas P, ed: New Perspectives in Wood Anatomy. 1982. ISBN 90-247-2526-7
Prins CFL, ed: Production, Marketing and Use of Finger-Jointed Sawnwood. 1982. ISBN 90-247-2569-0
Oldeman RAA, et al., eds: Tropical Hardwood Utilization: Practice and Prospects. 1982. ISBN 90-247-2581-X
Den Ouden P and Boom BK: Manual of Cultivated Conifers: Hardy in Cold and Warm-Temperate Zone. 1982. ISBN 90-247-2148-2
Bonga JM and Durzan DJ, eds: Tissue Culture in Forestry. 1982. ISBN 90-247-2660-3
Satoo T and Magwick HAI: Forest Biomass. 1982. ISBN 90-247-2710-3
Van Nao T, ed: Forest Fire Prevention and Control. 1982. ISBN 90-247-3050-3
Douglas J: A Re-appraisal of Forestry Development in Developing Countries. 1983. ISBN 90-247-2830-4
Gordon JC and Wheeler CT, eds: Biological Nitrogen Fixation in Forest Ecosystems: Foundations and Applications. 1983. ISBN 90-247-2849-5
Németh MV: The Virus-Mycoplasma and Rikettsia Disease of Fruit Trees. ISBN 90-247-2868-1
Duryea ML and Landis TD, eds: Forest Nursery Manual: Production of Bareroot Seedlings. 1984. ISBN 90-247-2913-0
Hummel FC, ed: Forest Policy: A Contribution to Resource Development. 1984. ISBN 90-247-2883-5
Manion PD, ed: Scleroderris Canker of Conifers. 1984. ISBN 90-247-2912-2
Duryea ML and Brown GN, eds: Seedling Physiology and Reforestation Success. 1984. ISBN 90-247-2949-1
Staaf KAG and Wiksten NA: Tree Harvesting Techniques. 1984. ISBN 90-247-2994-7
Boyd JD: Biophysical Control of Microfibril Orientation in Plant Cell Walls. 1985. ISBN 90-247-3101-1
Findlay WPK, ed: Preservation of Timber in the Tropics. 1985. ISBN 90-247-3112-7
Samset I: Winch and Cable Systems. 1985. ISBN 90-247-3205-0
Leary RA: Interaction Theory in Forest Ecology and Management. 1985. ISBN 90-247-3220-4
Gessel SP: Forest Site and Productivity. 1986. ISBN 90-247-3284-0
Hennessey TC, Dougherty PM, Kossuth SV and Johnson JD, eds: Stress Physiology and Forest Productivity. 1986. ISBN 90-247-3359-6
Shepherd KR: Plantation Silviculture. 1986. ISBN 90-247-3379-0
Sohlberg S and Sokolov VE, eds: Practical Application of Remote Sensing in Forestry. 1986. ISBN 90-247-3392-8
Bonga JM and Durzan DJ, eds: Cell and Tissue Culture in Forestry. Volume 1: General Principles and Biotechnology. 1987. ISBN 90-247-3430-4
Bonga JM and Durzan DJ, eds: Cell and Tissue Culture in Forestry. Volume 2: Specific Principles and Methods: Growth and Developments. 1987. ISBN 90-247-3431-2
Bonga JM and Durzan DJ, eds: Cell and Tissue Culture in Forestry. Volume 3: Case Histories: Gymnosperms, Angiosperms and Palms. 1987. ISBN 90-247-3432-0
(Set Vols. 1–3: ISBN 90-247-3433-9)

Cell and Tissue Culture in Forestry

Volume 2
Specific Principles and Methods: Growth and Developments

edited by

J.M. BONGA and DON J. DURZAN
Canadian Forestry Service / University of California

1987 **MARTINUS NIJHOFF PUBLISHERS**
a member of the KLUWER ACADEMIC PUBLISHERS GROUP
DORDRECHT / BOSTON / LANCASTER

Distributors

for the United States and Canada: Kluwer Academic Publishers, P.O. Box 358, Accord Station, Hingham, MA 02018-0358, USA
for the UK and Ireland: Kluwer Academic Publishers, MTP Press Limited, Falcon House, Queen Square, Lancaster LA1 1RN, UK
for all other countries: Kluwer Academic Publishers Group, Distribution Center, P.O. Box 322, 3300 AH Dordrecht, The Netherlands

Library of Congress Cataloging in Publication Data

Cell and tissue culture in forestry.

(Forestry sciences)
Rev. ed. of: Tissue culture in forestry. 1982.
Includes index.
Contents: v. 1. General principles and biotechnology -- v. 2. Specific principles and methods.
1. Plant cell culture--Collected works. 2. Plant tissue culture--Collected works. 3. Forests and forestry--Collected works. I. Bonga, J. M. II. Durzan, D. J. III. Tissue culture in forestry. IV. Series.
SD403.5.C45 1987 634.9'56 86-23767
ISBN 90-247-3433-9 (set)

ISBN 90-247-3431-2

Copyright

PRINTED IN THE NETHERLANDS

VOLUME 2

TABLE OF CONTENTS

CONTRIBUTORS

AITKEN-CHRISTIE J, Forest Research Institute, Private Bag, Rotorua, New Zealand
CHEN THH, Alberta Research Council, Biotechnology Department, Edmonton, Alta., Canada
CHEN Z, Academia Sinica, Institute of Genetics, Beijing, China
COUMANS M, Universite de Liège-Sart Tilman, Hormonologie Fondamentale et Appliquéé, Institute de Botanique (B22), B-4000 Liège, Belgium
DAVID A, Université de Bordeaux I, Laboratoire de Physiologie Végétale, F-33405 Talence, France
DINER AM, Michigan Technological University, Department of Forestry, BioSource Institute, Houghton, Mich. 49931, USA
DIXON RK, University of Minnesota, College of Forestry, Department of Forest Resources, 110 Green Hall, 1530 North Cleveland Avenue, St. Paul, Minn. 55108, USA
DRIVER JA, Plant Research Laboratory, Driver Nursery, 2737 North Avenue, Modesta, Calif. 95351, USA
DURZAN DJ, University of California, Davis, Department of Pomology, 1035 Wickson Hall, Davis, Calif. 95616, USA
EVERS PW, Research Institute for Forestry and Landscape Planning, "De Dorschkamp", P.O. Box 23, 6700 AA, Wageningen, The Netherlands
FOWLER MW, University of Sheffield, Wolfson Institute of Biotechnology, Sheffield S10 2TN, England
GASPAR Th, Université de Liège-Sart Tilman, Hormonologie Fondamentale et Appliquéé, Institute de Botanique (B22), B-4000 Liège, Belgium
GREY D, University of Sheffield, Wolfson Institute of Biotechnology, Sheffield S10 2TN, England
HO RH, Ontario Ministry of Natural Resources, Ontario Tree Improvement and Forest Biomass Institute, Maple, Ont., LOJ 1EO, Canada
KARNOSKY DF, Michigan Technological University, Department of Forestry, Houghton, Mich. 49931, USA
KARTHA KK, National Research Council of Canada, Plant Biotechnology Institute, Saskatoon, Sask., S7N OW9, Canada
LAKSHMI SITA G, Indian Institute of Science, Microbiology and Cell Biology Laboratory, Bangalore 560012, India
MARX DH, USDA Forest Service, Southeastern Forest Experiment Station, Forestry Sciences Laboratory, Athens, Ga. 30602, USA
MCCOWN BH, University of Wisconsin, Department of Horticulture, Madison, Wis. 53706, USA
PATEL KR, University of Calgary, Department of Biology, Calgary, Alberta, T2N 1N4, Canada
ROHR R, Université de Nancy I, Laboratoire de Botanique II, Boite Postale 239, 54506 Vandoeuvre les Nancy Cedex, France
ROMANI RJ, University of California, Davis, Department of Pomology, 1035 Wickson Hall, Davis, Calif. 95616, USA

RUSSELL JA, University of Wisconsin, Department of Horticulture, Madison, Wis. 53706, USA
SINGH AP, Forest Research Institute, Private Bag, Rotorua, New Zealand
STEPAN-SARKISSIAN G, University of Sheffield, Wolfson Institute of Biotechnology, Sheffield S10 2TN, England
SUTTLE GRL, Plant Research Laboratory, Driver Nursery, 2737 North Avenue, Modesta, Calif. 95351, USA
THORPE TA, University of Calgary, Department of Biology, Calgary, Alberta, T2N 1N4, Canada
TRAN THANH VAN K, Laboratoire de Phytotron, C.N.R.S., 91190 Gif-Sur-Yvette, France
TRINH TH, Laboratoire de Phytotron, C.N.R.S., 91190 Gif-Sur-Yvette, France
TULECKE W, Antioch College, Yellow Springs, Ohio 45387, USA
WICKER M, Université Pierre et Marie Curie, Laboratoire d'Oncogenese Végétale, 12 Rue Cuvier, 75005 Paris, France
YILMAZ-LENTZ D, Laboratoire de Phytotron, C.N.R.S., 91190 Gif-Sur-Yvette, France

1. INTRODUCTION

This is the second in a series of three new volumes dealing with cell and tissue culture of forest trees, a research area that has developed very rapidly since our 1982 Edition.

The first volume deals with general aspects of nutrient media and physical elements of the culture environment, and contains chapters on the genetics and physiology of clonal propagation. In addition, because of the increasing importance of the subjects, it contains several chapters discussing genetic variation, genetic engineering, and ultrastructure.

The second volume deals with specific topics, in particular in the field of growth and development. Protoplast culture, haploid and triploid culture, embryo rescue, and somatic embryogenesis are important elements in attempts to obtain genetically improved sexual and clonal stock. Progress in these areas is covered by several chapters. Various aspects of morphogenesis and organogenesis, including positional effects, are also discussed. Now that more species are being micropropagated every year, it is increasingly important to pay attention to problems with the finished products, e.g., plantlets, artificial seeds. Therefore, chapters on pathology, mycorrhizae, senescence, cold storage, germplasm preservation, and nursery practice are included.

In the third volume, case histories are presented for a wide variety of gymnosperm and angiosperm (including palm) species. These species represent a wide range of geographic areas. In some cases, micropropagation efforts have recently been started and large-scale field-testing has not yet been carried out. For others, the initial large-scale field-testing is completed and micropropagation is practiced on a commercial scale.

In summary, the three volumes cover a wide range of research activities in the field of cell and tissue culture of forest trees, attempt to identify future trends, and where possible, identify the appropriate biotechnologies.

2. CONIFER PROTOPLASTS

A DAVID

1. INTRODUCTION

Genetic modification of cells by somatic hybridization or by the introduction of organelles (chloroplasts, mitochondria) and DNA or RNA molecules (via liposomes) has a bright future, thanks to modern protoplast technology. However, it may be rather utopian as yet to wish to genetically modify conifers by isolation and culture of protoplasts because this group of trees is notoriously difficult to manipulate *in vitro*. Nevertheless, there are at least three good reasons to attempt culture of conifer protoplasts. First, the protoplast system permits reprogramming of the nucleus and cytoplasm, which could facilitate cellular rejuvenation and thus a return to the expression of totipotency. Second, it is convenient material for experimentation in genetic engineering of conifers. Third, cells are now available that have the capacity to form somatic embryos and these are logical sources for the production of protoplasts (cf. Tulecke and Durzan in this volume).

The first isolations of protoplasts were obtained from cultures of *Picea abies* (2) and *Pseudotsuga menziesii* (35) in 1973 and 1975 respectively, but these remained mitotically inactive. In this chapter I will review the progress made with conifer protoplast cultures up to March 1985. Attention is focussed in particular on the nature of the starting material, and the conditions required for isolation and culture of the protoplasts.

2. STARTING MATERIAL

2.1. *Organs*

2.1.1. *Cotyledons*. Viable protoplasts were obtained from surface sterilized cotyledons that were excised from 2-4 week old seedlings of Pseudotsuga menziesii (23). However, with Pinus pinaster, a smaller number of viable protoplasts was obtained from surface sterilized cotyledons than from cotyledons of seedlings grown aseptically from surface sterilized seeds (6×10^5 versus 20×10^5 protoplasts/g fresh weight of tissue) (9).

To obtain starting material that is physiologically as homogenous as possible, a pretreatment can be used. For example, culture of cotyledons of Pseudotsuga menziesii for 8-14 days on a semi-solid nutrient medium containing, amongst others, 500 μ M α-naphthaleneacetic acid (NAA) and 15 μ M N^6-benzyladenine (BA) prior to protoplast preparation, permitted production of 5.5×10^4 protoplasts/100 mg fresh weight of tissue (23). It was suggested that this pretreatment, which assures a rapid growth of the cotyledons, caused a change in the cell wall chemistry, thus rendering them more sensitive to the enzymes that are emploid in protoplast preparation. We have pretreated Pinus pinaster as follows. The roots were cut from small seedlings and discarded. The cut end of the hypocotyl was subsequently dipped for 7 days in a solution containing minerals and growth regulators. Protoplasts from cotyledons thus treated produced cells with a higher mitotic rate than cells from protoplasts from non-pretreated seedlings (7, 8, 9).

The stage of development of cotyledons at the moment of excision is important, in particular when no pretreatment was used. Not yet fully developed cotyledons of Pinus pinaster produce 3 times as many protoplasts as fully developed ones, and go through 3 division cycles in three weeks. Protoplasts from fully developed cotyledons only go through one cycle (9). Protoplasts from not yet fully developed cotyledons have also been obtained from aseptically grown seedlings of Biota orientalis (8), Pinus coulteri (26), Pinus sylvestris (1).

2.1.2. Needles. Needles excised from aseptic Pinus pinaster seedlings, or from shoots obtained from axillary buds in vitro, have provided protoplasts. Again it is important that the tissues are mitotically active when used for protoplast preparation (6).

2.1.3. Roots. Protoplasts have been isolated from 20mm long root segments of Pinus pinaster, taken 7 mm from the root tip in the region of differentiation immediately behind the elongation zone (12). In spite of the fact that these root sections were excised from seedlings grown from surface sterilized seeds in an aseptic environment, these root sections were contaminated and had to be surface sterilized in 0.1% mercuric chloride before aseptic protoplasts could be obtained. This treatment did not diminish protoplast viability, which was sometimes as high as 90%. Similar problems in obtaining protoplasts from root sections have been encountered with Glycine max for instance. With this species bacterial contamination of protoplasts was avoided by adding antibiotics to the enzyme solutions (36).

2.2. Callus and cell suspensions

Callus and cell suspension cultures have the advantage of providing cell populations that are relatively homogenous and undifferentiated, and that are easily accessible to enzyme action. Protoplasts have been prepared from hypocotyl callus of Picea excelsa (31), from cell suspensions derived from hypocotyls of Pinus taeda (33), from cell suspensions derived from embryos of Pinus contorta (15), and from cell suspensions obtained from phloem and cambium tissues excised from the trunk of mature Sequoia sempervirens (6).

A study of the cell division intensity of suspension cultures should be carried out to determine at what culture stage the highest number of viable protoplasts can be obtained. Several parameters could be useful to determine the optimal stage for protoplast preparation, e.g. the mitotic index, packed cell volume, or dry weight increase. With Pinus contorta (15, Fig. 1) the highest percentage of viable protoplasts was obtained from cell suspensions in the exponential growth phase.

2.3. Pollen

In spite of the interest in haploid cells, in particular because of their potential in somatic hybridization, there is only one reference that deals with pollen protoplasts. Duhoux (11) hydrated Cupressus arizonica pollen to burst their exine layer before removing the intine by enzyme digestion. However, hydration of pollen initiates the synthesis of polysaccharides used in the formation of the pollen tube wall. To overcome this active production of polysaccharides an incubation of more than 20 hr in a

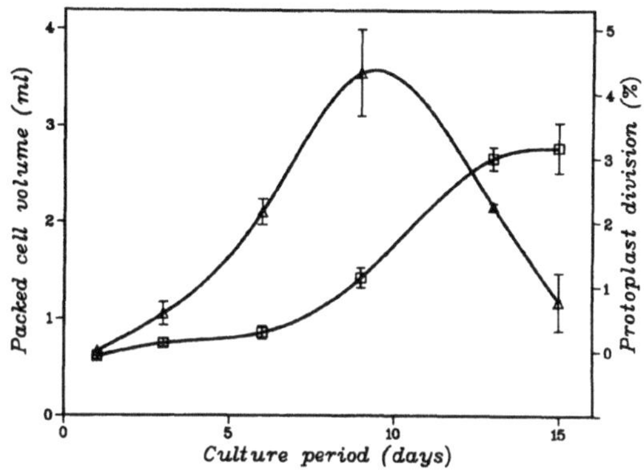

FIGURE 1. Relationship between growth phase of Pinus contorta cells in suspension culture frequency of isolated protoplasts. The points in the growth curve represent mean values from 4-8 different experiments (□). The division frequency of protoplasts after two days in culture are based on 3-4 different experiments (Δ). Vertical bars indicated standard errors. (From Hakman and Von Arnold (15), by permission of Plant Cell Reports).

solution with a high concentration (12.5% w/v) of enzyme is required. Unfortunately such a strong enzyme treatment is generally incompatible with survival of the protoplasts (11).

3. PROTOPLAST ISOLATION

3.1. Enzymatic solutions

The enzymatic solutions used for the hydrolysis of the cell wall generally contain a cellulase as the principal component, a hemicellulase and a pectinase. Each of these is added either in its commercially available form or in a more purified form (20). A contamination with nucleases has been noted in some commercial enzymes (4). Addition of bovine serum albumin to the enzyme solution has raised the survival rate of Pinus taeda protoplasts from 89% to 99.7% (33). The bovine serum albumin may have served as a substrate for proteases that are present as contaminants in the enzymes used. This may have largely inactivated the proteases and thus protected the proteins in the protoplast membranes. The mitotic activity of protoplasts of Pinus pinaster is higher if "desalted-prior-to-use" enzymes are used rather than non-desalted ones (9).

The enzyme concentration can affect the percentage of spontaneous fusions of protoplasts. A 16 hr incubation of Pinus pinaster cotyledons in a 2% w/v enzyme solution resulted in 83% mononucleate protoplasts, whereas in a 0.16% w/v solution the percentage was 98% (7).

The enzyme concentration that is required for protoplast isolation varies for the different organs that serve as protoplast source. Whereas a 0.2% w/v solution results in good protoplast formation if cotyledons of Pinus pinaster are used, the concentration has to be raised to 0.5% w/v to achieve the same with needles. This indicates a considerable difference in cell wall composition or density in these two organs (6).

To facilitate isolation of protoplasts, and to protect their integrity, the cells are plasmolyzed prior to or during enzyme treatment by exposing the explant to a metabolizable osmoticum (glucose) or a non- metabolizable one (sorbitol or mannitol). Some investigators have pretreated the tissues before their exposure to the enzyme solution. For example, tissues of Pinus taeda were plasmolyzed for 90 min in a solution containing 0.4 M mannitol and 0.2 M sorbitol (33). Root sections (20 mm long) of Pinus pinaster exposed for 1 hr to 0.7 M sorbitol and 30 mM cysteine before enzyme treatment, formed 3.4×10^3 protoplasts per section, while non-pretreated sections only yielded 1.3×10^3 protoplasts per section (12). L-cysteine was added to avoid inhibition of the enzymes by phenolics (30, 34). With both treatments about 85% of the protoplasts produced were viable.

Different plant materials require different osmoticum concentrations. For cotyledons of Pinus pinaster a 0.6 M sorbitol treatment is optimal (10), while for suspension cultures of Pinus contorta 0.4 M sorbitol gave the best results (15). This difference between cotyledons and suspension cultures indicates that a change in osmolarity occurs in the vacuole when a cell passes from an organized grouping of cells in a tissue, to their dissociated state in a suspension culture.

When osmoticum is added to the enzyme solution, the osmoticum and enzyme concentrations have to be mutually adjusted. Good production of viable Pseudotsuga menziesii protoplasts was obtained with 6% w/v enzyme plus 0.4M sorbitol, as well as with 2.5% w/v enzyme plus 0.6 M sorbitol, (23).

The enzymes and osmoticum are dissolved either in a "physiologically adjusted solution", i.e., in the liquid protoplast culture medium, or in a "physical-chemical solution" in which the stability of the pH is assured by adding MES, and that of the plasma membrane by adding extra calcium (up to 5.6 mM).

3.2. Isolation procedures

When organs are used, they should be cut into small fragments to facilitate enzyme penetration into the tissues. This is essential in particular when cotyledons or needles serve as protoplast source, because the cuticular waxes of these organs present effective barriers to the penetration of enzymes.

While the cell walls are being dissolved, the enzyme solution is agitated on a a shaker (40-70 rpm) and is maintained at a pH of 5.5-5.8 and a constant temperature, e.g. 20°C (15), or 30°C (26). The length of the treatment depends on the concentration of the enzymes; overnight treatment was used in enzyme concentrations of 0.2% w/v for Pinus pinaster cotyledons (7, 9) and 0.5% w/v and 1.5% w/v for Sequoia sempervirens (6) and Pinus contorta cell suspensions (15); treatment of a few hours in concentrations higher than 2% w/v was used in other cases (23, 26, 31, 33). Because much cellular debris is formed early during the enzyme treatment, cotyledons of Pseudotsuga menziesii were transferred after 30 min to a fresh enzyme solution and were kept in this fresh solution for an additional 6 hr (23).

3.3. Purification of protoplasts

Purification of conifer protoplasts follows a protocol that is similar to that used for many herbacious plants (5). After incubation the enzyme solution is removed by filtration followed by centrifugation and washing. The prospects of obtaining a high density of "good looking" protoplasts, i.e., protoplasts with spherical well dispersed organelles, is sometimes improved by

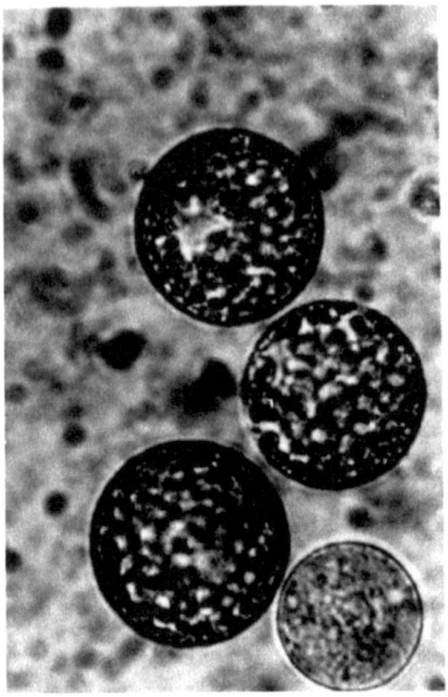 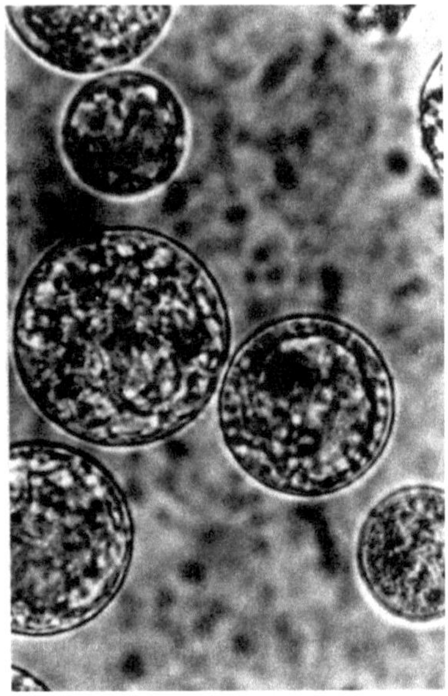

FIGURE 2. Pinus pinaster cotyledon proptoplasts.
FIGURE 3. Sequoia sempervirens cell suspension protoplasts.

centrifugation in a dextran gradient (31) or on a layer of 21% saccharose (26) (Fig. 2 and 3).

3.4. Yield and Viability

Whether or not true protoplasts were obtained is determined by testing for fluorescence in Calcofluor White ST (5, 25). Yield is determined by counting the number of rounded protoplasts with well dispersed chloroplasts with a haemocytometer. A yield of 5.5×10^5/g fresh weight was obtained from Pseudotsuga menziesii cotyledons (23) and 1.5×10^7/g fresh weight from Pinus pinaster (9) cotyledons. With cell suspension cultures much higher yields are possible. In fact, 91% of the cells of a suspension culture of Pinus taeda turned into protoplasts after 7hr of exposure to the enzymes. Cells in suspension cultures have thin, easily hydrolyzed walls during the exponential growth phase of the cultures (33), which could account for the high yields of protoplast that were obtained from these cultures.

Viability of the protoplasts, determined by vital staining in fluoresceine diacetate (29) or Evans blue (18), varied between 80 and 95%. The degree of penetration of vital stain is controlled by the degree of integrity of the plasma membrane, which in turn corresponds with the level of viability.

4. PROTOPLAST CULTURE

4.1. Culture conditions

Most protoplasts are cultured in 50-100 µl drops of medium (9, 23, 26) in densities varying from 10^4 to 5 x 10^5 ml^{-1}. The basal media that are generally used for protoplast culture are usually similar to those used for tissue and cell culture (3, 24), except that several carbohydrates, organic acids and vitamins are often added (19).

To stimulate mitosis in the cells obtained from protoplasts they are cultured on a fabric support composed of 0.5mm thick polyester fibre fragments saturated with liquid medium. The fabric improves gas exchange and expedites removal of inhibitory substances produced by the protoplasts and cells. With such a system cell division was maintained beyond the stage of clumps containing 20 cells (23).

A variety of light and temperature regimes has been used for culture, e.g. 18 hr/day of 2000 lux illumination at 25°C (23), 16 hr/day of 300 lux illumination at 23°C (33), or continuous darkness at 25°C for 16 hr during daytime and a lower 21°C for 8 hr during the night (9). In each case the choice of these regimes probably was determined more by the controlled environment equipment available than by any other consideration.

4.2. Protoplast behavior

4.2.1. Protoplast morphology. Under the light microscope the protoplasts appear spherical. About 80% of the protoplasts obtained from 5-7 mm long needles of *Pinus pinaster* have a diameter of 20-35 µm if kept at 610 mOsm/Kg H_2O (6). Electron microscope observations of protoplasts obtained from *Pinus pinaster* cotyledons confirm that walls are completely removed by the enzyme treatment (Fig. 4). These protoplasts showed a thin layer of peripheral cytoplasm around a large vacuole with tannins. The structure of the organelles is not notably changed by plasmolysis and removal of the cell wall.

4.2.2. Cell wall regeneration. Protoplasts will regenerate new walls if cultured in a suitable environment (see section 3.4). At least 35% of *Pinus pinaster* protoplasts will show evidence of wall formation after 24 hr (10). Protoplasts of *Pseudotsuga menziesii*

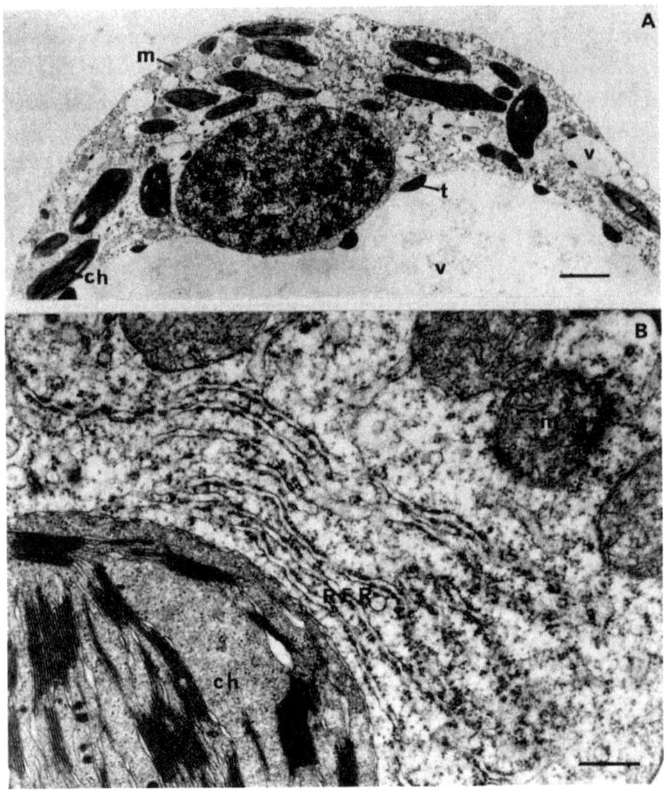

FIGURE 4. (A-B) Electron micrographs of protoplasts. (A) general morphology of a newly-formed protoplast and distribution of the cell organelles, X 3200. (B) enlargement of a protoplast showing part of a chloroplast with its typical thylakoid system and with rough ER arranged parallel to the chloroplast margin. X 25500. Scale bar in (A) = 3 μm; in (B) = 0.4 μm.
ch = chloroplast, m = mitochondrion, n = nucleus, RER = rough endoplasmic reticulum, t = tannin body, V = Vacuole. (From David, Jarlet and David (10), by permission of Physiologia Plantarum).

generally regenerate new walls within 72 hr (23), while protoplasts of Picea excelsa show fluorescence after 20 hr (31).

The first stages of cell wall regeneration have been studied with C^{14} glucose. Radioactive polysaccharides appeared on the protoplast surface after about 30 min. Acid hydrolysis of these polysaccharides, followed by chemical analysis of the hydrolysis products showed the presence of galactose, glucose, mannose, arabinose and xylose, which suggests that most enzymes involved in cell wall polysaccharide synthesis are activated soon after the protoplasts are removed from the enzyme solution (31).

4.2.3. Division of cells derived from protoplasts. From 10 to 15% of the cells divided once during the first week in culture (12, 15, 23). The frequency of division after that was strongly influenced by the osmolality of the medium. In protoplast cultures of Pseudotsuga menziesii the sorbitol level, which initially was 0.6 M, had to be progressively reduced to maintain mitotic activity (23). Similarly the osmolality of the medium of cell cultures derived from Pinus pinaster protoplasts had to be reduced from 610 to 225 mOsm/Kg H_2O after 40 days in culture (10). These results were similar to those otained with, for example, cultures of mesophyll protoplasts of peas (16). In each case the osmoticum concentrations probably had to be reduced to avoid prolonged osmotic shock with its consequent changes in protein synthesis. These changes in protein synthesis could be similar to those provoked by carbohydrate or oxygen depletion, and heat or other stresses that produce specific "stress" proteins (13).

The nature and concentration of nitrogen in the medium plays an important role. Cell colonies derived from protoplasts of cotyledons of Pseudotsuga menziesii required 10 mM L-glutamine to grow (21, 23). In cell suspensions and callus cultures derived from protoplasts of Pinus pinaster, cell division has been maintained for more than 3 months using 24-40 mM of glutamine as the only nitrogen source (10). In cell suspensions of Pseudotsuga menziesii, the addition of 50 mM glutamine to the medium has resulted in better growth, less vacuolization and denser cytoplasm (22). These results show that cells are incapable of assimilating inorganic nitrogen in sufficient amounts to sustain growth, and that an additional source of reduced organic nitrogen is required.

Protoplast stability is improved by the addition of Ca^{++} and Mg^{++} to the medium, with the extra calcium being needed for cell wall regeneration (28). In cell suspensions derived from protoplasts from cotyledons (10) or needles (6) of Pinus pinaster the calcium concentration had to be reduced from 5.6 to 1.5 mM after 20 days in culture to maintain growth (Fig. 5). The cell suspensions of needle origin have grown for 8 months without a reduction in mitotic rates.

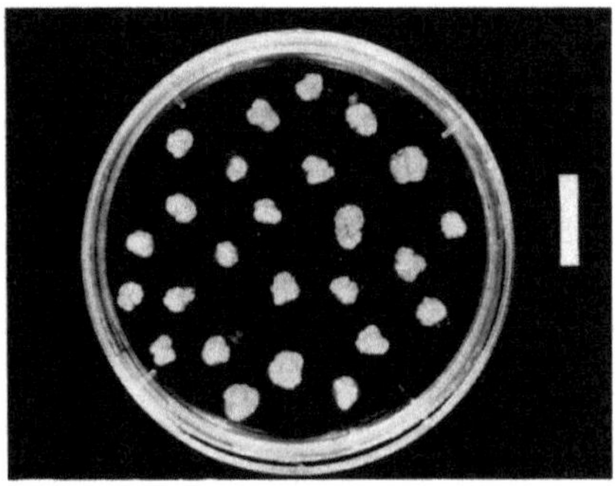

FIGURE 5. Callus from protoplasts of Pinus pinaster after growing for 3 months on an agar medium. Scale bar = 20 mm.

Studies of the ultrastructure of the regenerated cells show numerous organelles, including peroxysomes and plastids with starch, and numerous plasmodesmata between the cells (10).

Differentiation of tracheids and the formation of phenolics in the regenerated cells, presumably caused by an activation of phenylalanine-ammonia-lyase (PAL) in response to the reduction in nitrogen in the medium (14), inhibited further mitotic activity (10, 26).

5. CONCLUSION

Protoplast isolation, followed by cell wall regeneration and cell division has been achieved with explants from the following conifer species: Biota orientalis (8), Pinus contorta (15), Pinus coulteri (26), Pinus pinaster (7, 9, 10, 12), Pinus sylvestris

(1), Pinus taeda (33), Pseudotsuga menziesii (21, 23) and Sequoia sempervirens (6). In none of these cultures has mitotic activity been continuous. This leads one to ask why conifer protoplasts show such a lack of responsiveness in comparison to, for example, those of tobacco. Does this indicate that there are profound differences between protoplasts of perennials and conifers or that the nutrient media for conifer protoplasts still are deficient in some important aspects?

In protoplast cultures of many herbacious species, excessive phenol synthesis is one of the factors leading to mitotic arrest. The same appears to be the case in protoplast cultures of conifers. To overcome this problem the following avenues could be explored. 1) Addition of activated charcoal, because this has stimulated division in cells derived from Marchantia polymorpha protoplasts (32). 2) A reduction in the concentration of sucrose, or an increase in the concentration of nitrogen, because high sucrose and low nitrogen favored overproduction of phenols in cell cultures of Acer pseudoplatanus (27). 3) The synthesis and activation of PAL could possibly be inhibited by lower light, higher auxin, lower cytokinin and avoidance of deconditioning of the medium by frequent transfers or dilutions (17).

The first few cell divisions are probably the most important ones for organogenesis, and future research should focus on this aspect. Preferably, this research should involve protoplasts from a wide variety of organs. Finally, differences in protein synthesis between callus obtained directly from organs, and protoplasts obtained from the same organs, should be studied because such differences could indicate what physiological changes were induced by the passage through the protoplast stage.

6. REFERENCES

1. BORNMAN CH Personal Communication
2. CHALUPA V 1973 Tissue cultures of conifers and their potential use in tree breeding. In Proc 1st Colloq Plant Tissue Cult Genet Breeding, pp 181-191
3. CHENG TY 1977 Factors affecting adventitious bud formation of cotyledon culture of Douglas Fir. Plant Sci Lett 9: 179-187
4. COCKING EC 1972 Plant cell protoplast isolation and development. Annu Rev Plant Physiol 23: 29-50

5. CONSTABEL F 1975 Isolation and culture of plant protoplasts. In OL Gamborg, LR Wetter, eds, Tissue Culture Methods. Nat Res Counc, Saskatoon, pp 11-21
6. DAVID A Nonpublished results
7. DAVID A, H DAVID 1979 Isolation and callus formation from cotyledon protoplasts of pine (Pinus pinaster). Z Pflanzenphysiol 94: 173-177
8. DAVID H, A DAVID, T MATEILLE 1981 Isolement et culture des protoplastes de deux Gymnospermes: Pinus pinaster, Biota orientalis. In Proc IUFRO Sect S2 01 5. Int Workshop "In Vitro" Cultivation For Tree Species, Fontainebleau, France, pp 338-347
9. DAVID H, A DAVID, T MATEILLE 1982 Evaluation of parameters affecting the yield, viability and cell division of Pinus pinaster protoplasts. Physiol Plant 56: 108-113
10. DAVID H, E JARLET, A DAVID 1984 Effects of nitrogen source, calcium concentration and osmotic stress on protoplasts and protoplast - derived cell cultures of Pinus pinaster cotyledons. Physiol Plant 61: 477-482
11. DUHOUX E 1980 Protoplast isolation of Gymnosperm pollen. Z Pflanzenphysiol 99: 207-214
12. FAYE M, A DAVID 1983 Isolation and culture of gymnosperm root protoplasts (Pinus pinaster). Physiol Plant 59: 359-362
13. FLECK J, A DURR, C FRITSCH, T VERNET, L HIRTH 1982 Osmotic-shock 'stress proteins' in protoplasts of Nicotiana sylvestris. Plant Sci Lett 26: 159-165
14. HAHLBROCK K 1974 Correlation between nitrate uptake, growth and changes in metabolic activities of cultured plant cells. In HE Street, ed, Tissue Culture and Plant Science. Academic Press, London, New York, pp 363-378
15. HAKMAN IC, S VON ARNOLD 1983 Isolation and Growth of protoplasts from cell suspensions of Pinus contorta Dougl. ex Loud. Plant Cell Rep 2: 92-94
16. JIA SR 1982 Factors affecting the division frequency of pea mesophyll protoplasts. Can J Bot 60: 2192-2196
17. JONES DH 1984 Phenylalanine ammonia-lyase: regulation of its induction and its role in plant development. Phytochem 23: 1349-1359
18. KANAI R, GE EDWARDS 1973 Purification of enzymatically isolated mesophyll protoplasts from C_3, C_4 and crassulacean acid metabolism plants using an aqueous dextran-polyethylene glycol two-phase system. Plant Physiol 52: 484-490
19. KAO KN 1977 Chromosomal behavior in somatic hybrids of soybean - Nicotiana glauca. Mol Gen Genet 150: 225-230
20. KAO KN, OL GAMBORG, RA MILLER 1971 Cell divisions in cells regenerated from protoplasts of soybean and Haplopappus gracilis. Nature New Biol 232: 124-130
21. KIRBY EG 1980 Factors affecting proliferation of protoplasts and cell cultures of Douglas-Fir. In F Sala, B Parisi, R Cella, O Ciferri, eds, Plant Cell Cultures: Results and Perspectives. Elsevier/North-Holland Biomed Press, pp 289-293
22. KIRBY EG 1982 The effects of organic nitrogen sources on growth of cell cultures of Douglas-fir. Physiol Plant 56: 114-117
23. KIRBY EG, T CHENG 1979 Colony formation from protoplasts derived from Douglas fir cotyledons. Plant Sci Lett 14: 145-154

24. MURASHIGE T, F SKOOG 1962 A revised medium for rapid growth and bioassays with tobacco tissue cultures. Physiol Plant 15: 473-497
25. NAGATA T, I TAKEBE 1970 Cell wall regeneration and cell division in isolated tabacco mesophyll protoplasts. Planta 92: 301-308
26. PATEL KR, NS SHEKHAWAT, GP BERLYN, TA THORPE 1984 Isolation and culture of protoplasts from cotyledons of Pinus coulteri. Plant Cell Tissue Organ Cult 3: 85-90
27. PHILLIPS R, G HENSHAW 1977 The regulations of synthesis of phenolics in stationary phase cell cultures of Acer pseudoplatanus. J Exp Bot 28: 785-794
28. ROSE R 1980 Factors that influence the yield, stability in culture and cell wall regeneration of spinach mesophyll protoplasts. Aust J Plant Physiol 7: 713-725
29. ROTMAN BR, RO PAPERMASTER 1966 Membrane properties of living cells as studied by enzymatic hydrolysis of fluorogenic esters. Proc Nat Acad Sci USA 55: 134-141
30. SENN A, PE PILET 1980 Isolation and some morphological properties of maize root protoplasts. Z Pflanzenphysiol 100: 299-310
31. STRMEN J, M CIERNA 1981 Cell wall regeneration of the spruce tissue culture protoplasts (Picea excelsa). In Proc IUFRO Sect S2 01 5. Int Workshop "In Vitro" Cultivation For Tree Species, Fontainebleau, France, pp 355-360
32. SUGAWARA Y, K MORI, H MATSUSHIMA, M TAKEUCHI 1983 Enhancement of cell division in Marchantia protoplast culture by activated charcoal. Z Pflanzenphysiol 109: 275-278
33. TEASDALE RD, E RUGINI 1983 Preparation of viable protoplasts from suspension-cultured loblolly pine (Pinus taeda) cells and subsequent regeneration to callus. Plant Cell Tissue Organ Cult 2: 253-261
34. WALLIN A, K GLIMELIUS, T ERIKSSON 1977 Pretreatment of cell suspensions as a method to increase the protoplast yield of Haplopapus gracilis. Physiol Plant 40: 307-311
35. WINTON LL, RA PARHAM, HM KAUSTINEN 1975 Isolation of conifer protoplasts. Genet Physiol Notes No 20, Inst Pap Chem, Appleton, Wisconsin: 1-7
36. XU ZH, MR DAVEY, EC COCKING 1982 Callus formation from root protoplasts of Glycine max. Plant Sci Lett 24: 111-115

3. PROTOPLAST CULTURE OF HARDWOODS

B.H. MCCOWN AND J.A. RUSSELL

1. INTRODUCTION

Hardwood trees, like most woody plants, present difficult challenges to the plant breeder. Not only are the plants unwieldy and thus difficult to manipulate in large numbers, but the life cycle with the attendant juvenile period of years to decades strongly limits the rapidity of improvement. In addition, many hardwood tree species are genetically diverse (heterozygous) and thus complex breeding strategies must be employed to stabilize a genetic character in a seed-propagated population. These characteristics make breeding hardwood trees both time-consuming and expensive. Since most of the hardwood species are not the focus of large forestry programs, such genetic improvement programs are usually not undertaken on an extensive basis.

The potential of manipulating genes by non-traditional methods (the new biotechnology including genetic engineering) can reduce the disadvantages of working with trees as crop organisms. In particular, the application of cellular-based gene transfers will eliminate the need to grow the parents to sexual maturity at every breeding cycle. Gene cloning and transfer allows the insertion of a few characters such as herbicide resistance into an otherwise superior genotype without necessarily having to engage in a time-consuming classical selection/backcross program. Thus the application of biotechnology to forest crops has immense potential (43).

A mainstay technique of cellular-based gene transfer in plants is protoplast culture. In the past, protoplasts were essential for most molecularly-based genetic manipulations. However, recent advances have shown that other tissue can be the host for gene transfer methodology, especially where such vector systems as *Agrobacterium tumefaciens* are utilized (12). Nevertheless, protoplasts do offer the potential of a plethora of alternatives for cellular gene manipulations including protoplast fusion, electroporation, microinjection, genetic transformation, and somaclonal variation. It is conceivable that when working with difficult to manage crops like deciduous trees, multiple manipulations would be performed at the cellular level before moving into selection and improvement at the whole plant level.

The realization of the above potential for hardwood trees depends on our ability to induce protoplasts to form normal plants. The history of research in this area, although not extensively explored by any measure, has not been encouraging. Although the isolation of viable protoplasts from hardwood species appears to be relatively successful, the continued division of the protoplasts into multicellular units (calli) has been much less common. Even more remote has been the differentiation of protoplast-derived calli for there are only four cases where the latter has been reported for hardwood species.

During the last five years, our laboratory has been involved in defining the major biological problems in protoplast culture of hardwood species (34,44,45,46). This research has focused on the genera *Betula*, *Rosa*, and *Rhododendron*. Recently, in conjunction with the USDA-Forest Service Biotechnology Program headed by the North Central Forest Experiment Station, we have been focusing on the genus *Populus*. This chapter will discuss some of these findings in light of the past work.

2. REQUIREMENTS OF A SUCCESSFUL PROTOPLAST SYSTEM

For a protoplast culture system to be fully usable, three general requirements must be met:

1. A uniform population of protoplasts showing high viability and very high purity (absence of cells and debris) must be obtained. The isolation system must be reproducible over time.

2. The protoplasts must progress in an orderly manner through early development eventually culminating in calli or suspension cultures.
3. The protoplast-derived cells must be capable of differentiating autotrophic and complete plants.

2.1. Isolation

Many reports are available that describe the isolation of protoplasts from hardwoods and related plants (Table 1). However, the purity of the preparation and the reproducibility of the protocol is usually not documented and quite often the viability is also not assessed.

Experience indicates that there are two variables of critical importance when working with hardwood species - source tissue and uniform release of the protoplasts from the source tissue.

2.1.1. Source tissue. The most successful source tissue for protoplasts of woody plants is juvenile tissue. Sources involving seedling tissues are often used, however the applicability of this approach to general tree improvement where adult parents must be used is questionable. Sources using callus/suspension cultures are probably also using juvenile cells, although this is presently impossible to measure and verify. Shoot cultures are also relatively juvenile when stabilized (see Chapter by D.D. McCown and B.H. McCown).

In addition to juvenility, other aspects of the physiological condition of the source tissue can be very important. For example, the environment under which the stock plants are grown can influence both protoplast yield and viability (15,20,41). Even under identical environmental conditions as obtained in shoot cultures, the physiological condition of the source tissue can have profound effects on the yield and viability of the resultant protoplasts (45).

An important consideration as to choice of source tissue for hardwood tree research is the general availability, uniformity and reproducibility of the source. A source that requires complex and time-consuming manipulations for its production and where the specifics of the production may be highly genotype-dependent (as

Table 1. Reports of the *in vitro* culture of hardwood trees and related plants where the isolation of protoplasts, the early protoplast division, the formation of protoplast-derived calli, and the regeneration of plants from protoplasts has been indicated. An individual report is listed only in the category of most advanced protoplast development mentioned in that report. Revised from Smith (44).

SPECIES	SOURCE TISSUE	REFERENCE
PROTOPLAST ISOLATION		
Betula x 'Boerner'	shoot cultures	44
Betula papyrifera	shoot cultures	44
Corylus avellana	leaves	28
Datura candida	shoot cultures	38
Datura sanguinea	shoot cultures	38
Kalmia latifolia	shoot cultures	44
Malus domestica 'Jonathan'	leaves	18
Morus alba	leaves	21
Paulownia fortunei	seedlings	35
Paulownia taiwaniana	seedlings	36,37
Populus x *euramericana*	seedlings	36,37
Populus hybrid TT32	suspension cultures	11
Populus nigra x *P. laurifolia* (NC-5272)	shoot cultures	33
Populus maximowiczii x *P. trichocarpa* (NC-11390)	shoot cultures	33
Populus tremuloides	seedlings	58
Quercus petraea	leaves	3
Rhododendron x 'Boule de Neige'	callus	44
Rhododendron x 'Gilbraltar'	shoot cultures	45
Santalum album	leaves	42
Skimmia japonica	leaves, callus	32
Ulmus americana	meiotic cells	31
Vaccinium sp.	shoot cultures	44
PROTOPLAST DIVISION		
Alnus incana	seedlings	13
Betula x 'Boerner'	callus	44
Betula papyrifera	callus	44
Betula platyphylla szechuanica	shoot cultures	45
Citrus sp.	cotyledons	7
Fagus sylvatica 'Atropunicea'	leaves	4
Fagus sylvatica 'Zlatia'	leaves	4
Malus domestica 'Golden Delicious' 'McIntosh' 'Mutsu'	callus, suspension cultures	14

(Table 1 continued)

SPECIES	SOURCE TISSUE	REFERENCE
Populus tremula	leaves	1,2
Populus tremula 'Erecta'	shoot cultures	33,44
Populus tremuloides	leaves	2
Rhododendron x 'Boule de Neige'	shoot cultures	44
Rosa sp. 'Paul's Scarlett'	suspension cultures	27
Ulmus americana	leaves	10
	cotyledons, callus, suspension cultures	30
Ulmus campestris	leaves	10
Ulmus x campestris	leaves	10
Ulmus levis	leaves	10
Ulmus parvifolia	cotyledons, callus	30
Ulmus pumula	cotyledons, callus, suspension cultures	30
CALLUS FROM PROTOPLASTS		
Acer pseudoplatanus	callus	6
Alnus glutinosa	seedlings	13
Alnus incana	suspension cultures	53
Citrus sinensis 'Valencia'	cotyledons	8
Coffea arabica	callus	47
	leaves	23
Coffea canephora	leaves	23
Copaifera multijuga	seedlings	57
Leucaena leucocephala	seedlings	57
Malus domestica 'Jonathan'	callus, suspension cultures	18
Prunus dulcis 'Nonpareil'	suspension cultures	59
Rosa sp. 'Paul's Scarlett'	suspension cultures	49
Rosa sp. 'Soraya'	callus	19
Santalum album	callus	42
Sapium sebiferum	seedlings	57
Sorbus aucuparia	shoot cultures	5
Ulmus x 'Homestead'	suspension cultures	48,49
Ulmus x 'Regal'	shoot cultures	33
REGENERATION FROM PROTOPLASTS		
Citrus sp.	nucellar callus	16,55,56
Populus alba x P. grandidentata (NC-5339)	shoot cultures	33
Populus nigra 'Betulifolia' x P. trichocarpa (NC-5331)	shoot cultures	33
Santalum album	suspension cultures	25,29
Ulmus x 'Pioneer'	suspension cultures	48

is often the case with suspension cultures) complicates immensely the research and the subsequent reproducibility and general applicability. We have found that stabilized shoot cultures are an ideal source tissue for protoplast generation since shoot cultures are easily maintained in large numbers in a uniform condition at any time of the year, are sterile thus eliminating the injurious and nonuniform effects of chemical sterilization, and are relatively juvenile. Shoot cultures keep tissue in a differentiated state and thus the genetic uniformity is high and the original capacity of the tissue is maintained. In addition, the data gained while establishing the shoot cultures provides information about tissue sensitivities and growth requirements useful in the design of protoplast/cell culture experiments (39). Finally, and most importantly, the methodologies relying on shoot cultures are the least species and genotype-specific and thus the most generally applicable.

2.1.2. Protoplast release from the source tissue. The uniform release of protoplasts from a source tissue is dependent on the uniformity of the cells in the tissue and the uniformity of exposure of the cells to the cell-wall degrading enzymes. Leaves provide a large population of relatively uniform cells-the mesophyll cells-and thus leaves, especially when obtained from shoot cultures, are a logical and convenient source for protoplast research. However, the enzymes commonly used for protoplast release do not readily penetrate the leaves of most hardwood species. Since the epidermis of the leaves of most hardwood plants cannot be peeled to expose the mesophyll cells, slicing or cutting of the leaf tissue before enzyme treatment is usually necessary. However, microscopic studies done in our laboratory indicate that with *Betula* and *Populus*, enzyme penetration into the leaf is still limited and only those cells on the cut surfaces are readily attacked. Thus with time, linear arrays of cells are attacked and released as protoplasts as the leaf is slowly digested from the cut surfaces inward. This results in a population varying from overdigested protoplasts (thus suffering from long-term exposure to the toxic enzyme preparation) to protoplasts with incompletely-digested cell walls. Only a small percentage of the protoplasts is in optimum condition and thus the yield of viable protoplasts is low (usually under 2 million protoplasts per g source tissue). We have solved this problem by employing a leaf-grinding procedure that both strips-away the epidermal layers and cuts the leaf into fine sections (34). Thus many cells are uniformly exposed to the enzymes at the same time. The result is a high yield (up to 20 million protoplasts per g source tissue) using a digestion time (4 hours), only 1/2 that used previously.

By employing both the shoot culture and the leaf preparation techniques and by using purification procedures adapted from work done with potato (40,41), we can routinely obtain large populations of protoplasts of diverse genotypes of Populus that have viabilities of 85% or greater at isolation and have few if any contaminating cells.

2.2. Protoplast plating/culture

The limited number of cases of successful culture of woody plant protoplasts (Table 1) and the diverse methodologies used in these reports prohibit the selection of a single protocol for successful culture. Whether such diversity of methodologies is needed or whether it is just the result of many laboratories working independently on diverse species and thus not using similar protocols is not known. We have found that with Populus, some of the requirements for protoplast development are also commonly cited in various reports using other plants. In particular, these requirements involve eliminating toxicities that lead to inhibiting protoplast development more than adding factors that stimulate growth. Three such requirements are the elimination of ammonium from the culture medium (22,41,54,60), the need to avoid exudate build-up and other medium changes (9,26,51), and avoidance of the inhibitory effects of agar (17).

Protoplast plating density is another commonly cited factor in the culture of protoplasts. There apparently exists an optimal population density below which protoplast division is limited and above which protoplast division is suppressed, as demonstrated by the work of Sticklen et al. (49) using Ulmus suspension-culture derived protoplasts. We have observed a similar phenomenon with Populus protoplasts where plating densities of 10^5 and above are stressful to newly plated protoplasts as evidenced by increased budding, clearing of cell contents, and death. Although the basis for the stressful nature of high protoplast plating densities is not clear, the release of inhibitory compounds from the cells is a likely factor.

An interesting stimulatory factor observed for protoplasts of Populus that has not been generally reported for other species (although may occur unintentionally) is physical contact with a solid surface. Protoplasts placed in liquid drops or wells float to the surface, survive for long periods (weeks), but do not divide. However, protoplasts that have adhered to a solid surface (for example to the bottom of a plastic plate or to a floating disc) can undergo extensive division and form large cell colonies. Whether or not such physical contact is a general stimulatory factor in protoplast development of other hardwoods is not known.

2.3. Early development of protoplasts.

The number of reported cases where protoplasts from hardwoods and related species have formed cell walls, initiated division, and developed into self-perpetuating cell colonies is limited. Although the number of reports of protoplasts showing some division, usually from 2 to 16 cell colonies, is relatively large (Table 1), continued division is often not observed or reported. However, a significant enough number of reports of callus formation are available (Table 1) to indicate that protoplast-derived callus formation is certainly possible with hardwood trees.

Our experience with both woody and herbaceous plant protoplasts suggests there may be two pathways of early protoplast development. Protoplasts from herbaceous plant tissues and from undifferentiated/meristematic tissues of woody plants generally exhibit classical, bipolar division in 50% or greater of the population within the first week of culture. However, we have observed a slow, staged developmental sequence with protoplasts isolated from leaf tissue (Fig. 1). Less than 10% of the original protoplast population show such staged development and eventually form colonies in a period of 6 weeks or more. Why two quite distinct developmental pathways should occur is not clear. We do see protoplasts undergoing each pathway in the same protoplast preparation. It would be intriguing to determine how many distinct subpopulations of protoplasts occur in a preparation from an organ as complex as the leaf and if these populations differ in their cellular characteristics (24,52), in their cultural requirements, in their totipotency, and/or in their developmental pattern. The slow developmental sequence may be a result of the highly differentiated nature of the original cell or merely the consequence of stress factors still present in the culture system that particularly influence a subpopulation of the original cells.

One advantage of a slow rate of development of some hardwood protoplasts is that the stages in the developmental sequence can be clearly discerned (Fig. 1). Interestingly, if we closely observe such rapidly developing leaf protoplast systems as potato, we see the same stages, however the time each stage is prevalent may be only hours or days, not weeks as with the woody species. This may make woody plants preferential subjects for research involving detailed studies of the anatomical/biochemical events associated with the protoplast developmental sequence.

2.4. Regeneration of shoots.

The number of reports of shoot regeneration from protoplasts of hardwoods is limited to four (Table 1). Again one should note that the source tissue (nucellar callus) for the *Citrus* example is highly juvenile and probably limited in application to *Citrus* because of its specialized origin. It is interesting to note also that utilizing other source tissue for *Citrus* (such as leaves or cotyledons) has not produced successful results (7,8). The reliance on suspension cultures as with *Santalum* and *Ulmus* also limits the general applicability since these suspension-culture procedures can be quite species/genotype specific.

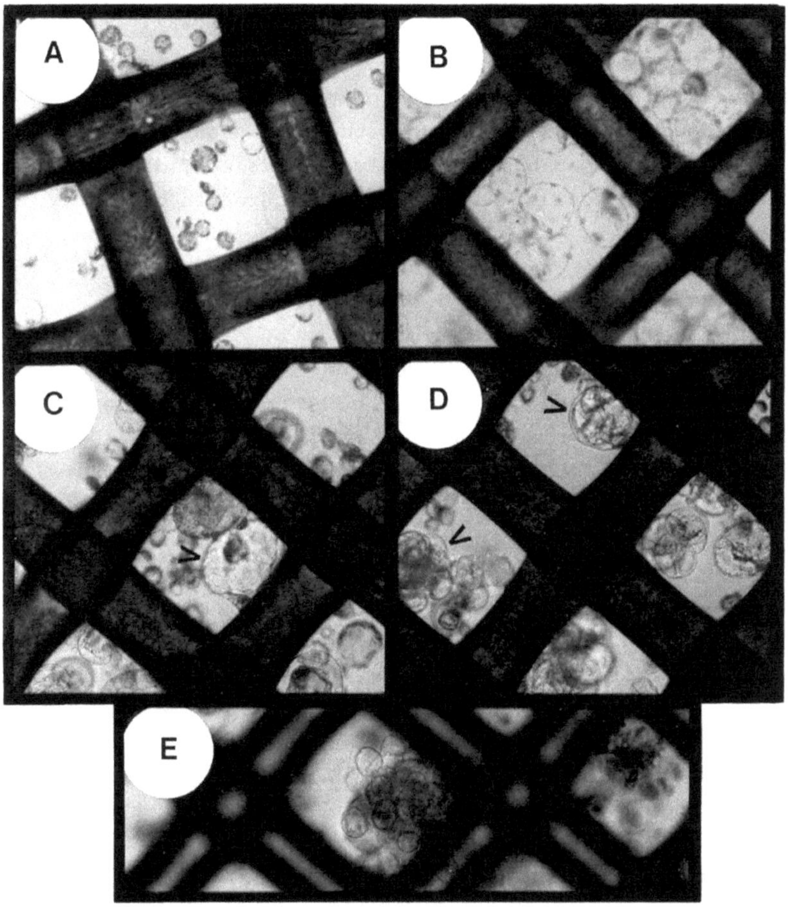

FIGURE 1. A developmental sequence of protoplasts of Populus. a) Newly isolated protoplasts. b) After cell wall formation and swelling are mostly completed, about 2 weeks later than a. c) The "systrophe" stage (arrow), about 2 to 3 weeks after b. The cell denoted by the arrow is in both systrophe and the early stages of "intensification" where a marked increase in cytoplasmic content is apparent. d) The "fracturing" stage, where cell division within the boundaries of the original cell wall occurs, about 1 week after c. e) The "break-out" stage where the cells formed in "e" divide beyond the original cell boundary and develop into colonies. Note that the new cells are now close to the same size as the original isolated protoplasts as seen in "a". Square is 150μ I.D.

Thus one has to conclude that regeneration of plants from protoplast-derived cells of hardwoods is the major problem. Although getting cell colonies from protoplasts is not yet easy, there is enough success to indicate that the problem is entirely within our present capability to solve. Such confidence is not warranted with protoplast-based shoot regeneration of hardwoods in general.

3. CONCLUDING REMARKS

We have been working with protoplasts of hardwoods for more than 6 years; at the same time, we have also utilized herbaceous crops. It is intriguing that with all the countless combinations of media/environment/tissue that have been tested, the most successful systems for hardwoods are some of the simplest. In addition, most of the critical variables that have been discovered fall into the category of removing inhibition to development, not of stimulating development. This is in contrast to the herbaceous species where one has the distinct feeling that the culture techniques in large part control development. This attitude is enforced by the large differences in rates of development of the two plant types. The culture of woody plants seems to function by processes much less influenced by the culture conditions than the culture of herbaceous plants. It may be that the endogenous controls on cellular development for woody species are much tighter (less plastic?) than with herbaceous plants. Indeed, anyone growing the crops for production of these two plant types soon comes to a similar conclusion. With the dynamic broadening of our understanding of gene action and control that is occurring at present, it will be fascinating if such loose generalizations turn out to be based in fact.

4. REFERENCES

1. AHUJA MR 1983 Developmental potential of mega and normal protoplasts in *Populus*. *In* I Potrykus, CT Harms, A Hinnen, R Hutter, PJ King, RD Shillito, eds, 6th Int Protoplast Symp, Basel, Experientia Suppl 45:28-29
2. AHUJA MR 1983 Isolation and culture of mega and normal protoplasts in Aspen. Silvae Genet 32:5-6

3. AHUJA MR 1984 Protoplast research in woody plants. Silvae Genet 33:32-37
4. AHUJA MR 1984 Isolation and culture of mesophyll protoplasts from mature beech trees. Silvae Genet 33:37-39
5. BINDING H, J JORGENSEN, G KRUMBIEGEL-SCHROEREN, J FINGER, G MORDHORST, G SUCHWIAT 1983 Culture of apical protoplasts from shoot cultures in the orders Fabales, Rosales, and Caryophyllales. *In* I Potrykus, CT Harms, A Hinnen, R Hutter, PJ King, RD Shillito, eds, 6th Int Protoplast Symp, Basel, Experientia Suppl 45:34-35
6. BRIGHT SWJ, DH NORTHCOTE 1974 Protoplast regeneration from normal and bromodeoxuridine-resistant sycamore cells. J Cell Sci 16:445-463
7. BURGER DW, WP HACKETT 1981 Protoplast culture of several citrus tissues. Hortscience 16:417
8. BURGER DW, WP HACKETT 1982 The isolation, culture, and division of protoplasts from citrus cotyledons. Physiol Plant 56:324-328
9. CARLBERG I, K GLIMELIUS, T ERIKSSON 1983 Improved culture ability of potato protoplasts by use of activated charcoal. Plant Cell Rep 2:223-225
10. DORION N, B GODIN, C BIGOT 1983 Isolation and culture of leaf protoplasts from *Ulmus* sp. *In* I Potrykus, CT Harms, A Hinnen, R Hutter, PJ King, RD Shillito, eds, 6th Int Protoplast Symp, Basel, Experientia Suppl 45:8-9
11. DOUGLAS G 1982 Protoplast isolation from totipotent cell-cultures of *Populus* hybrid TT32. *In* A Fujiwara, ed, Plant Tissue Culture 1982, Proc 5th Int Cong Plant Tissue Cell Culture, Tokyo, pp 605-606
12. HORSCH RB, JE FRY, NL HOFFMANN, D EICHHOLTZ, SG ROGERS, RT FRALEY 1985 A simple and general method for transferring genes into plants. Science 277:1229-1231
13. HUHTINEN O, J HONKANEN, LK SIMOLA 1982/83 Ornithine and putrescine-supported divisions and cell colony formation in leaf protoplasts of alders (*Alnus glutinosa* and *A. incana*). Plant Sci Lett 28:3-9
14. HURWITZ CD, GN AGRIOS 1984 Isolation and culture of protoplasts from apple callus and cell suspension cultures. J Am Soc Hortic Sci 109:348-350
15. KARUNARATNE SM, KJ SCOTT 1981 Mitotic activity in protoplasts isolated from *Sorghum bicolor* leaves. Plant Sci Lett 23:11-16
16. KOBAYASHI S, H UCHIMIY, I IKEDA 1983 Plant regeneration from 'Trovita' orange protoplasts. Jpn J Breed 33:119-122
17. KOHLENBACH HW, W WERNICKE 1978 Investigations on the inhibitory effect of agar and the function of active carbon in anther culture. Z Pflanzenphysiol 86:463-472
18. KOUIDER M, R HAUPTMANN, JM WIDHOLM, RM SKIRVIN, SS KORBAN 1984 Callus formation from *Malus* x *domestica* cv. 'Jonathan' protoplasts. Plant Cell Rep 3:142-145
19. KRISHNAMURTHY KV, RR HENDRE, DA GODBOLE, VM KULKARNI, AF MASCARENHAS, V JAGANNATHAN 1979 Isolation and regeneration of rose bud callus protoplasts (*Rosa* sp. cv. Soraya). Plant Sci Lett 15:135-137

20. MORENO V, L ZUBELDIA, LA ROIG 1984 A method for obtaining callus cultures from mesophyll protoplasts of melon (Cucumis melo L.). Plant Sci Lett 34:195-201
21. OHYAMA K, S OKA 1975 Isolation of protoplasts from mesophyll cells of Morus alba. Proc Crop Sci Soc Jpn 44:121-122
22. OKAMURA M, T HAYASHI, S MIYAZAKI 1984 Inhibiting effect of ammonium ion in protoplast culture of some Asteraceae plants. Plant Cell Physiol 25:281-286
23. OROZCO FJ, O SCHIEDER 1983 Isolation and culture of coffee leaf protoplasts. In I Potrykus, CT Harms, A Hinnen, R Hutter, PJ King, RD Shillito, eds, 6th Int Protoplast Symp, Basel, Experientia Suppl 45:52-53
24. OZEKI Y, A KOMAMINE 1981 Induction of anthocyanin synthesis in relation to embryogenesis in a carrot suspension culture: Correlation of metabolic differentiation with morphological differentiation. Physiol Plant 53:570-577
25. OZIAS-AKINS P, PS RAO, O SCHIEDER 1984 Plant regeneration from embryonic suspension-derived protoplasts of sandalwood (Santalum album). In RH Henke, KW Hughes, MJ Constantin, A Hollaender, eds, Tissue Culture in Forestry and Agriculture, Proceedings of the Third Tennessee Symposium on Plant Cell and Tissue Culture, Plenum Press, New York, pp 338-339
26. PATEL KR, NS SHEKHAWAT, GP BERLYN, TA THORPE 1984 Isolation and culture of protoplasts from cotyledons of Pinus coulteri D.Don. Plant Cell Tissue Organ Culture 3:85-90
27. PEARCE RS, EC COCKING 1973 Behaviour in culture of isolated protoplasts from "Paul's Scarlett" rose suspension culture cells. Protoplasma 77:165-180
28. RADOJEVIC L, A KOVOOR 1978 Characterization and estimation of newly synthesized DNA in higher plant protoplasts during the initial period of culture. J Exp Bot 29:963-968
29. RAO PS, P OZIAS-AKINS 1985 Plant regeneration through somatic embryogenesis in protoplast cultures of sandalwood (Santalum album L.) Protoplasma 124:80-86
30. REDENBAUGH K, DF KARNOSKY, RD WESTFALL 1981 Protoplast isolation and fusion in three Ulmus species. Can J Bot 59:1436-1443
31. REDENBAUGH MK, RD WESTFALL, DF KARNOSKY 1980 Protoplast isolation from Ulmus americana pollen mother cells, tetrads and microspores. Can J For Res 10:284-289
32. ROBENEK H, E PEVELING 1977 Ultrastructure of the cell wall regeneration of isolated protoplasts of Skimmia japonica Thunb. Planta 136:135-145
33. RUSSELL JA, BH MCCOWN, in preparation
34. RUSSELL JA, EL ZELDIN, BH MCCOWN 1985 Leaf grinding improves release of mesophyll protoplasts. Hortscience 20:571
35. SAITO A 1976 Isolation of protoplasts from mesophyll cells of Paulownia fortunei Hemsl. and Populus euramericana cv. I-45/51. J Jpn For Sci 58:301-305
36. SAITO A 1980 Isolation of protoplasts from mesophyll cells of Paulownia and Populus. Bulletin of the Forestry and Forest Products Research Institute (Japan) 309:1-6

37. SAITO A 1980 Fusion of protoplasts isolated from somatic cells of tree species. Bulletin of the Forestry and Forest Products Research Institute (Japan) 309:7-12
38. SCHIEDER O 1980 Somatic hybrids between a herbaceous and two tree Datura species. Z Pflanzenphysiol 98:119-127
39. SELLMER JC, JA RUSSELL, EL ZELDIN, BH MCCOWN 1985 Utilization of cytokinin response curves in tissue evaluation of Populus for biotechnology research. Hortscience 20:593
40. SHEPARD JF 1980 Mutant selection and plant regeneration from potato mesophyll protoplasts. In B Gengenbach, RL Phillips, CE Green, eds, Genetic Improvement of Crops: Emergent Techniques, Univ Minnesota press, Minneapolis, pp 185-219
41. SHEPARD JF, RE TOTTEN 1977 Mesophyll cell protoplasts of potato. Isolation, proliferation, and plant regeneration. Plant Physiol 60:313-316
42. SITA GL, BS RANI 1983 Preliminary studies on isolation and culture of protoplasts from sandalwood (Santalum album). In I Potrykus, CT Harms, A Hinnen, R Hutter, PJ King, RD Shillito, eds, 6th Int Protoplast Symp, Basel, Experientia Suppl 45:4-5
43. SKELSEY AF 1984 Biotechnology in agriculture - New tools for the oldest science. Joint Council on Food and Agricultural Science, USDA, 36 pp
44. SMITH MAL 1984 Application of Microculture and Protoplast Culture Techniques to Physiological Analysis of Woody Crops, PhD Thesis, Univ Wisconsin, Madison
45. SMITH MAL, BH MCCOWN 1982/83 A comparison of source tissue for protoplast isolation from three woody plant species. Plant Sci Lett 28:149-156
46. SMITH MAL, JP PALTA, BH MCCOWN 1984 The measurement of isotonicity and maintenance of osmotic balance in plant protoplast manipulations. Plant Sci Lett 33:249-258
47. SONDAHL MR, MS CHAPMAN, WR SHARP 1980 Protoplast liberation, cell wall reconstitution, and callus proliferation in Coffea arabica L. callus tissues. Turrialba 30:161-165
48. STICKLEN MB, RD LINEBERGER, SC DOMIR 1985 Isolation and culture of Ulmus x 'Homestead' protoplasts. Hortscience 20:571
49. STICKLEN MB, RD LINEBURGER, SC DOMIR 1985 Isolation and culture of protoplasts of Ulmus x 'Homestead'. Plant Sci 41:117-120
50. STRAUSS A, I POTRYKUS 1980 Callus formation from protoplasts of cell suspension cultures of Rosa 'Paul's Scarlett'. Physiol Plant 48:15-20
51. SUGAWARA Y, K MORI, H MATSUSHIMA, M TAKEUCHI 1983 Enhancement of cell division in Marchantia protoplast culture by activated charcoal. Z Pflanzenphysiol 109:275-278
52. TAKEDA J, M SENDA 1984 Relationship of the electrophysiological property of tobacco cultured cells to their regeneration ability. Plant and Cell Physiology 25:619-624
53. TREMBLAY FM, JB POWER, M LALONDE 1985 Callus regeneration from Alnus incana protoplasts isolated from cell suspensions. Plant Sci 41:211-216

54. UPADHYA M 1975 Isolation and culture of mesophyll protoplasts of potato (Solanum tuberosum L.). Potato Res 18:438-445
55. VARDI A, P SPEIGEL-ROY, E GALUN 1975 Citrus cell culture: Isolation of protoplasts, plating densities, effect of mutagens and regeneration of embryos. Plant Sci Lett 4:231-236
56. VARDI A, P SPEIGEL-ROY, E GALUN 1982 Plant regeneration from Citrus protoplasts: variability in methodological requirements among cultivars and species. Theor Appl Genet 62:171-176
57. VENKETESWARAN S, V GANDHI 1980 Protoplast isolation and culture of tree genera for biomass production. Eur J Cell Biol 22:501
58. VERMA DC, SR WANN 1983 Isolation of high yields of viable protoplasts from quaking aspen seedlings and cultured loblolly pine cell suspensions. In I Potrykus, CT Harms, A Hinnen, R Hutter, PJ King, RD Shillito, eds, 6th Int Protoplast Symp, Basel, Experientia Suppl 45:10-11
59. WU SC, AH KUNIYUK 1985 Isolation and culture of almond protoplasts. Plant Sci 41:55-60
60. ZAPATA FJ, KC SINK, EC COCKING 1981 Callus formation from leaf mesophyll protoplasts of three Lycopersicon species: L. esculentum, cv. Walter, L. pimpinillifolium and L. hirsutum, f.glabratum. Plant Sci Lett 23:41-46

4. BIOCHEMISTRY OF FOREST TREE SPECIES IN CULTURE

D. GREY, G. STEPAN-SARKISSIAN and M. W. FOWLER

1. INTRODUCTION

While tissue culture systems represent an area of major potential in tree biology, cell culture systems are ideally suited for investigations aimed at achieving a greater understanding of the cellular physiology and biochemistry of tree species. Over the years a number of tree species have been established in cell culture and a variety of information has been obtained from such sources. We now have a greater understanding of the influence of various factors on growth rate, mechanisms of resistance against microbial attack and secondary product biosynthesis.

Since a tree is a highly integrated multicellular organism, there is a need for care when extrapolating observations from the cellular to the multicellular level. In spite of this there is a base of literature from studies with cell cultures of forest species which is highly relevant to the longer term development of arboriculture and silviculture.

Early studies with cell cultures of tree species were little concerned with the potential improvement of the parent plant. Rather they were directed towards studying general aspects of plant biochemistry and physiology (51). Much of the early work was carried out with cell cultures derived from the sycamore tree (*Acer pseudoplatanus* L.) (52). Since then many other species have also been investigated, a development accompanied by a change in research emphasis away from the primary metabolism and towards areas more directly related to commercial exploitation. Much of this work has however been carried out with callus cultures and has only been referred to in passing, this review focussing specifically on cell cultures.

2. BIOCHEMISTRY

Various aspects of the biochemistry of plant cell cultures have been investigated, ranging from studies of basic biochemistry and primary metabolism to the more complex pathways of secondary metabolite synthesis. Although strictly speaking growth kinetics do not fall within the confines of biochemistry, they are included in the present discussion since biochemical parameters are generally described in terms of the growth phases through which cell cultures are deemed to pass (47).

2.1. Metabolic activities and growth patterns

Much of the early work on growth kinetics of a forest tree species in culture was carried out with *A. pseudoplatanus* (9, 34, 39, 81, 92). The choice of sycamore cells as experimental material was dictated by the fact that they form fine cell suspension cultures with a high density of isolated single cells (55).

Cultures of *A. pseudoplatanus* cells have been used to study cell growth in both batch and continuous culture (51, 52, 86, 103). In batch culture the pattern of growth in terms of cell density is ideally described by a sigmoid curve (51). Several stages in the curve describe the metabolic events occurring in the cell population. When introduced into fresh medium plant cells undergo a lag phase during which they take up fresh nutrients and use them to sustain intense biosynthetic activity (25, 104). This activity declines at the start of the next stage, i.e. the exponential phase, which is characterized by cell division. During the exponential phase metabolic activity is directed towards polymerization (104). The final stage of the growth cycle is described as a stationary phase during which the cell number remains constant but average cell size expands and dry weight per cell increases. The exponential phase of growth is relatively short and consists of 2-4 cell divisions each lasting an average of 60-70h (4, 51). Shorter generation times (48h) (56) and longer periods of exponential phase (7-8 days) have however been reported for sycamore cells (55).

Manipulation of medium components has significant effects on the rate of growth of A. pseudoplatanus cell suspension cultures. For example, doubling the concentration of 2,4-dichlorophenoxyacetic acid (2,4-D) results in a reduction of 20h in the mean cell doubling time (51), although this effect is observed only within a certain range of auxin concentrations (59). In contrast various concentrations of auxin and cytokinin have little effect on the growth of poplar (Populus) cells in culture (63). Growth is, however, found to be markedly affected when the 'standard' carbon source (glucose or sucrose) is replaced by other sugars or non-sugar sources. In particular, the generation time of sycamore cells is increased by 20% when sucrose grown cells are introduced into a glycerol containing medium (79). Parallel studies produced similar observations (36). The glycerol grown cells however took twice as long to reach their maximum biomass. This suggests that the metabolic pathways responsible for glycerol catabolism are able, in this instance at least, to support adequate biosynthetic activity.

Simpkins and co-workers (81) experimenting with carbon sources other than sucrose have reported that sycamore cells grow equally well on glucose, fructose, galactose, maltose and starch. Equimolar concentrations of glucose and fructose also support good growth. The same workers have also investigated the effect of initial sucrose concentrations (ranging from 1-10%) on the growth rate and final biomass yield of sycamore cells. The highest biomass yields in terms of dry weight were obtained between 4% and 6% initial sucrose concentrations. In the case of poplar cells equal levels of growth have been observed on sucrose, glucose and fructose but biomass accumulation drops to 50% or less when cells are grown on soluble starch, raffinose, galactose, melezitose, trehalose, maltose and lactose (64, 65). Levels of sucrose (ranging from 2-5%) do not appear to influence the growth rate but cell dry weight increases by 60% when the sucrose concentration is raised from 2% to 3% (64). Further increases in sucrose concentration to 4% and 5% do not affect biomass accumulation significantly. Such data concerning the growth of plant cells with a variety of carbon sources and concentrations should however only be taken as indicative. In most cases information on the stability of these cultures and constancy of response over several passages is lacking.

There have been relatively few reports of the establishment of suspension cultures from other forest trees coupled with studies of their physiology and biochemistry. An early report by Reinert (75) described a liquid culture initiated from dissociating callus of spruce (*Picea glauca*), but no quantitative growth data were provided. Further work with cell suspension cultures of spruce was reported by Durzan and co-workers (10, 11, 20, 23), who provided preliminary information on growth patterns and biomass accumulation as well as origins and development of tannin bodies in such systems. The same workers also established cell suspension cultures of jack pine (*Pinus banksiana*) and studied the effect of light and darkness on their growth patterns (21) as well as the distribution of nitrogeneous compounds in the cells and growth medium (22). Cell suspension cultures of lodgepole pine (*Pinus contorta*) were found difficult to maintain mainly because of the high levels of oxidation of phenolic acids associated with cell lysis (100). An initial report (105) on the establishment of Douglas-fir [*Pseudotsuga menziesii* (Mirb.) Franco] cell cultures has not so far been followed by reports of their characterization. Other workers have used suspension cultures of Douglas-fir and loblolly pine (*Pinus taeda* L.) to study the site of tannin synthesis in plant cells (72, 73), and cultures of eucalyptus (*Eucalyptus globulus*) to study antimicrobial activities (49). Green suspension cultures of Douglas-fir have been used to study the structure and synthetic patterns of condensed tannins (83, 84, 85, 86), but little detailed information is available regarding the growth and biochemistry of these cultures.

2.2. Nutrient uptake

Data regarding the uptake of nutrients from culture media by suspension cultures of tree species are scarce.

Cells of A. pseudoplatanus L. preferentially absorb glucose when glucose and fructose are supplied in the culture medium in equimolar amounts (81). This phenomenon has also been observed when sycamore cells are inoculated into medium containing 2% sucrose (composed of glucose and fructose) as carbon source. Although very little glucose or fructose is present at subculture, their levels rise first and fall later as the sucrose is depleted. This suggests that the cells possess surface-localized hydrolytic enzymes (invertases), a hypothesis supported by many other studies with plant cell cultures (90). The level of fructose remains above that of glucose throughout the growth cycle.

Simpkins and Street (82) observed an interesting interaction between growth regulators and the carbon source. High kinetin levels (2.5mg/l) were found to depress cell yield when sycamore cultures were grown with sucrose or its constituent monosaccharides as the carbon source. The effect was much greater with glucose or glucose + fructose than with sucrose. The reduced growth could not be attributed to a slower uptake of carbon since by day 21 of culture, only traces of sugar remained in the medium.

To gain information as to the nature of cell wall hydrolysis Copping and Street (12) measured the levels of 'acid' and 'neutral' invertase activity in cell wall and 'soluble' fractions from A. pseudoplatanus cells. While acid invertase activity in the cell wall increased early in the growth cycle, activity of neutral invertase declined. A reciprocal relationship was observed with enzyme activity in the 'soluble' fraction. This behaviour indicates a movement from the protoplast to the cell wall of acid invertase and a reverse movement of neutral invertase. It is thought that acid invertase is concerned with sucrose uptake from the medium and the removal of stored sugar from the vacuole. Neutral invertase is responsible for sucrose hydrolysis in the cytoplasm. However, unless invertase is considered to be a constitutive enzyme it is difficult to explain its function in cells grown exclusively on glucose, since no sucrose is detected in the medium.

The rate of sucrose disappearance from the medium, may be linked to the level and nature of the nitrogen source (42). At equimolar nitrogen concentration the rate of sucrose uptake is consistently higher in nitrate-grown sycamore culture compared with those grown on glutamate. Sucrose exhaustion is complete on 14mM nitrate after 336h, but it is a further 24h before it is completely removed in the presence of 14mM glutamate.

Another carbohydrate upon which attention has been focussed is myo-inositol. When grown on standard Murashige and Skoog (70) liquid medium (3% sucrose, 1 mg/l 6-benzylaminopurine, 0.1 mg/l 2,4-D) A. pseudoplatanus cell suspensions remove tritiated myo-inositol up to 100 mg/l in 24h with up to 20% being used for pectin biosynthesis. When the level of 6-benzylaminopurine is increased 10-fold, myo-inositol uptake is reduced drastically and very little is available for pectin biosynthesis (99). Although a similar increase in the level of 2,4-D has no effect on myo-inositol uptake, it does reduce the amount available for utilization. However the uptake of the cyclitol is drastically reduced by 2,4-D in the presence of 3% sucrose. The distribution of myo-inositol in the cell wall constituents is similar to that in untreated cells.

Nitrogen utilization varies with the nature of the nitrogen source. Nitrate supplied at a concentration of 14mM to sycamore cultures is depleted from the medium by mid-log phase (42). Glutamate uptake in contrast occurs at a slower rate and appears dependent upon the initial concentration. Nitrogen depletion corresponds with the entry of the culture into stationary phase. For some as yet unexplained reason the intracellular nitrate pools seem to be utilized at a slower rate than the nitrate in the medium.

Sycamore cultures absorb urea more rapidly than nitrate (104). The difference is due possibly to the operation of distinct and separate uptake systems with characteristic kinetics for each nitrogen source. Alternatively there may be an effect by the nitrogen source on cell metabolism which then directly or indirectly determines subsequent demands for nitrogen from the medium. The increase in medium pH which occurs during the uptake of both NH_4^+ and NO_3^- may also affect the uptake and utilization of other ions (31).

Cell suspension cultures of sycamore have been shown to take up and incorporate [2-^{14}C]-uridine in all stages of the growth cycle (13). The rate of uptake per 10^6 cells is highest at inoculation and declines steeply throughout the growth of the culture. Although labelled uridine is rapidly taken up by sycamore cells, it is several hours before the acid soluble pool becomes saturated with this nucleoside. This implies either a substantial pool expansion or a slow equilibration with the pre-existing pool. Apparently pool expansion has little effect on the kinetics of [^{14}C]-uridine uptake. It is thought that large pools of UMP and UDP-glucose exist in sycamore cells since the time taken for the pools to become saturated with radioactivity was comparatively long.

2.3. Primary metabolism

Space limitation prevents a detailed discussion of primary metabolism. Comments are confined to the utilization of two key macroelements, carbon and nitrogen. In cell cultures their level together with the form in which they are supplied exerts a major influence on the metabolic activity, production of biomass, secondary metabolite systems as well as on cellular organization and development. An understanding of the interactions between these two nutrient factors in cultured cells should provide valuable insights into the metabolism of carbon and nitrogen in plant tissues (19).

2.3.1. Carbohydrate metabolism. The rationale behind early work on carbohydrate utilization in plant cell cultures was their potential as a tool for the study of biochemical differentiation in plants. As a first step enzymes acting on, and pathways originating from, the carbon source were identified and investigated.

The enzyme invertase (EC 3.2.1.26) catalyzes the hydrolytic cleavage of the sucrose molecule. Based on the pH optimum for activity, acid, neutral and alkaline forms of this enzymes have been reported from a wide range of plant cultures (63, 89). Depending on the species, or even on cultures derived from the same species, invertase can be present in one, two or all three forms. For example, in cultured carrot cells (77) only an acid invertase exists. Sycamore cells show acid and neutral invertase activity (12). Sugar cane cells possess all three forms of the enzyme (63). Amongst forest tree species invertase has been studied in tissue cultures of incense cedar (Libocedrus decurrens Torr.) and cypress (Cupressus funebris Endl.) (89), and in cell cultures of sycamore (12). In cells of A. pseudoplatanus acid invertase activity is equally distributed between the cell wall and 'soluble' fractions, whereas a significant proportion of neutral invertase is associated with the cell wall. This strong association of invertase activity with the cell wall, also reported for other species in culture (26, 63, 88), indicate the important role played by the cell wall in sucrose hydrolysis.

Upon the entry of glucose into the cell, either as a product of sucrose hydrolysis or by direct uptake from a glucose-supplemented medium, it becomes a substrate for two principal functions; i) provision of energy, and ii) supply of carbon skeletons for cell components. For glucose to become available to the pathways of intermediary metabolism a phosphorylation step to glucose-6-phosphate is required. The reaction is catalyzed by the enzyme hexokinase (EC 2.7.1.1) which has been reported in batch (29) and continuous (28) cultures of A. pseudoplatanus cells. In batch cultures two pH optima (6.5 and 8.5) are reported for hexokinase and are attributed to the presence of two isoenzymes associated with the particulate (mitochondrial) and 'soluble' fractions. The instability of hexokinase during purification has prevented more extensive investigation of its physiological role in plant cells growing in batch culture. An insight, albeit in a limited form, into the role of hexokinase is provided by continuous cultures of sycamore cells. These cultures have an advantage over batch cultures in that they can be maintained in steady states in selected growth phases and at specific growth rates (103). In continuous cultures of A. pseudoplatanus cells (28) the extractable activity of hexokinase remains low at various growth rates in comparison with the activities of other enzymes of carbohydrate oxidation. Further supportive evidence has been provided by Jessup and Fowler (43) using sycamore cells grown in batch cultures. This feature of hexokinase suggests that the phosphorylation of glucose may be the rate limiting step of hexose oxidation and therefore an important control point of carbon flux through glycolysis and the pentose phosphate pathway.

Detailed studies of these two pathways have been carried out with sycamore cells. These studies were based on measurements of enzyme activities and intracellular metabolite levels (25, 28, 43, 80). Early work with the glycolytic and the pentose phosphate pathways (24, 25) had indicated significant changes in the total capacity and in the relative activities of their key enzymes. During the early period of cell growth in batch culture the increase in the capacity of the pentose phosphate pathway was greater than in glycolytic capacity, presumably because in the period of 3-4 days subsequent to subculturing the metabolism of the cells is targeted towards biosynthetic activities. Subsequent work with continuous (chemostat) cultures of sycamore cells (28) revealed that this increased capacity of the pentose phosphate pathway may not be at the expense of the glycolytic flux but rather due to a net increase in metabolite entry into the respiratory pathways. Increased levels of oxygen uptake in sycamore batch (34, 80) and continuous (28) cultures may all be ascribed to this phenomenon. Glycolysis and the pentose phosphate pathway are thought to have distinct roles within the framework of intermediary metabolism. Thus glycolysis provides the cells with ATP and precursors for tricarboxylic acid cycle intermediates, whereas the main role of the pentose phosphate pathway appears to be the supply of reducing equivalents in the form of NADPH and pentose precursors for nucleic acid synthesis (54).

The shift in carbohydrate oxidation through the pentose phosphate pathway in sycamore cells has been attributed to a higher demand for NADPH (24, 25, 28). This reducing equivalent can donate electrons to a variety of biosynthetic reactions. Of these, the reductive assimilation of nitrate to ammonia and the removal of TCA cycle intermediates for biosynthesis have received some attention in sycamore cells.

Jessup and Fowler (14) have reported that in the presence of nitrate-nitrogen the estimated relative flux of the two hexose oxidation routes indicates an increase in glucose metabolism via the pentose phosphate pathway. The involvement of this pathway with biosynthetic activity via NADPH has also been postulated by Wilson (104). He reported a high level of cyanide-insensitive respiratory activity in mitochondrial preparations from sycamore cells in the lag phase of growth. This phase is known to be a period of biosynthetic activity that precedes the phase of cell division during which the balance of metabolism changes to polymerization. Further support for this pattern of metabolic activity has been provided by Shimizu and her colleagues (80) who reported high levels of NAD, NADPH, ATP and energy charge in sycamore cells before they begin dividing. Following the onset of cell division, the rate of oxygen uptake, the level of NADPH and energy charge begin to decline indicating a decrease in demand for biosynthetic reducing equivalents. In parallel with the increase in pentose phosphate pathway activity for the provision of reducing power, there should be other routes for the removal of C-5 moieties to prevent their accumulation and possible feedback inhibition of the pathway. One such route is the incorporation of pentoses into nucleotides.

Brown and Short (5) reported a significant increase in the nucleotide content of sycamore cells in the lag phase. This increase was mainly due to a large rise in UDP-glucose and ATP concentrations. High levels of nucleoside-glycosyl intermediates in the lag phase would assist the biosynthesis of polysaccharide cell wall material during the ensuing phase of rapid cell division.

Despite the relatively wide range of compounds used as carbon sources for tree species in culture (Section 2.1.), very little attention has been paid to the biochemistry of the pathways involved in their mobilization. A notable exception is the work of Scala and Semersky (79) with sycamore cells growing on glycerol. This substrate may enter the later stages of gluconeogenesis as an important precursor for hexose and pentose synthesis, both of which are required for growth. Scala and Semersky (79) confined their studies to the induction kinetics of fructose-1,6-bisphosphatase (EC 3.1.3.11), a regulatory enzyme of gluconeogenesis. Unlike most other plant fructose-1,6-bisphosphatases, the sycamore enzyme was found to be unaffected by AMP and inhibited by EDTA perhaps because of the presence of a metal ligand. The enzyme was found to be induced immediately upon the introduction of cells into glycerol-containing medium and the activity continuously increased before reaching a plateau after 20 days.

2.3.2. Nitrogen metabolism. Another important aspect of cellular primary metabolism is the incorporation of nitrogen into amino acids and ultimately proteins. In boreal forest soils the principal limiting nutrient is nitrogen and its deficiency often curtails the productivity of forest tree species (19).

Nitrogen is predominantly supplied to plant cell cultures in the form of a nitrate salt, although in the field it is available to most forest trees in a reduced form, namely ammonia (19). The use of ammonia as a sole source of nitrogen has not been successful with plant cell cultures. Sycamore (27) and poplar (64) cells supplied with ammonia fail to grow mainly because of a drop in medium pH following ammonia uptake (31). However since ammonia is the primary route through which nitrogen enters organic compounds it is necessary for the cells to effect its provision via the enzymes of nitrogen assimilation. This pathway comprises the enzymes nitrate reductase (EC 1.6.6.1), nitrite reductase (EC 1.6.6.4), glutamine synthetase (EC 6.3.1.2) and glutamate synthase (EC 1.4.1.13). With the exception of the glutamine synthetase step all the other reactions are reductive processes. Nitrate reductase and glutamate synthase are thought to utilize NADH as a source of reducing power generated by the reactions of glycolysis. Although the electron donor for nitrite reductase in non-photosynthetic cells has not been identified, there is strong evidence to suggest the involvement of NADPH. In chlorophyllous plant tissues it has been shown that reducing equivalents from NADPH are transferred to nitrite via the one-electron donor ferredoxin (40, 45, 62). It is likely therefore that a similar feature will be present in the green tissues of forest trees. However to date no work has been reported involving the use of photosynthetic cell suspension cultures and as yet no evidence has been found for the existence of ferredoxin in non-green cells (38, 67).

The source of reducing power for nitrate assimilation has been investigated by Jessup and Fowler (43) using non-photosynthetic sycamore cells. Cultures were grown on nitrate and glutamate, and the carbon flux through the pentose phosphate pathway and glycolysis was estimated. As mentioned above (Section 2.3.1.) the increased capacity of the pentose phosphate pathway in the presence of nitrate may be in response to a higher demand for reducing equivalents for the conversion of nitrite to ammonia. The NADP-malic enzyme (EC 1.1.1.40) is another potential source of reductant for nitrite reductase.

Upon formation of glutamate the amino nitrogen group may be transferred to the appropriate keto acid receptor to yield amino acids. These reactions are catalyzed by glutamate-pyruvate transaminase (EC 2.6.1.2) and glutamate-oxaloacetate transaminase (EC 2.6.1.1) to give serine and aspartate respectively. These amino acids may then undergo incorporation into proteins.

It was thought at one time that glutamate dehydrogenase (EC 1.4.1.3), which has both aminating and deaminating activities, was the route by which nitrogen entered organic compounds. Although the presence of glutamine synthetase in higher plants had been known for a number of years (93), it was not until the discovery of glutamate synthase (GOGAT) (30) that an alternative pathway of ammonia assimilation was proposed.

The operation of glutamine synthetase in the nitrate assimilation pathway has been verified by experiments involving the growth of cell suspension cultures on different nitrogen regimes and parallel measurements of associated enzyme activities and metabolite levels.

The patterns of in vivo activities of nitrate and nitrite reductases from batch sycamore suspension cultures differ greatly depending on the nitrogen source present in the medium (41). Although nitrate reductase activity is not generally detected in stationary phase cells, cultures rapidly develop nitrate reductase upon inoculation into fresh medium. Nitrite reductase activity follows a similar pattern except that there is often considerable activity of this enzyme at inoculation. However when the cells are grown on medium containing glutamate as sole nitrogen source the levels of both these reductases are very different. The level of nitrate reductase remains very low throughout the 14-day growth period and nitrite reductase activity slowly declines from its initial level at day 0 to undetectable levels by day 14.

Observations of Jessup (41) are in agreement with those of Young (106) who measured various enzyme activities and amino acid levels in sycamore suspensions in continuous culture. Under steady state conditions the levels of nitrate reductase remained constant, but the enzyme had the lowest level of activity of all the enzymes measured. From this observation Young (106) suggested that nitrate reduction is the rate limiting step in nitrate assimilation under nitrogen-limited conditions. The enzyme activities measured were sufficient to account for the rate of nitrogen assimilation and therefore the culture biomass doubled in response to a two-fold increase in medium nitrate. When nitrate was replaced by glutamate the specific activity of nitrate reductase remained reasonably constant. The activity levels of glutamine synthetase, glutamate-oxaloacetate transaminase and glutamate-pyruvate transaminase increased. In batch grown sycamore cells (41) all three enzymes maintain the same level of activity irrespective of nitrogen source, suggesting that the amount of glutamate synthesized via nitrate reductase and nitrite reductase is sufficient to meet metabolic requirements. However the observations reported from continuous cultures (106) suggest that the amount of glutamate available to glutamine synthetase is limited by the activity of either nitrate or nitrite reductase when nitrate is supplied as nitrogen source. The change in nitrogen source of the culture (106) also causes a change in catalytic activity of glutamate dehydrogenase resulting in an increase in aminating activity and a decrease in deaminating activity. This pattern is difficult to explain since it is not understood why the activity of the enzyme should be diverted towards glutamate formation when this substrate is the sole source of nitrogen. Fowler and Barker (27) have studied the levels of glutamate dehydrogenase (aminating) in sycamore cells grown in batch culture in the presence of nitrate and ammonia. At its peak glutamate dehydrogenase activity is 25-fold higher in ammonia + nitrate grown cells than in those grown on nitrate alone at the same period. The enzymes glutamine synthetase and glutamate synthase exhibit similar activity profiles during growth on nitrate plus ammonia. It is likely that the increase in glutamate dehydrogenase activity may be in response to high levels of ammonia. When nitrate grown cells are transferred to medium containing only ammonia no growth occurs and the enzyme activities decrease with time. The usual levels of ammonia present in plant tissues of 4-5 x 10^{-5}M (35) would be insufficient to support the activity of glutamate dehydrogenase, which has been reported to require an ammonia concentration of 0.1M for optimal activity.

Analysis of amino acid levels has revealed that when the nitrogen source is switched from nitrate to glutamate, there is a 6-fold increase in alanine content compared to an only 2-fold increase in aspartate levels (106). This may be due to either a higher level of glutamate-pyruvate transaminase activity under these conditions or to the ability of the cells to provide pyruvate as opposed to oxaloacetate for transmination. The incorporation of amino acids into more complex nitrogenous compounds in cultured sycamore cells has been reported by a number of workers (34, 82). In both cases the cells attained their maximum content of insoluble nitrogen during the first 3 days prior to or by the beginning of the exponential phase and decline after about a week, which would account for the fall in cellular nitrogen reported by Givan and Collin (34). These workers simultaneously measured the changes in respiration rate during the culture period and found that it corresponded closely with the increased synthesis of insoluble nitrogen compounds which were presumed to be mainly proteins. They suggested that oxygen consumption is closely linked to nitrogen metabolism through ADP and inorganic levels of enzymes and coenzymes.

Attempts have also been made to grow plant cell cultures on media containing organic nitrogen, mainly in the form of urea or as a mixture of amino acids. With sycamore cultures neither nutrient appears able to support growth as sole nitrogen source although their inclusion in nitrate-containing medium results in enhanced growth compared with cells growing on nitrate only (81). A similar phenomenon has been observed with cell suspension cultures of spruce (11). The analysis of data concerned with nitrogen metabolism from cell cultures supplied with urea or amino acids can be problematic because of conflicting information about the contribution of their carbon moieties to intermediary metabolism (31). Further systematic research in this area with cell cultures is required to elucidate the metabolism and ultimate fate of urea in plants. Urea-based fertilizers have been known to improve wood deposition in spruce trees (37). There are indications that urea can act as a precursor for a series of nitrogenous compounds (17, 18) such as biotin which is a cofactor for enzymes catalyzing the non-photosynthetic fixation of carbon dioxide (19). This process, which is widespread in plant cells (97), may supplement the level of TCA cycle intermediates used for amino acid synthesis. A requirement for CO_2 in the initiation of growth has been demonstrated in sycamore cell cultures (33).

2.4. Cell wall metabolism

A major product of plant cell metabolic activity is the cell wall, which has an added dimension in tree species through an enhanced skeletal/reinforcing function.

Early work by Lamport and Northcote (56, 57) suggested that the composition of the cambial cell walls of the intact sycamore tree corresponded closely to that of the cultured suspension cells. Available data suggest that the cell walls of cultured cells are composed of cellulose microfibrils embedded in a cross-linked matrix of xyloglucan, pectic polymer, arabinogalactans and hydroxyproline-rich glycoprotein (3, 6, 48, 94).

Studies by Lamport and Northcote (56, 57) identified pectin, hemicellulose and x-cellulose as major constituents of cultured plant cell walls. Darvill and colleagues (15) have reported that the primary cell walls of suspension cultured sycamore cells are composed of about 34% pectin, 24% hemicellulose, 23% cellulose and 19% hydroxyproline-rich glycoprotein. The pectic polymers are composed of neutral arabinan (10%), neutral galactan (8%) and acidic rhamnogalacturonan (16%) (94). Two different forms of rhamnogalacturonan, designated I and II, have been identified by various workers (14, 66). It is estimated that rhamnogalacturonan I contributes 7% to the mass of the cell wall and has a molecular weight of 200,000, while rhamnogalacturonan II contributes approximately 3-4% by weight. Both pectins are extremely complex molecules yielding ten different monosaccharides including L-rhamnosyl, D-galacturonosyl, L-arabinosyl and D-galactosyl residues. They also contain the rarely observed sugar apiose, 2-0-methylxylose and 2-0-methylfucose.

The initial stages of the biosynthetic pathway of pectic polymers have been characterized in cell suspensions of *A. pseudoplatanus*. The initial reaction is catalyzed by D-glucose-6-phosphate cycloaldolase and involves the cyclization of D-glucose-6-phosphate to 1-L-myo-inositol-1-phosphate (60, 61). Myo-inositol is then formed by hydrolysis of its monophosphate and undergoes a series of reactions culminating in the incorporation of pentoses into the pectic polymers of the cell wall.

A number of workers (3, 102) have suggested that hemicelluloses may have a structural function by inter-connecting the cellulose fibrils with the pectic polysaccharides of the wall via non-covalent and covalent bonds. The two hemicelluloses identified in the cell walls of sycamore cells in suspension are xyloglucan and glucuronoarabinoxylan which are estimated to contribute 19% and 5% of the cell wall respectively (16). They have been shown to possess many of the structural characteristics of analogous polysaccharides that have been isolated previously (15). The structure of xyloglucan is based on a repeating heptasaccharide unit consisting of four residues of glucose and three residues of xylose. Glucuroarabinoxylan, on the other hand, has a backbone with both acidic and neutral side chains attached along its length and terminating in glucuronosyl or 4-O-methylglucuronosyl and arabinosyl and/or xylosyl residues respectively.

A detailed investigation into the specific nature of plant cell walls during development was undertaken by Nevins and his co-workers (71). They found that there were changes in the polysaccharide composition during development of sycamore seedlings. Such changes were also observed in sycamore cell cultures grown on different carbon sources. Arabinose was the most abundant sugar released from the cell wall of cultured cells. When sugar levels of seedling cell walls were analyzed xylose was present in the highest concentration, whereas it is present in only small amounts in cultured cell walls.

Cells grown on various carbon sources exhibited the same pattern of sugar yield after acid hydrolysis of their walls. The only exception was with galactose grown cells, which contained larger amounts of this sugar. However the percentage yield of total sugars in the cell wall material as a whole varies markedly depending on the carbon source in the medium. For example when the cells were grown on sucrose, the hydrolyzed sugars accounted for 62% of the total wall material, whereas when the same cells were grown on mannose the percentage contribution was only 24%. The workers involved (71) suggest that in the cell wall the ratio of polysaccharides is more strictly controlled than their absolute amounts. It is reasonable to assume that *in vivo* the cells synthesize walls from a specific mixture of metabolites. This cellular control, however, can be influenced by the constituents of the culture medium. Rubery and Northcote (78) observed that 2,4-D increased the balance of incorporation of L-[^{14}C]-arabinose from 58% to 73% of total radioactivity taken up into the neutral sugars.

As mentioned previously cell walls of sycamore cells contain a large amount of hydroxyproline-rich protein. Of the total hydroxyproline of the cells nearly all is found in the cell wall, while less than 0.1% appears in hydrolysates of the cell contents (57). The hydroxyproline content varies slightly under different culture conditions and the protein-containing hydroxyproline is an integral part of the cell wall of actively growing cells. Sycamore cells have also been shown to secrete hydroxyproline-rich protein into the culture medium (68). One advantage of plant cell cultures is that they also secrete soluble non-cellulosic polysaccharides into the medium (2, 3, 4, 68). This allows rapid and effective isolation of polymeric material from the medium without the problem of accompanying and possibly interfering non-carbohydrate substances (4).

Many enzymes catalyzing a variety of reactions are located in the cell wall of living cells. *A. pseudoplatanus* cell wall preparations have been shown to contain the enzyme acid phosphatase (EC 3.1.3.2) (56) and several glycosidases (47). Keegstra and Albersheim (47) studied four of these enzymes, namely, β-glucosidase (EC 3.2.1.21), N-acetyl-β-glucoseaminidase (EC 3.2.1.30), α-galactosidase (EC 3.2.1.22) and β-galactosidase (EC 3.2.1.23). β-Glucosidase was located primarily on the outside of the cell wall, with only a small fraction of its activity detected inside the cell. Approximately 50% of the β-galactosidase activity was located on the outside of the cell wall, while the rest of the activity was cellular in origin but attributed to an isoenzyme. A degree of the α-galactosidase activity was also found in the cell wall. N-acetyl-β-glucosaminidase was only located in the cytoplasm. The activities of β-glucosidase and at least that of one of the β-galactosidases increased during the lag phase and decreased as the growth of the culture ceased. These findings and reports by other workers (46, 53, 58), suggest that glycosidases may be able to degrade plant cell walls and could potentially be involved in facilitating cell growth.

Copping and Street (12) suggested that invertase may also have a role in cell growth in addition to its usual function in sucrose hydrolysis. As the cells prepare to divide there is a rise in the specific activity of the cellular invertases, which continues for most of the period of active cell growth. The activity starts to decline only when the cells enter progressive deceleration and stationary phase.

No evidence for the presence of secondary thickening in cell suspension cultures of tree species has been found (9). However secondary thickening in sycamore, birch (Betual platyphylla), pine (Pinus ponderosa) and ash (Fraxinus elatos) trees has been shown to involve deposition of α-cellulose, hemicellulose and lignin (95). The amount of pectic substances per cell remains constant during formation of sapwood cells but in heartwood cells there is an almost complete absence of this fraction. However the cambial cells of each species are very similar. In a subsequent paper Thornber and Northcote (96) also reported that xylans formed during secondary thickening appeared to contain greater amounts of uronic anhydride than that deposited in the primary cell wall. They found that pectic substances are lost during the development of the cambial cells to heartwood in the angiosperms. However as with other features the nature of carbohydrates found in cambial cells differs little from species to species.

2.5. Secondary metabolism

The use of plant cell cultures as a tool to study primary metabolism has been extended in recent years to include the vast array of biosynthetic pathways culminating in secondary metabolites. However, this additional application of plant cell cultures is not without problems. Although some suspension cultures produce - albeit at a reduced yield - secondary products characteristic of the parent plant (7, 76) others fail to do so to any significant degree (69). Certain plant cell systems have however been reported to produce novel products not present in the plant from which they were derived (32, 87). Plant cell cultures nevertheless possess certain desirable characteristics for the study of secondary metabolism. These features are:

a) independence of cells from intricate inter-relations in the parent plant,

b) rapid uptake and metabolism of precursor substances, and,

c) control of nutrition and environmental conditions with the possibility of short term alterations in selected parameters.

The most extensively studied pathway of secondary metabolism in forest tree species is that of phenolic synthesis. In their work with sycamore cell cultures Carceller and her colleagues (9) observed the appearance of lignin-like substances in the culture medium and investigated the effect of sucrose and hormone levels on their production. These workers reported that lignin content of the cells and its level in the medium were enhanced when the initial sucrose concentration was increased from 1% to 15%. Raising the concentration of sucrose in the medium from 2% to 4% resulted in a three-fold stimulation of phenolic production in sycamore cells (101). This cooperative effect of the carbon source on phenolic accumulation is also observed when sucrose-depleted sycamore cultures in stationary phase are supplemented with 2% sucrose (74). The presence of lignin-containing cells has also been observed in suspension cultures of lodgepole pine (P. contorta) (101).

In a series of experiments Matsumoto and co-workers (65) investigated the effect of sucrose and other carbon sources on anthocyanin formation by cell cultures of a poplar hybrid (Populus maximowiczii x P. nigra). Increasing the sucrose concentration from 0.3% to 5% resulted in a 20-fold stimulation of anthocyanin production. Of the other carbon sources tested, glucose, fructose and raffinose were found to support anthocyanin formation, whereas lactose, galactose, sorbitol, maltose, trehalose, melezitose and starch were ineffective.

A close positive relationship between stimulation of lignin production and increased levels of auxin (2,4-D) and cytokinin have also been observed (9). No attempt was made however to compare the structure of cell culture lignin with that of the sycamore tree. Despite the production of lignin by sycamore cell suspensions they lack a clearly defined secondary wall and tracheary elements (9, 98). It may be argued that a certain degree of aggregation and cytodifferentiation would be necessary before such structures are synthesized. Carceller and co-workers (9) have observed that even the short-lived phase of sycamore aggregation was an important factor in triggering lignin biosynthesis.

In studies with plant cell cultures auxin and nitrogen levels have emerged as the two major media factors influencing secondary metabolism. Matsumoto and colleagues (65) reported that naphthalene acetic acid and indole acetic acid were both effective in stimulating anthocyanin formation in poplar cells. Unfortunately their studies did not extend to cover the relationship between auxin level and anthocyanin synthesis. In steady-state cultures of sycamore cells a reduction in the level of auxin (2,4-D) results in an enhancement of phenolic production (50). Similarly exhaustion of nitrogen from the medium of sycamore cells coincides with the onset of tannin production (101). The major class of phenolics produced by A. pseudoplatanus cells are condensed tannins consisting of flavonoid monomers (+)-catechin, (-)-epicatechin and leucocyanin. The negative effect of nitrogen metabolism on phenolic biosynthesis was further confirmed when the doubling of medium nitrogen delayed the production of phenolics by 6 days (101).

The cooperative and antagonistic effects of sucrose and nitrogen respectively on secondary metabolism might by explained in terms of competition for a common precursor (74). Presumably in the presence of a nitrogen source cultured cells would use carbon skeletons for protein synthesis. However, upon depletion of medium nitrogen and the consequent cessation of protein synthesis carbon precursors are possibly diverted towards secondary product formation.

In view of the importance of phenylalanine ammonia lyase (PAL) in the context of phenolic metabolism in cultures of forest tree species, it is surprising that this enzyme has not been studied more extensively. Phenylalanine ammonia lyase (EC 4.1.1.5) catalyzes the deamination of L-phenylalanine yielding ammonia and trans-cinnamic acid which then becomes the precursor for the biosynthesis of phenylpropanoid compounds in plants. This central position of PAL as a bridge between the primary and secondary domains of plant metabolism has made it the most extensively studied enzyme of secondary metabolism in recent years (8, 44). In cultured poplar cells the profile of PAL activity parallels that of anthocyanin synthesis although the latter lags two days behind the former (65). PAL is in turn affected by light which causes a two fold increase in the extractable activity of the enzyme. A series of parallel experiments with monochromatic light have indicated an effective promotion of anthocyanin production by red light. However, no clear data are provided on the induction by red light of PAL in poplar cells as reported for a PAL isoenzyme in mung bean seedlings (1). The photoregulation of PAL has been widely observed in a variety of plant tissues and the available evidence has been reviewed (107).

A positive correlation between PAL activity and production of phenolics has also been reported for sycamore cells (101). The delaying effect of medium nitrogen on phenolic production, discussed earlier, was also observed for the peak of PAL activity which appeared six days later when the medium nitrogen content was doubled. Westcott and Henshaw (101) also reported that the extractable activity of PAL even at reduced level following its peak was sufficient to account for the rate of phenolic biosynthesis. The apparent interrelationship between nitrogen metabolism and carbon utilization for the biosynthesis of secondary products raises the question of regulatory mechanisms involved between PAL activity and phenolic production on the one hand and protein biosynthesis on the other and warrants further investigation.

3. CONCLUSION

While the literature on the physiology of cell cultures of tree species is relatively sparse, it is obvious from the work discussed in this review that much is to be learned in this area from plant cells grown in culture. Although tissue cultures are better suited for regeneration and propagation of forest tree species than cell cultures, the potential of cell cultures is of a different nature and may provide much useful information which later may be used to advantage.

4. REFERENCES

1. AHMED SI, T SWAIN 1970 The effect of light on the activity of enzymes of the aromatic pathway in peas and mung bean. Phytochemistry 9: 2287-2290
2. ASPINALL GO, JA MOLLOY, JWT CRAIG 1969 Extracellular polysaccharides from suspension-cultured sycamore cells. Can J Biochem 47: 1063-1070
3. BAUER WD, KW TALMADGE, K KEEGSTRA, P ALBERSHEIM 1973 The structure of plant cell walls. II. The hemicellulose of the walls of suspension-cultured sycamore cells. Plant Physiol 51: 174-187
4. BECKER GE, PA HUI, P ALBERSHEIM 1964 Synthesis of extra-cellular polysaccharides by suspensions of <u>Acer pseudoplatanus</u> cells. Plant Physiol 39: 913-920
5. BROWN EG, KC SHORT 1969 The changing nucleotide pattern of sycamore cells during culture in suspension. Phytochemistry 8: 1365-1372
6. BURKE D, P KAUFMAN, M MCNEIL, P ALBERSHEIM 1974 The structure of plant cell walls. V1. A survey of the walls of suspension-cultured monocots. Plant Physiol 54: 109-115

7. BUTCHER DN 1977 Secondary products in tissue culture. In J. Reinert, YPS Bajaj, eds, Applied and Fundamental Aspects of Plant Cell Tissue and Organ Culture. Springer-Verlag, Berlin, pp 668-693
8. CAMM EL, GHN TOWERS 1973 Phenylalanine ammonia lyase. Phytochemistry 12: 961-973
9. CARCELLER M, MR DAVEY, MW FOWLER, HE STREET 1971 The influence of sucrose, 2,4-D, and kinetin on the growth, fine structure, and lignin content of cultured sycamore cells. Protoplasma 73: 367-385
10. CHAFE SC, DJ DURZAN 1973 Tannin inclusions in cell suspension cultures of white spruce. Planta 113: 251-262
11. CHALUPA V, DJ DURZAN 1973 Growth of Norway Spruce [*Picea abies* (L.) Karst] tissue and cell cultures. Commun Inst For Cech 8: 111-125
12. COPPING LG, HE STREET 1972 Properties of the invertase of cultured sycamore cells and changes in their activity during culture growth. Physiol Plant 26: 346-354
13. COX BJ, G TURNOCK, HE STREET 1973 Studies on the growth in culture of plant cells. XV. Uptake and utilization of uridine during the growth of *Acer pseudoplatanus* L. cells in suspension culture. J Exp Bot 24: 159-174
14. DARVILL AG, M MCNEIL, P ALBERSHEIM 1978 Structure of plant cell walls. VIII. A new pectic polysaccharide. Plant Physiol 62: 418-422
15. DARVILL AG, M MCNEIL, P ALBERSHEIM, DD DELMER 1980 The primary cell walls of flowering plants. In NE Talbert, ed, The Plant Cell. Academic Press, New York, pp 91-162
16. DARVILL JE, M MCNEIL, AG DARVILL, P ALBERSHEIM 1980 Structure of plant cell walls. XI. Glucoronoarabinoxylan, a second hemicellulose in the primary cell walls of suspension-cultured sycamore cells. Plant Physiol 66: 1135-1139
17. DURZAN DJ 1973 Nitrogen metabolism of *Picea glauca*. V. Metabolism of uniformly labelled ^{14}C-L-proline and ^{14}C-L-glutamine by dormant buds in late fall. Can J Bot 51: 359-369
18. DURZAN DJ 1973 The incorporation of tritiated water into amino acids in the presence of urea by white spruce seedlings in light and darkness. Can J Bot 51: 351-358
19. DURZAN DJ 1976 Biochemical changes during gymnosperm development. Acta Hortic 56: 183-194
20. DURZAN DJ, SC CHAFE, SM LOPUSHANSKI 1973 Effects of environmental changes on sugars, tannins and organized growth in cell suspension cultures of white spruce. Planta 113: 241-249
21. DURZAN DJ, V CHALUPA 1976 Growth and metabolism of cells and tissue of jack pine (*Pinus banksiana*). 3. Growth of cells in liquid suspension cultures in light and darkness. Can J Bot 54: 456-467
22. DURZAN DJ, V CHALUPA 1976 Growth and metabolism of cells and tissue of jack pine (*Pinus banksiana*. 6. Free nitrogenous compounds in cell suspension cultures of jack pine as affected by light and darkness. Can J Bot 54: 496-506

23. DURZAN DJ, FC STEWARD 1971 Morphogenesis in cell cultures of gymnosperms: some growth patterns. In Workshop on Sexual Reproduction in Forest Trees. IUFRO, Varparanta (Finland), pp 16-17
24. FOWLER MW 1971 Studies on the growth in culture of plant cells. XIV. Carbohydrate oxidation during the growth of Acer pseudoplatanus L. cells in suspension culture. J Exp Bot 22: 715-721
25. FOWLER MW 1972 The pathways of carbohydrate oxidation in growing and dividing cells of Acer pseudoplatanus L. (sycamore) grown in batch culture. FEBS Symp 24: 169-173
26. FOWLER MW 1978 Regulation of carbohydrate metabolism in cell suspension cultures. In TA Thorpe, ed, Frontiers of Plant Tissue Culture 1978. Int Assoc Plant Tissue Cult, Calgary, pp 443-452
27. FOWLER MW, RJD BARKER 1979 Assimilation of ammonia in non-chlorophyllous tissue. In EJ Hewitt, CV Cutting, eds, Nitrogen Assimilation of Plants. Academic Press, London, pp 484-500
28. FOWLER MW, A CLIFTON 1974 Activities of enzymes of carbohydrate metabolism in cells of Acer pseudoplatanus L. maintained in continuous (chemostat) culture. Eur J Biochem 45: 445-450
29. FOWLER MW, A CLIFTON 1975 Hexokinase activity in cultured sycamore cells. New Phytol 75: 533-538
30. FOWLER MW, W JESSUP, GS SARKISSIAN 1974 Glutamate synthetase type activity in higher plants. FEBS Lett 46: 340-342
31. FOWLER MW, R WATSON, I LYONS 1982 Substrate utilization, carbon and nitrogen, by suspension cultured plant cells. In A Fujiwara, ed, Plant Tissue Culture 1982. Proc 5th Int Cong Plant Tissue Cell Cult, Tokyo, pp 225-228
32. GAMBORG OL, F CONSTABEL, TAG LA RUE, RA MILLER, W STECK 1971 The influence of hormones on secondary metabolite formation in plant cell cultures. Colloq Int CNRS 193: 335
33. GATHERCOLE RWE, KJ MANSFIELD, HE STREET 1976 Carbon dioxide as an essential requirement for cultured sycamore cells. Physiol Plant 37: 213-217
34. GIVAN CV, HA COLLIN 1967 Studies on the growth in culture of plant cells. II. Changes in respiration rate and nitrogen content associated with the growth of Acer pseudoplatanus cells in suspension culture. J Exp Bot 18: 321-331
35. GOOD NE 1960 Activation of the Hill reaction by amines. Biochim Biophys Acta 40: 502-517
36. GROUT BWW, KW CHAN, I SIMPKIN 1976 Aspects of growth and metabolism in a suspension culture of Acer pseudoplatanus (L.) grown on a glycerol carbon source. J Exp Bot 27: 77-86
37. HAGNER S, J SORASTER, B JOHANSSON, A AHGREN 1966 Timber production by forest fertilization. Sver Skogsvardsfoerb Tidskr, Halfe 2. Swedish Cellulose Company, pp 171
38. HASLETT BG, R CAMMACK, FR WHATLEY 1973 Quantitative studies on ferredoxin in greening bean leaves. Biochem J 136: 697-703

39. HENSHAW GG, KK JHA, AR MEHTA, OJ SHAKESHAFT, HE STREET 1966 Studies on the growth in culture of plant cells. I. Growth patterns in batch propagated suspension culture. J Exp Bot 17: 362-377
40. HEWITT EJ, GF BETTS 1963 The reduction of nitrate and hydroxylamine by ferredoxin and chloroplast grana from Cucurbita pepo. Biochem J 89: 20P
41. JESSUP W 1977 Nitrate assimilation and carbohydrate oxidation in cultured plant cells. PhD Thesis, Univ Sheffield, UK pp 46-64
42. JESSUP W, MW FOWLER 1976 Interrelationship between carbohydrate metabolism and nitrogen assimilation in cultured plant cells. II. Effect of the nitrogen source and concentration on nutrient uptake and respiratory activity in cultured sycamore cells. Planta 132: 125-129
43. JESSUP W, MW FOWLER 1977 Interrelationships between carbohydrate metabolism and nitrogen assimilation in cultured plant cells. III. Effects of the nitrogen source on the pattern of carbohydrate oxidation in cells of Acer pseudoplatanus L. grown in culture. Planta 137: 71-76
44. JONES DH 1984 Phenylalanine ammonia-lyase : regulation of its induction, and its role in plant development. Phytochemistry 23: 1349-1359
45. JOY KW, RH HAGEMAN 1966 The purification and properties of nitrate reductase from higher plants and its dependence on ferredoxin. Biochem J 100: 263-273
46. KATZ M, L ORDIN 1967 A cell wall polysaccharide hydrolyzing enzyme system in Avena sativa L. coleoptiles. Biochim Biophys Acta 141: 126-134
47. KEEGSTRA K, P ALBERSHEIM 1970 The involvement of glycosidases in the cell wall metabolism of suspension-cultured Acer pseudoplatanus cells. Plant Physiol 45: 675-678
48. KEEGSTRA K, KW TALMADGE, WD BAUER, P ALBERSHEIM 1973 The structure of plant cell walls. III. A model of the walls of suspension-cultured sycamore cells based on the interconnections of the macromolecular components. Plant Physiol 51: 188-197
49. KHANNA P, EJ STABA 1968 Antimicrobials from plant tissue cultures. Lloydia 31: 180-189
50. KING PJ 1976 Studies on the growth in culture of plant cells. XX. Utilization of 2,4-dichlorophenoxyacetic acid by steady-state cell cultures of Acer pseudoplatanus L. J Exp Bot 27: 1053-1072
51. KING PJ, KJ MANSFIELD, HE STREET 1973 Control of growth and cell division in plant cell suspension cultures. Can J Bot 51: 1807-1823
52. KING PJ, HE STREET 1977 Growth patterns in cell cultures. In HE Street, ed, Plant Tissue and Cell Cultures, 2nd ed. Blackwell Scientific Publications, Oxford, pp 307-387
53. KIVILAAN A, S LEE 1966 Autolysis of plant cell walls in vitro. Plant Physiol 41: lxix - lxx
54. KREBS HA, LV EGGLESTON 1974 The regulation of the pentose phosphate cycle in rat liver. Adv Enz Regul 12: 421-434

55. LAMPORT DTA 1964 Cell suspension cultures of higher plants: isolation and growth energetics. Exp Cell Res 33: 195-206
56. LAMPORT DTA, DH NORTHCOTE 1960 The use of tissue cultures for the study of plant-cell walls. Biochem J 76: 52P
57. LAMPORT DTA, DH NORTHCOTE 1960 Hydroxyproline in primary cell walls of higher plants. Nature 188: 665-666
58. LEE S, A KIVILAAN, RS BANDURSKI 1967 In vitro autolysis of plant cell walls. Plant Physiol 42: 968-972
59. LEGUAY J-J, J GUERN 1975 Quantitative effects of 2,4-dichlorophenoxyacetic acid on growth of suspension-cultured Acer pseudoplatanus cells. Plant Physiol 56: 356-359
60. LOEWUS MW, F LOEWUS 1971 The isolation and characterization of D-glucose-6-phosphate cycloaldolase (NAD-dependent) from Acer pseudoplatanus L. cell cultures. Plant Physiol 48: 255-260
61. LOEWUS MW, F LOEWUS 1973 D-Glucose 6-phosphate cycloaldolase : inhibition studies and aldolase function. Plant Physiol 51: 263-266
62. LOSADA M, A PANEQUE, JM RAMIREZ, FF DEL CAMPO 1963 Mechanism of nitrite reductase in chloroplast. Biochem Biophys Res Comm 10: 298-303
63. MARETZKI A, M THOM, LG NICKELL 1974 Utilization and metabolism of carbohydrates in cell and callus cultures. In HE Street, ed, Tissue Culture and Plant Science 1974. Academic Press, London, pp 329-361
64. MATSUMOTO T, K OKUNISHI, K NISHIDA, M NOGUCHI, E TAMAKI 1971 Studies on the culture conditions of higher plant cells in suspension culture. Part II. Effect of nutritional factors on the growth. Agr Biol Chem 35: 543-551
65. MATSUMOTO T, K NISHIDA, M NOGUCHI, E TANAKI 1973 Some factors affecting the anthocyanin formation by Populus cells in suspension culture. Agr Biol Chem 37: 561-567
66. MCNEIL M, AG DARVILL, P ALBERSHEIM 1980 Structure of plant cell walls. X. Rhamnogalacturonan I, a structurally complex pectic polysaccharide in the walls of suspension-cultured sycamore cells. Plant Physiol 66: 1128-1134
67. MELANDRI BA, A BACCARINI, G FORTI 1967 Light-induced formation of ferredoxin in bean leaves. Physiol Veg 5: 337-339
68. MORRIS MR, DH NORTHCOTE 1977 Influence of cations at the plasma membrane in controlling polysaccharide secretion from sycamore suspension cells. Biochem J 166: 603-618
69. MORRIS P, MW FOWLER 1980 Growth and alkaloid content of cell suspenison cultures of Papaver somniferum. Planta Med 39: 284-285
70. MURASHIGE T, F SKOOG 1962 A revised medium for rapid growth and bioassays with tobacco tissue culture. Physiol Plant 15: 473-497
71. NEVINS DJ, PD ENGLISH, P ALBERSHEIM 1967 The specific nature of plant cell wall polysaccharides. Plant Physiol 42: 900-906

72. PARHAM RA, HM KAUSTINEN 1976 Differential staining of tannin in sections of epoxy-embodied plant cells. Stain Technol 51: 237-240

73. PARHAM RA, HM KAUSTINEN 1977 On the site of tannin synthesis in plant cells. Bot Gaz 138: 465-467

74. PHILLIPS R, GG HENSHAW 1977 The regulation of synthesis of phenolics in stationary phase cell cultures of Acer pseudoplatanus L. J Exp Bot 28: 785-794

75. REINERT J. 1956 Dissociation of cultures from Picea glauca into small tissue fragments and single cells. Science 123: 457-458

76. REINHARD E, G CORDUAN, OH VOLK 1968 Uber Gewebekulturen von Ruta graveolens. Planta Med 16: 8-16

77. RICARDO CPP, T APREES, WA FULLER 1972 Effects of sugars on invertase activity of carrot cells. Phytochemistry 11: 2435-2436

78. RUBERY PH, DH NORTHCOTE 1970 The effect of auxin (2,4-dichlorophenoxyacetic acid) on the synthesis of cell wall polysaccharides in cultured sycamore cells. Biochim Biophys Acta 222: 95-108

79. SCALA J, FE SEMERSKY 1971 An induced fructose-1,6-diphosphatase from cultured cells of Acer pseudoplatanus (English sycamore). Phytochemistry 10: 567-570

80. SHIMIZU T, A CLIFTON, A KOMAMINE, MW FOWLER 1977 Changes in metabolite levels during growth of Acer pseudoplatanus (sycamore) cells in batch suspension culture. Physiol Plant 40: 125-129

81. SIMPKINS I, HA COLLIN, HE STREET 1970 The growth of Acer pseudoplatanus cells in a synthetic liquid medium : response to the carbohydrate, nitrogenous and growth hormone constituents. Physiol Plant 23: 385-396

82. SIMPKINS I, HE STREET 1970 Studies on the growth in culture of plant cells. VII. Effects of kinetin on the carbohydrate and nitrogen metabolism of Acer pseudoplatanus, L. cells grown in suspension culture. J Exp Bot 21: 170-185

83. STAFFORD HA, TY CHENG 1980 The procyanidins of Douglas-fir seedlings, callus and cell suspension cultures derived from cotyledons. Phytochemistry 19: 131-135

84. STAFFORD HA, HH LESTER 1980 Procyanidins (condensed tannins) in green cell suspension cultures of Douglas-fir compared with those in strawberry and avocado leaves by means of C_{18}-reversed-phase chromatography. Plant Physiol 66: 1085-1090

85. STAFFORD HA, HH LESTER 1981 Proanthocyanidins and potential precursors in needles of Douglas-fir and in cell suspension cultures derived from seedling shoot tissue. Plant Physiol 68: 1035-1040

86. STAFFORD HA, M SHIMAMOTO, HH LESTER 1982 Incorporation of [^{14}C]phenylalanine into flavan-3-ols and procyanidins in cell suspension cultures of Douglas-fir. Plant Physiol 69: 1055-1059

87. STECK W, BK BAILEY, JP SHYLUK, OL GAMBORG 1971 Coumarins and alkaloids from cell cultures of Ruta graveolens. Phytochemistry 10: 191-198

88. STEPAN-SARKISSIAN G, MW FOWLER 1986 The metabolism and utilization of carbohydrates by suspension cultures of plant cells. In M. Morgan, ed, Carbohydrate Metabolism in Cultured Cells. Plenum Publishing Company, London, (in press)
89. STRAUS J 1962 Invertase in cell walls of plant tissue cultures. Plant Physiol 37: 342-348
90. STREET HE 1966 The nutrition and metabolism of plant tissue and organ cultures. In EN Willmer, ed, Cells and Tissues in Culture. Academic Press, London, 3: 533-629
91. STREET HE 1977 Cell (suspension) cultures - techniques. In HE Street, ed, Plant Tissue and Cell Cultures, 2nd edition. Blackwell Scientific Publications, Oxford, pp 61-102
92. STUART R, HE STREET 1969 Studies on the growth in culture of plant cells. IV. The initiation of division in suspensions of stationary phase cells of Acer pseudoplatanus, L. J Exp Bot 20: 556-571
93. STUMPF PK, WD LOOMIS 1950 Observations of a plant amide system requiring manganese and phosphate. Arch Biochem 25: 451-453
94. TALMADGE KW, K KEEGSTRA, WD BAUER, P ALBERSHEIM 1973 The structure of plant cell walls. I. The macromolecular components of the walls of suspension-cultured sycamore cells with a detailed analysis of the pectic polysaccharides. Plant Physiol 51: 158-173
95. THORNBER JP, DH NORTHCOTE 1961 Changes in the chemical composition of a cambial cell during its differentiation into xylem and phloem tissue in trees. 1. Main components. Biochem J 81: 449-455
96. THORNBER JP, DH NORTHCOTE 1961 Changes in the chemical composition of a cambial cell during its differentiation into xylem and phloem tissue in trees. 2. Carbohydrate constituents of each main component. Biochem J 81: 455-464
97. TING IP 1971 Non-autotrophic CO_2 fixation and crassulacean acid metabolism. In MD Hatch, CB Osmund, RO Slayter, eds. Photosynthesis and Photorespiration. John Wiley & Sons Inc., New York, pp 169-183
98. TORREY JG 1968 Hormonal control of cytodifferentiation in agar and cell suspension cultures. In F Wightman, G Setterfield, eds. Biochemistry and Physiology of Plant Growth Substances. Range Press Ltd , Ottawa, pp 843
99. VERMA DC, J TAVARES, FA LOEWUS 1976 Effect of benzyladenine, 2,4-dichlorophenoxyacetic acid, and D-glucose on myo-inositol metabolism in Acer pseudoplatanus L. cells grown in suspension cultures. Plant Physiol 57: 241-244
100. WEBB KJ 1981 Growth and cytodifferentiation in cell suspension cultures of Pinus contorta. In Colloque International sur la Culture 'in vitro' des Essences Forestieres, IUFRO, International Workshop, Fontainbleau, France. AFOCEL, Nangis, France. pp 217-224
101. WESTCOTT RJ, GG HENSHAW 1976 Phenolic synthesis and phenylalanine ammonia-lyase activity in suspension cultures of Acer pseudoplatanus L. Planta 131: 67-73

102. WILDER BM, P ALBERSHEIM 1973 The structure of plant cell walls. IV. A structural comparison of the wall hemicellulose of cell suspension cultures of sycamore (Acer pseudoplatanus) and of red kidney bean (Phaseolus vulgaris). Plant Physiol 51: 889-893
103. WILSON G 1980 Continuous culture of plant cells using the chemostat principle. Adv Biochem Eng 16: 1-25
104. WILSON SB 1971 Studies of the growth in culture of plant cells. XIII. Properties of mitochondria isolated from batch cultures of Acer pseudoplatanus cells. J Exp Bot 22: 725-34
105. WINTON LL 1972 Callus and cell cultures of Douglas-fir. For Sci 18: 151-154
106. YOUNG M 1973 Studies on the growth in culture of plant cells. XVI. Nitrogen assimilation during nitrogen-limited growth of Acer pseudoplatanus L. cells in chemostat culture. J Exp Bot 24: 1172-85
107. ZUCKER M 1972 Light and enzymes. Ann Rev Plant Physiol 23: 133-156

5. SOMATIC EMBRYOGENESIS IN WOODY PERENNIALS

W TULECKE

1. INTRODUCTION

1.1. Extent of this review

Since the first reported induction of somatic embryogenesis in carrot cell cultures more than 25 years ago (131, 151), there has been significant progress and numerous reviews (3-6, 8, 17, 25, 35, 40, 41, 48, 51, 98, 103, 123, 126, 128, 135, 145, 166, 167). Consequently, this review will emphasize the more recent work on woody perennials, including conifers and woody monocotyledonous plants. Reviewed elsewhere in this series are somatic embryogenesis in endosperm (Lakshmi Sita), haploids (Rohr, Chen), and the role of ammonia in somatic embryogenesis (Durzan).

1.2. Definition of terms

Somatic embryogenesis has been defined clearly by Haccius (46) as a non-sexual developmental process which produces a bipolar embryo from somatic tissue. Developmental stages similar to normal embryogenesis occur and yield an embryo with no vascular connection to the parent tissue. Sharp *et al*. (136) distinguished two types of somatic embryogenesis, direct and indirect. Direct somatic embryogenesis refers to the development of an embryo directly from the original explant tissue. Indirect somatic embryogenesis is the formation of embryos from callus or cell suspension, or from cells or groups of cells of somatic embryos. The latter process is called "repetitive somatic embryogenesis" (3, 4). Embryos derived in this manner are sometimes described as adventive or secondary. Embryos from tissues such as the megagametophyte of conifers would be formed by the process of gametophytic embryogenesis. Any cells which can develop into somatic embryos are said to possess embryogenic competence. Whether these cells are target cells which respond to special signals or whether most cells have this capability is not yet known. The selection of specific developmental stages of explant material, conditioning media, sequential transfers and appropriate environmental conditions are generally necessary for successful embryogenesis. The primary events required for cells to enter the developmental program of embryogenesis are unknown, but the techniques are available to begin to address this question.

1.3. Woody perennials

The introduction of somatic embryogenesis in woody perennials dates from the early work of LaRue (82) and research with *Biota orientalis* Engl. (69) and *Zamia integrifolia* Ait. (105). Somatic embryogenesis has since been induced in more than 25 families, 44 genera, 60 species and numerous cultivars of trees (Table 1). Among characteristics of woody plants which make them more intractable for studies of somatic embryogenesis are: 1) the short seasonal period of time when any particular tissue or stage of development is available for culture, 2) the long period required for regeneration, 3) the frequent production of phenolic compounds in browning reactions, and 4) the long term committment required for productive research. A singular advantage, however, is the long life of the individual plant. The potential for propagating new germplasm from protoplast fusion products, embryogenic lines derived from wide crosses, somatic variants, and bioengineered creations might well be justified in forest crops (96) orchard trees or other woody species.

2. METHODS

2.1. Explant source

Initiating the developmental program for somatic embryogenesis from a cell or a group of cells frequently depends on the nature of the explant source. This refers to the conditions under which the source plant was grown and the stage of development of the plant part from which the explant was taken. Juvenile tissues of certain types appear to be the most suitable for induction of somatic embryogenesis. Mullins and Srinivasan (97), who first reported somatic embryogenesis from the nucellus of grape, referred to that tissue as an island of juvenility. The characteristics of juvenile and adult tissues have been discussed by Bonga (16, 18) and Durzan (33).

The integrated physiology of the whole plant controls the embryogenic potential of each cell in the plant. Cells with embryogenic potential may be separated from neighboring cells by a thick cell wall (157) or culture conditions might favor the development of certain "inducible cells" (77) or limit the growth of other cells. Furthermore, metabolic products, produced in vitro, might stimulate embryogenesis but limit the differentiation of other cells. Among the tissues used to induce somatic embryos are the cotyledons, ovule, leaf, megagametophyte, embryo, hypocotyl, cell suspensions (Table 1), as well as anthers, pollen, and endosperm.

2.2. General requirements

The induction of somatic embryos depends on selecting the optimum stage of development of the appropriate tissue, using an effective conditioning medium and providing suitable environmental conditions. Thus, after transfer from the conditioning to the inductive medium, embryogenic centers may develop and these can be used to develop an embryogenic line. Alternatively, callus cultures may be directed to embryogenesis by the selection of usually nodular, densely cytoplasmic, small-celled tissue masses and transferring these to inductive media. On the inducing medium embryogenesis generally progresses from the globular to the torpedo, cotyledonary and mature somatic embryo stages of development. For later stages of embryo development (cotyledonary and mature) another medium may sometimes be required.

2.3. Quantitation

Quantitation methods have been difficult to apply to embryogenic systems because of the lack of homogeneity of cell populations or of stages of development. Controlling the cell cycle, sieving and collecting clusters of cells of certain size classes (131) and using abscisic acid (ABA) to accumulate mature ungerminated somatic embryos (2) are steps toward effective quantitation. In addition, better quantitation of the following is required: the frequency of induction from the original explant, the frequency of occurrence in an embryogenic line, the criteria used for selection, the rate of successful movement of somatic embryos through their development and including germination, plantlet growth, transplanting to soil, acclimatization, and growth in the field.

2.4. Media requirements

The nutritional requirements for somatic embryogenesis are not well understood. They are neither specific nor exclusive, since various recipes produce results. The most commonly used medium is that of Murashige and Skoog (99), either in its original form or modified, but other media work as well or better for specific tissues (30). The conditioning medium usually contains plant growth substances and is understood to generate the embryogenic potential of some cells in the explant or the tissue culture. The inductive medium is often a simple basal medium lacking growth substances and provides for the expression of embryogenic potential. The two main components of conditioning media are growth regulators such as 2,4-dichlorophenoxyacetic acid (2,4-D) and benzylaminopurine (BAP), and reduced nitrogen sources such as ammonium or l-glutamine (see Durzan, Kirby, this publication). Carbohydrates (22, 62, 64, 65, 67) and osmoticum are useful in some embryogenic systems. The possible role of other nutrients has been reviewed by Kohlenback (68). Work with Liquidambar (137) showed that with this species no clear hormone combination was required for embryogenesis, while with Pterocarya (162) a single basal medium was adequate for conditioning and induction of embryogenesis. The transfer of the primary explant from the conditioning medium to the simpler basal medium is a general procedure that is followed to assure maturation and to prevent abnormal development of the somatic embryos. The timing of this event appears to be important, even though the role of media sequences in the formation of somatic embryos is not known.

2.5. Media effects

Many studies have been concerned with the supplementation of basal media with substances for conditioning, for induction, or for growth and development of somatic embryos. Many of these have been discussed in the reviews mentioned earlier. Only a few examples will be given here. In one report, Kononowicz and Janick

(71) showed that the kind and amount of carbohydrate influenced the rate of embryogensis in cotyledon tissue of Theobroma cacao. Other studies showed that high sucrose increased the concentration of anthocyanin (115, 121), alkaloids (110) and fatty acids (116, 170). These substances are formed in normal zygotic embryos and thus indicate normal maturation in somatic embryos. Cotyledons of somatic embryos grown on glucose media contained more triglycerides than those grown on sucrose media, which contained more phospho- and glycolipids. The studies of Ben Hayyim and Neumann (12) on Citrus showed a greater conversion of nucellar callus to somatic embryos on glycerol media (280 mM) than on sucrose (70 mM). On glycerol media most of the callus was converted to somatic embryos which were green. In other studies (67) galactose and lactose stimulated embryogenesis in Citrus 6-12 fold over media with sucrose.

2.6. Inhibitors and promoters

Substances which stimulate or inhibit cell growth, cell differentiation or gene function have been reported to occur in the tissues of somatic embryos as well as in their culture environment (159, 160). Among the substance reported are ethylene, ethanol, gibberellic acid (GA), auxin, phenolics, tannins, polyamines, ABA and carbon dioxide. Substances such as dimethylamino succinamic acid (alar), ethephon, ABA (63) and inhibitors of auxin synthesis (63) have been reported to stimulate embryogenesis. Additional factors which were correlated with increased embryogenesis were low levels of sucrose (62, 68), irradiation (143, 164), or charcoal (86) and the absence of ageing (62).

2.7. Other factors

Certain physical and environmental conditions may be critical to the process of embryogenesis in some systems. The variables include the rate of gas exchange in a culture vessel (ethylene buildup or dissipation), the volume of medium per explant (dilution and conditioning of the medium), light quality, quantity and duration, the periods and interruption of dark, the frequency of transfer (riddance of phenolics and inhibitors, renewal of nutrients), the selection of tissues (nodular, color, or other criteria), temperature, and characteristics of the explant. For example, prolonged culture in the dark was required for embryogeneis in Coffea (140) but light was better than dark for Theobroma (70). Other conditions which may be important are the effect of cold storage, media sequences and gradients.

3. SOMATIC EMBRYO DEVELOPMENT

3.1. Origin and development

The question of the origin of somatic embryos has been under discussion ever since Steward et al. (151) suggested that single isolated cells of carrot express a totipotency which mimicks zygote ontogeny during somatic embryogenesis. Street (152) likewise emphasized the single cell origin of somatic embryos and Kato (59) demonstrated it with serial photographs. For woody perennials there have reports for the single cell origin of somatic embryos from Citrus (24), single epidermal cells of Anacardium occidentale (118), Paulownia (121), Phoenix dactylifera (157), the spongy mesophyll of Coffea (141), from cotyledon tissue of Coffea (141), Anacardium (118) and Juglans (162) and stem callus of Vitis (76, Fig. 4). Histological and developmental studies of the origin of somatic embryos suggest that more than one pattern of cellular organization can lead to normal somatic embryogenesis. Proembryonal cell complexes (10) suggest a multicellular origin of at least some somatic embryos as compared to the single cell origin mentioned above. Ontogenetic studies, including scanning electron microscope photographs, of cell clusters of Coffea (150), Ilex (52) and Citrus (40) showed them to be the origin of globular embryos. The cell clusters consisted of cells which were small, densely cytoplasmic, rich in ribosomes, and contained numerous starch grains. About 50-100 somatic embryos were formed per culture from callus tissue from leaf mesophyll which was grown for 70 days on a conditioning medium and 40 days on a basal medium. Complete plants were grown from these somatic embryos. In the case of Citrus (23) the nucellar callus tissue formed nodules, proembryos, heart-shaped and cotyledonary embryos. At any of these stages of development additional nodules were formed from the hypocotyl and developed into proembryos. In callus cultures of Phoenix dactylifera (157), the early two-celled somatic embryos were separated from the surrounding tissue by a thick wall. This development took place in nodular white callus but starch grain accumulation was not a feature of any stage of embryo development.

Adventive somatic embryos have been reported to occur from the cotyledons, roots, and hypocotyls of somatic embryos, as shown for Carica (Fig. 2), Mangifera (Fig. 5) and Juglans (Fig. 7 and 8). Pence et al. (114, 115) described the development of somatic embryos from the internal tissue of cotyledons and from the tip of glandular hairlike structures of Theobroma cacao. Other work with Picea abies (47, 75), Pinus lambertiana (45), Pseudotsuga menziesii (45) and Larix decidua (101, Fig. 3) indicates an origin of somatic embryos from suspensor or suspensorlike cells. These studies call attention to the importance of the physical dimensions of the embryogenic process, including gradients and electrical polarity (21).

The importance of the "position effect" on embryogenesis is illustrated by the results of Stamp and Henshaw (148, 149) with Manihot esculenta Crantz (cassava) cultures. Only the adaxial surface of the cotyledon explants were embryogenically competent and formed somatic embryos. The stage of development and the medium were critical in the embryogenic process, which was described as resembling cleavage polyembryony. Abnormalities were present, but sixty percent of the somatic embryos developed into normal plants. The total time from an explant to a plant in the greenhouse was 140-180 days. Stamp and Henshaw noted that the larger more well-developed somatic embryos were more likely to develop into plants.

3.2. Protoplasts and cell suspensions

Only a few examples of somatic embryogenesis from protoplasts or from cell suspensions are reported for woody perennials. Work with Citrus (164) demonstrated an enhancement of embryogenesis when protoplasts were irradiated with x-rays and then plated at high densities. The surviving cells served as feeders for the mitotically competent embryogenic cells. Protoplasts were obtained from the nucellus of eight Citrus species (165) and plants were regenerated from each. The Citrus system appears reliable and potentially useful for protoplast fusion and for the regeneration of plants with novel germplasm (144).

Experiments with Santalum (131) showed that protoplasts isolated from 3-year-old callus derived from a 20-year-old tree, could be induced to form embryogenic cell aggregates, globular and cotyledonary somatic embryos (Fig. 1) and plantlets. The somatic embryos were numerous, well developed and formed adventive embryos. The callus was grown as a cell suspension and transferred weekly. Small cell clusters were repeatedly selected, finally yielding a rapidly growing culture. Every 4-5 days, 6-8 ml of cell suspension was transferred to 14 ml of fresh medium in a 100 ml Erlenmeyer flask. This method gave nutrient renewal, maximum aeration, and a high cell density. The growth of somatic embryos from cell suspensions of Liquidambar (137) and Carica (86) has also been reported.

3.3. Abnormal development

Abnormal forms of somatic embryos are frequently found in most embryogenic cultures, expecially in those systems which do not use cell suspensions as the source of embryos. The somatic embryos appear to be influenced by the neighboring structures around them. The solid medium and microenvironments may contribute to the formation of fused embryos, fasciated parts, multiple shoots, many-rooted embryos, hypertrophied tissue and callus. Where there is ambiguity in the early stages of embryo formation, the structures have been called proembryos, embryoids, embryo-like, meristemoids, sphaeroblasts, or neomorphs.

4. PLANT DEVELOPMENT

4.1. Selection of somatic embryos for propagation

Plant development from somatic embryos has been reported in about 50 percent of those species whose tissues have been observed to undergo somatic embryogenesis (Table 1). Some of the criteria used for the selection of somatic embryos capable of normal plant development are: bipolarity, the development of apical and root meristems, an opaque white color indicating the storage of starch (162), an appearance of vigor, cotyledon growth, greening in the dark in the case of conifers, a lack of anthocyanin (51) or its presence (115). If selection is undertaken at the globular stage of development or from suspended cells, the procedures may include the use of selective media and survival, followed by regeneration via somatic embryogenesis. An example of the latter technique was the selection of Shamouti orange cells or cell clusters for salt tolerance (11, 66). The plants from these cells were later reported to be stable and true variants (109). Isozyme studies were made of embryogenic and non-embryogenic lines of Citrus ovular callus (10) and they showed an increase in peroxidase activity with embryogenicity. In Citrus nucellar callus gene expression was correlated with ten enzyme systems, suggesting that this approach might be useful in separating interspecific hybrids from non-hybrids in populations of plants derived from fused protoplasts.

4.2. Dormancy

Some woody perennials produce zygotic embryos which are dormant and require a cold treatment (2-6°C) before germination will take place. Somatic embryos from these same plants also pass through a dormant stage and require a similar cold treatment, for example, in Vitis (78) and Juglans (162). The work with Vitis (124) showed that ABA increased during the development of somatic embryos and that chilling markedly reduced the endogenous levels. Further studies by Takeno et al. (154) determined that the levels of gibberellic acid and GA precursors in unchilled somatic embryos of Vitis increased during the chilling process as the ABA level decreased.

4.3. Germination

Following cold treatment (2-6°C) or appropriate use of GA or ABA, the germination of somatic embryos sometimes requires a series of coordinated treatments at the correct stage of development in order to obtain growth into plants. The selection of vigorous embryos is a first step, but other factors must be considered. With Ilex (50, 58), it was necessary to remove the somatic embryo from the parent tissue to obtain elongation of the radicle. Wang and Janick (168) found that leaching the somatic embryos of cacao gave precocious germination, presumably by removing inhibitors. Precocious germination is often undesirable since it may results in abnormal plants. Further development of somatic embryos is sometimes promoted by using more dilute media with lowered salts, less sucrose, and a support system such as filter paper bridges, vermiculite or peat plugs (162).

4.4. Transfer to soil

Once sufficient root development occurs and a few small leaves emerge beyond the cotyledons, the plants may be removed from the sterile environment, washed thoroughly to get rid of nutrients, sucrose, agar and exfoliated cells and planted in sterile soils or other support system. Mixtures of equal parts soil, sand, peat and vermiculite work well. The change from high relative humidity of the culture vessel to the ambient fluctuations of moisture of the atmosphere is usually accomplished by enclosing the young plant in a container (polyethylene bag or plastic cup) to restrict water loss. Carbon dioxide for carbon assimilation may be provided by microbial activity in the soil organic matter. The container is gradually perforated over a period of 3-4 weeks and later removed. In similar ways the plants are acclimatized to light and cool or hot temperatures. Fungicides are recommended to prevent wilt and to increase plant survival. Many of these procedures are similar to those used for acclimatizing stem cuttings from micropropogation (see Driver, this volume).

5. PHYSIOLOGY

5.1. Fatty acids

Studies have been done on the somatic embryos of those woody perennials which use fats rather than carbohydrates as the major energy storage compounds. In embryogenic lines of tissue from the oil palm (163) there was an accummulation of lipid in the very early stages of somatic embryo formation. This accummulation was correlated with an increase in acetyl-Co-A carboxylase activity. Enzymatic activity was then used to measure the embryogenic potential of different cell lines. Other studies comparing somatic and zygotic embryos of cacao (116) showed an effect of high sucrose concentrations (up to 27% w/v) on fatty acid composition. As the sucrose concentration was increased the stearic and oleic acid content of embryos increased and linoleic decreased. The amount of fatty acid varied with temperature (10-35°C) and somatic embryos grown at 26 C had a fatty acid composition which resembled that of cocoabutter produced by mature zygotic embryos grown in culture (55). Plant hormones (ABA, GA, BAP, Naphthaleneacetic acid [NAA]) or nutrients (coconut water, casein hydrolysate, K, P, N) had no effect. The high sucrose also increased the anthocyanin content (115) and the caffeine and theobromine level in somatic embryos (110). Similarly, in cultures of Simmondsia (jojoba), Lee, Thomas and Price (83) found that the wax constituents of somatic embryos resembled those produced by seeds.

5.2. Plant growth regulators

Plant growth regulators are commonly used in conditioning media for explants being induced to form somatic embryos. Studies on the effect of these substances on embryogenesis (30, 175) or on somatic embryo development (39) or germination (2) are also reported (see also Minocha, this volume). For example, GA was effective in overcoming the dormancy requirement in Corylus (117) but was without effect in Paulownia (121). Takeno et al. (154) showed that the germination of somatic embryos of Vitis was controlled by the changing levels of GA-like substances and ABA. The high level of ABA in unchilled embryos decreased during chilling and the pool of free GA increased. The embryogenic competence of clones of callus of Theobroma (72) was stimulated by GA and Amo 1618, but was depressed by 2-chloroethyl-trimethyl ammonium chloride (CCC). In Citrus ovular callus (63), GA suppressed embryogenesis and ABA, Alar, and CCC stimulated the process. It was also possible to select Citrus sinensis ovular callus for tolerance to 2,4-D (146).

5.3. Phenolics and polyamines

Tissues of woody perennials which are placed in culture sometimes produce phenolic compounds, turn brown and appear non-viable. This browning reaction is common in plants such as Coffea (140) and the use of antioxidants such as ascorbic acid or cysteine has been suggested to reduce phenolic oxidation (95, 138). Incubation of tissues in the dark and the use of activated charcoal have been recommended (87, 88). The absence of light would reduce the oxidation of chlorogenic acid, caffeic acid, ferulic and other phenolics to dark quinones. However, these brown reaction products are not necessarily inhibitory to the process of somatic embryogenesis. Somatic embryos have been derived from brown tissue masses of Vitis (76), Elaeis (107), Coffea (138), Paulownia (121), Bambusa (93), and Juglans (162). In embryogenic callus derived from ovules of Paulownia, Radojevic (121) found that the placental tissue became necrotic but was essential to the initiation of embryogenesis. Krul and Worley (76) noted the presence of necrotic cells adjacent to embryogenic tissue of Vitis, and Hanower and Hanower (49) showed that phloroglucinol and phloridzin stimulated embryogenesis in Elaeis. It is thus apparent that the substances involved in the browning reaction deserve more attention. See also the discussion of polyamines (Bagni and Biondi) in volume one, and senescence (Romani) in this volume.

6. DISCUSSION

6.1. Questions

A number of questions need to be answered to provide more understanding of the process of somatic embryogenesis. For example, what is the basis of embryogenic competence? Is it an intrinsic feature of most plant cells (122), a characteristic of more juvenile tissue (18), or a certain receptivity of target cells at a particular stage of a cell cycle? Are special signals and receptors essential to call forth the morphogenetic response called somatic embryogenesis? Which steps are critical for conditioning an explant tissue? How important are the sequences of environmental, nutrient and hormonal factors for the optimal response of a given tissue? Answers to these questions should come from some of the new techniques and materials now available.

6.2. Developmental patterns

The developmental patterns for somatic embryogenesis have been investigated in order to develop models of the entire process from the inception of the embryos to the whole plant in soil. In some cases, the initial cell or group of cells has been identified and protocols for obtaining somatic embryos from explants (142) or callus (131) have been described. The literature on somatic embryogenesis suggests that there may be only a few patterns of development which produce somatic embryos, while there may be several ways to induce the pattern. Inadequate protocols, poor timing or choice of explant may yield few results, abnormal development, or callus. Similar problems also exist for controlling somatic embryo maturation, germination and growth into a plant (Table 2).

6.3. Theories

The stage of development of an explant (the phenocritical time), (28) has been considered important for regeneration in culture. Christianson's theory of somatic embryogenesis (27) is based on his work with soybean and information from other research. The theory suggests that cells which are metabolically competent to form somatic embryos undergo "permissive" induction (see also Street, 152). These cells are receptive to a stimulus which will elicit the embryogenic developmental program. To be able to differentiate between one state of permissive determination leading to a particular morphogenetic pattern (a somatic embryo) and another permissive state which leads to leaves, roots, shoots or callus requires sufficient knowledge of the metabolism of a tissue system. Christianson suggests graphing several key sets of metabolic data in three dimensions to give a "metabolic response surface" (Durzan this volume, and 36). Determining these surfaces at certain times, for example during the permissive inductive period or conditioning, might help identify the signal for a developmental pattern. Presumably, receptor sites, altered cytoskeletons, feedback mechanisms, and other systems would have to be studied to understand the critical events of the stage specific sensitivities which lead to a particular pattern of development.

The concept of target cells in plant differentiation has been reviewed by Osborne (109). The concept suggests that target cells or cells in a permissive state are those which are competent to recognize a specific signal (or they have a sufficient number of receptor sites to be activated by the signal). The theory assumes that a particular complement of proteins forms the basis of competence for differentiation. Hence, it should be possible to detect some development associated, specific changes in protein and to identify which of those proteins relate to specific signals. An example of a target cell system would be the petiolar abscission cells which respond to ethylene; another would be the cells which are induced to form somatic embryos. It is also possible, as Tran Thanh Van has suggested (161) that there is cellular complementarity and that groups of cells, rather than single cells are the target. Furthermore, each targeted cell may pass on part of the signal it has received. This view would emphasize the network of interrelations between cells.

Markers of embryogenic tissue of oil palm cultures have been identified by Ferdinando, Hulme and Hughes (42). A protein fraction from zygotic embryos was used to obtain antibodies from rabbits; the antibodies were incubated with whole cell extracts of non-embryogenic palm callus to remove non-embryogenic species proteins. This competing out method gave an antibody fraction which was used for immunorecognition of proteins specific for zygotic embryos. These antibodies were also used to distinguish early stages of somatic embryo formation in oil palm callus. This provided a method for direct identification of proembryonic cells or stage specific markers. One- and two-dimensional electrophoretograms, immunoblot and immunofluorescent techniques were used to verify the specificity of these proteins.

6.4. Clones or variants

Somatic embryos are clonal material (60) compared to zygotic embryos which bring new gene combinations into existence. In some cases, the clonal nature of somatic embryos and the plants derived from them needs to be verified by detailed morphological studies, progeny tests, isozyme analysis, cytological studies, DNA base sequencing and other methods. The nature and extent of the variation in plants derived from tissue culture systems has been reviewed extensively (31, 94, 108, 133). Plants of Santalum (129) derived from tissue culture showed a high incidence of variants (up to 90 percent) and some of these showed exceptional vigor under field conditions. Most variants, however, are not new to the plant breeder familiar with the material. It remains to be determined whether most of a genome will remain stable and genetically true-to-type when somatic variants are selected for such characters as herbicide or disease resistance, vigor, or other traits.

6.5. Uses

Among the uses suggested for somatic embryos are the selection and propagation of elite plants or deriving plants from unusual genomes such as from endosperm tissue. Early selection at the single cell or protoplast stage, followed by somatic embryogensis is another useful approach. Somatic embryos could also be used to propagate cell lines produced by protoplast fusion, as has been suggested for *Citrus* (145) and as a means for multiplying zygotic embryos rescued from wide crosses, perhaps a good approach for *Carica* (86) and *Juglans* (162). For *Vitis* (78), these methods may provide virus-free clonal material for vineyards. Other work with leaf discs of *Coffea* (119) suggests that somatic embryos may be useful in cocultivation experiments in which callus cells from leaf discs could be transformed by specific desirable genes and markers, using the plasmids of *Agrobacterium* as vectors. Following regeneration by somatic embryogenesis, it should be possible to screen for somatic embryos with the inserted marker and gene. The demonstrated embryogenic competence of mature leaves of *Cocos*, *Malus*, *Cinchona* and stems of *Citrus* and *Santalum* (Table 1) should make this approach feasible for presently grown cultivars.

7. CONCLUSION

Compared to sexually generated embryos which have provided plant breeders with an array of gene combinations for selection, somatic embryos have been regarded as a potentially stable, true-to-type source of plant propagants. However, variation in plants derived from somatic embryos does occur (31, 94). This variation has been attributed to mutation, mitotic crossing over, transposable genes, variation in explants, methods of regeneration and the conditions of culture. Street (152) has pointed out that none of the wide array of environments used to initiate and grow somatic embryos precisely resembles the environment of the ovule in the intact plant. It is assumed that culture conditions exert a major influence on the very early stages of somatic embryo development. In subsequent development adjacent tissues start to exert additional effects. Conditions which support somatic embryo maturation, germination and plant development also play a role in the selection, vigor and survival of plants derived from the original population of embryogenic cells. It thus seems possible, through the use of both early and late selection systems, to use the process of somatic embryogenesis for the production of useful variants of woody perennials.

8. ACKNOWLEDGEMENTS

Several bibliographical data bases for plant tissue and cell culture were consulted, including IRIS, Information Retrieval Interactive System. I wish to thank Dr. Y. Tateno, Life Science Research Information Section, Institute of Physical and Chemical Research, 2-1 Hirosawa, Wako-Shu, Saitama-ken 351, Japan, for his assistance.

9. REFERENCES

1. ABO EL-NIL M 1980 Embryogenesis of gymnosperm forest trees. US Patent 4, 217, 730
2. AMMIRATO PV 1974 The effects of abscisic acid on the development of somatic embyros from caraway (Carum carvi L.). Bot Gaz 135: 328-337
3. AMMIRATO PV 1983 Embryogenesis. In DA Evans, WR Sharp, PV Ammirato, Y Yamada, eds, Handbook of Plant Cell Culture, Vol 1. MacMillan Pub, New York, pp 32-123
4. AMMIRATO PV 1984 Induction, maintenance, and manipulation of development in embryogenic cell suspension cultures. In IK Vasil, ed, Cell Culture and Somatic Cell Genetics of Plants, Vol 1. Academic Press, pp 139-151
5. AMMIRATO PV 1985 Patterns of development in culture. In RR Henke, KW Hughes, MJ Constantin, A Hollaender, CM Wilson, Tissue Culture in Forestry and Agriculture. Plenum Press, New York, pp 9-29
6. AMMIRATO PV, DA EVANS, WR SHARP, Y YAMADA 1984 Handbook of Plant Cell Culture, Vol 3. MacMillan, New York
7. BAPAT VA, RAO PS 1979 Somatic embryogenesis and plantlet formation in tissue cultures of sandalwood (Santalum album L.). Ann Bot 44: 629-630
8. BAPAT VA, PS RAO 1984 Regulatory factors for in vitro multiplication of sandalwood tree (Santalum album Linn.) I. Shoot bud regeneration and somatic embryogenesis in hypocotyl cultures. Proc Indian Acad Sci 93: 19-27
9. BANKS MS 1979 Plant regeneration from callus from two growth phases of English ivy, Hedera helix L. Z Pflanzenphysiol 92: 349-353

10. BEN-HAYYIM, GA SHANI, A VARDI 1982 Evaluation of isozyme systems in *Citrus* to facilitate identification of fusion products. Theor Appl Genet 64: 1-5
11. BEN-HAYYIM G, J KOCHBA 1983 Aspects of salt tolerance in a NaCl-selected stable cell line of *Citrus sinensis*. Plant Physiol 72: 685-690
12. BEN-HAYYIM G, H NEUMANN 1983 Stimulatory effect of glycerol on growth and somatic embryogenesis in *Citrus* callus cultures. Z Pflanzenphysiol 110: 331-338
13. BHANSALI R, HC ARYA 1978 Differentiation in explant of *Citrus paradisi* Macf. (grapefruit) grown in culture. Indian J Exp Biol 16: 409-411
14. BONGA JM 1971 Formation of holdfasts, callus, embryoids and haustorial cells in *in vitro* cultures of dwarf mistletoe *Arceuthobium pusillum*. Phytomorphology 21: 140-153
15. BONGA JM 1977 Organogenesis in *in vitro* cultures of embryonic shoots of *Abies balsamea* (balsam fir). In Vitro 13: 41-48
16. BONGA JM 1981 Organogenesis *in vitro* of tissues from mature conifers. In Vitro 17: 511-518
17. BONGA JM 1981 Vegetative propagation of mature trees by tissue culture. *In* AN Rao, ed, Tissue Culture of Economically Important Plants. COSTED and ANBS, Singapore, pp 191-196
18. BONGA JM 1982 Vegetative propagation in relation to juvenility, maturity and rejuvenation. *In* JM Bonga, DJ Durzan, eds, Tissue Culture in Forestry. Martinus Nijhoff/Dr W Junk Publ, The Hague, pp 387-412
19. BRADLEY PM, F EL FIKI, KL GILES 1984 Polyamines and arginine affect somatic embryogenesis of *Daucus carota*. Plant Sci Lett 34: 397-401
20. BRANTON RL, J BLAKE 1983 Development of organized structures in callus derived from explants of *Cocos nucifera* L. Ann Bot 52: 673-678
21. BRAWLEY SH, DF WETHERELL, KR ROBINSON 1984 Electrical polarity in embryos of wild carrot precedes cotyledon differentiation. Proc Natl Acad Sci 81: 6064-6057
22. BUTTON J 1978 The effects of some carbohydratres on the growth and organization of *Citrus* ovular callus. Z Pflanzenphysol 88: 61-68
23. BUTTON J, J KOCHBA, CH BORNMAN 1974 Fine structure of and embryoid development from embryogenic ovular callus of Shamouti orange (*Citrus sinensis* Osb.). J Exp Bot 25: 446-457
24. BUTTON J, CE BOTHA 1975 Enzymatic maceration of *Citrus* callus and the regeneration of plants from single cells. J Exp Bot 6: 723-729
25. BUTTON J, J KOCHBA 1977 Tissue culture in the *Citrus* industry. *In* J Reinert, YPS Bajaj, eds, Applied and Fundamental Aspects of Plant Cell, Tissue and Organ Culture. Springer Verlag, Berlin, pp 70-92
26. CARRON MP, F ENJALRIC 1982 Studies on vegetative micropropagation of *Hevea brasiliensis* by somatic embyogenesis and *in vitro* microcutting. *In* A Fujiwara, ed, Plant Tissue Culture 1982. Proc 5th Int Cong Plant Tissue Cell Cult, Tokyo, pp 751-752
27. CHRISTIANSON ML 1985 An embryogenic culture of soybean: towards a general theory of somatic embryogenesis. *In* RR

Henke, KW Hughes, MJ Constantin, A Hollaender, CM Wilson, Tissue Culture in Forestry and Agriculture. Plenum Press, New York, pp 83-103
28. CHRISTIANSON ML, DA WARNICK 1984 Phenocritical times in the process of in vitro shoot organogenesis. Dev Biol 101: 382-390
29. CONGER BV 1981 Cloning agricultural crops via in vitro techniques. CRC Press, Boca Raton
30. DEBRUIJNE E, E DELANGHE, R VAN RIJK 1974 Action of hormones and embryoid formation in callus cultures of Carica papaya. Meded Fac Landbouwett Rijksuniv Ghent 39: 637-645
31. DURIEU H, M BARBIER 1982 High frequency of genetic variant plants regenerated from cotyledons of tobacco. In ED Earle, Y Demarley, eds, Variability in Plants Regenerated from Tissue Culture. Praeger, New York, pp 211-227
32. DURZAN D 1982 Somatic embryogenesis and sphaeroblasts in conifer cell suspensions white spruce (Picea glauca) Douglas fir (Pseudotsuga menziesii). In A Fujiwara, ed, Plant Tissue Culture. Proc 5th Int Cong Plant Tissue Cell Culture, Tokyo, pp 113-114
33. DURZAN DJ 1984 Special problems: adult vs juvenile explants. In WR Sharp, DA Evans, PV Ammirato, Y Yamada, Handbook of Plant Cell Culture. MacMillan, New York, Vol 2, pp 471-500
34. DURZAN DJ 1985 Tissue culture and improvement of woody perennials: an overview. In RR Henke, KW Hughes, MJ Constantin, A Hollaender, CM Wilson, eds, Tissue Culture in Forestry and Agriculture. Plenum, pp 233-256
35. DURZAN DJ 1985 Potential for genetic manipulation of forest trees: totipotency, somaclonal aberration, and trueness to type. In J Hanower, D Karnosky, D Keathley, eds, Proc Int Symp Recent Adv For Biotechnol, Traverse City, Michigan, pp 104-125
36. DURZAN DJ 1985 Metabolic networks in developing pistachio embryos (Pistachia vera cv Kerman). Proc Int Soc Dev Biol, Tenth Int Cong Los Angeles, CA 1985 (in press).
37. DURZAN DJ, FC STEWARD 1968 Cell and tissue culture of white spruce and jack pine. Bi-Monthly Res Notes Dept Fisheries, Ottawa 24: 30
38. EICHHOLTZ DA, HA ROBITAILLE, PM HASEGAWA 1979 Adventive embryony in apple. HortScience 14: 699-700
39. EPSTEIN E, J KOCHBA, H NEUMANN 1977 Metabolism of indoleacetic acid by embryogenic and non-embryogenic callus lines of 'Shamouti' orange (Citrus sinensis Osb.). Z Pflanzenphysiol 85: 263-268
40. ESAN EB 1973 A detailed study of adventive embryogenesis in the Rutaceae. PhD Thesis, Univ Cal, Riverside
41. EVANS DA, WR SHARP, CL FLICK 1981 Plant regeneration from cell cultures. Hortic Rev 3: 214-314
42. FERDINANDO D, J HULME, WA HUGHES 1984 Oil palm embryogenesis. A biochemical and morphological study. In Wye International Symposium, Ovule Tissues: Their Micro-manipulation, Tissue Culture and Physiology
43. GHARYAL PK, SC MAHESHWARI 1981 In vitro differentiation of somatic embryoids in a leguminous tree, Albizzia lebbeck L. Naturwissenschaften 68: 379-380

44. GUPTA PK, SK KANDURKAR, VM KULKARNI, MV SHIRGURKAR, AF MASCARENHAS 1984 Somatic embryogenesis and plants from zygotic embryos of coconut (Cocos nucifera L.) in vitro. Plant Cell Rep 3: 222-225
45. GUPTA PK, DJ DURZAN 1985 (pers comm)
46. HACCIUS B 1978 Question of unicellular origin on non-zygotic embryos in callus cultures. Phytomorphology 28: 74-81
47. HAKMAN I, LC FOWKE, S VON ARNOLD, T ERIKSSON 1985 The development of somatic embryos in tissue cultures initiated from immature embryos of Picea abies (Norway spruce). Plant Sci Lett 38: 53-59
48. HAN HU, S QIQUAN 1981 Advances in plant cell and tissue culture in China. Adv Agron 34: 1-13
49. HANOWER J, P HANOWER 1984 Inhibition et stimulation en culture in vitro de l'embryogénèse des souches issues d'explant foliares de palmier à huile. CR Acad Sci, Paris 298: 45-48
50. HU CY, IM SUSSEX 1971 In vitro development of embryoids on cotyledons of Ilex aquifolium. Phytomorphology 21: 103-107
51. HU CY, JD OCHS, FM MANCINI 1978 Further observations on Ilex embryoid production. Z Pflanzenphysiol 89: 41-49
52. HENKE RR, KW HUGHES, MJ CONSTANTIN, A HOLLAENDER, CM WILSON, eds, 1985 Tissue Culture in Forestry and Agriculture. Plenum, New York
53. JAMES DJ, AJ PASEY, DC DEEMING 1984 Adventitious embryogenesis and the in vitro culture of apple seed parts. J Plant Physiol 115: 217-229
54. JANICK J 1982 Adventive embryony in pear. Acta Hortic 124: 37-41
55. JANICK J, DC WRIGHT, PM HASEGAWA 1982 In vitro production of cacao seed lipids. J Amer Soc Hortic Sci 107: 919-922
56. JOHRI BM, YPS BAJAJ 1965 Growth responses of globular proembryos of Dendrophthoe falcata (Lf.) Etting in culture. Phytomorphology 15: 292
57. JONES LH 1974 Propagation of clonal oil palms by tissue culture. Oil Palm News 17: 1-9
58. JORDAN M, I CORTES, G MONTENEGRO 1983 Regeneration of plantlets by embryogenesis from callus cultures of Carica candamarcensis. Plant Sci Lett 28: 321-326
59. KATO H 1968 The serial observations of the adventive embryogenesis in microculture of carrot tissue. Sci Pap College Gen Educ, Tokyo 18: 191-197
60. KESTER DH 1983 The clone in horticulture. HortScience 18: 831-837
61. KOCHBA J, P SPIEGEL-ROY, H SAFRAN 1972 Adventive plants from ovules and nucelli of Citrus. Planta 106: 237-245
62. KOCHBA J, J BUTTON 1974 The stimulation of embryogenesis and embryoid development in habituated ovular callus from the 'Shamouti" orange (Citrus sinensis) as affected by tissue age and sucrose concentration. Z Pflanzenphysiol 73: 415-421
63. KOCHBA J, P SPIEGEL-ROY, H NEUMANN, S SAAD 1978 Stimulation of embryogenesis in Citrus ovular callus by ABA, ethephon, CCC and Alar and its suppression by GA. Z Pflanzenphysiol 89: 427-432

64. KOCHBA J, P SPIEGEL-ROY, S SAAD, H NEUMANN 1978 Stimulation of embryogenesis in Citrus tissue culture by galactose. Naturwissenschaften 65: 261
65. KOCHBA J, P SPIEGEL-ROY, S SAAD, H NEUMANN 1978 Tissue culture studies with Citrus: 1) The effect of several sugars on embryogenesis and 2) application of Citrus tissue cultures for selection of mutants. In AW Alfermann, E Reinhard, eds, Production of Natural Compounds by Cell Culture Methods. Proc Int Symp Plant Tissue Cult, Munchen, pp 223-233
66. KOCHBA J, P SPIEGEL-ROY, S SAAD 1980 Selection for tolerance to sodium chloride (NaCl) and 2,4-dichlorophenoxyacetic acid (2,4-D) in ovular callus lines of Citrus sinensis. In F Sala, B Parisi, R Calla, O Ciferri, eds, Plant Cell Cultures: Results and Perspectives. Elsevier North Holland Biomed Press, Amsterdam, pp 187-192
67. KOCHBA J, P SPIEGEL-ROY, H NEUMANN, S SAAD 1982 Effect of carbohydrate on somatic embryogenesis in subcultured nucellar callus of Citrus cultivars. Z Pflanzenphysiol 105: 358-368
68. KOHLENBACK HW 1978 Comparative somatic embryogenesis. In TA Thorpe, ed, Frontiers of Plant Tissue Culture 1978. Int Assoc Plant Tissue Cult, Univ Calgary, pp 59-66
69. KONAR RN, YP OBEROI 1965 In vitro development of embryoids on the cotyledons of Biota orientalis. Phytomorphology 15: 137-140
70. KONG LS, AN RAO 1981 Induction of callus and organogenesis in cocoa tissues. In AN Rao, ed, Tissue Culture of Economically Important Plants. COSTED and ANBS, Singapore, pp 107-112
71. KONONOWICZ AK, J JANICK 1984 The influence of carbon on the growth and development of asexual embryos of Theobroma cacao. Physiol Plant 61: 155-162
72. KONONOWICZ H, J JANICK 1984 Response of embryogenic callus of Theobroma cacao L. to gibberellic acid and inhibitors of gibberellic acid synthesis. Z Pflanzenphysiol 113: 359-366
73. KONONOWICZ H. AK KONONOWICZ, J JANICK 1984 Asexual embryogenesis via callus of Theobroma cacao L. Z Pflanzenphysiol 113: 347-358
74. KRIKORIAN AD, M SINGH, CE QUINN 1981 Aseptic micropropagation of Cinchona: prospects and problems. In AN Rao, ed, Tissue Culture of Economically Important Plants. COSTED and ANBS, Singapore, pp 167-174
75. KROGSTRUP P 1984 Micropropagation of conifers. PhD Thesis Royal Vet Agric Univ, Copenhagen, 161 pp (in Danish)
76. KRUL WR, JF WORLEY 1977 Formation of adventitious embryos in callus cultures of 'Seyval', a French hybrid grape. J Am Soc Hortic Sci 102: 360-363
77. KRUL WR, J MYERSON 1980 In vitro propagation of grape. In R Zimmerman, ed, Proc Conf Nursery Prod Fruit Plants Through Tissue Cult Appl. USDA Sci Educ Admin Agric Res Results, ARR-NE-11
78. KRUL WR, GH MOWBRAY 1984 Grapes. In WR Sharp, DA Evans, PV Ammirato, Y Yamada, eds, Handbook of Plant Cell Culture, Vol 2. MacMillan, New York, pp 396-434
79. LAKSHMI SITA G, NV RAGHAVA RAM, CS VAIDYANATHAN 1979 Differentiation of embryoids and plantlets from shoot callus of sandalwood. Plant Sci Lett 15: 265-270

80. LAKSHMI SITA G, NV RAGHAVA RAM, CS VAIDYANATHAN 1980 Differentiation of embryoids and plantlets from shoot callus of sandalwood by experimental embryogensis. Plant Sci Lett 20: 63-69
81. LAKSHMI SITA G 1981 Tissue culture of Eucalyptus species. In AN Rao, ed, Tissue Culture of Economically Important Plants. COSTED and ANBS, Singapore, pp 180-184
82. LARUE CD 1954 Studies on growth and regeneration in gametophytes and sporophytes of gymnosperms. Brookhaven Symp Biol 6: 187-208
83. LEE CW, JC THOMAS, RL PRICE 1982 Jojoba oil production from embryo cultures. In A Elias-Cesnick, ed, Proc 5th Int Conf Jojoba and Its Uses. Tucson pp 9-20
84. LITZ RE 1984 In vitro responses of adventitious embryos of two polyembryonic Eugenia species. HortScience 19: 720-722
85. LITZ RE 1984 In vitro somatic embryogenesis from nucellar callus of mono-embryonic mango. HortScience 19: 715-717
86. LITZ RE 1984 Papaya. In WR Sharp, DA Evans, PV Ammirato, Y Yamada, eds, Handbook of Plant Cell Culture. MacMillan, New York, Vol 2, pp 349-368
87. LITZ RE 1985 Somatic embryogenesis in tropical fruit trees. In RR Henke, KW Hughes, MJ Constantin, A Hollaender, CM Wilson, eds, Tissue Culture in Forestry and Agriculture. Plenum, New York, pp 179-193
88. LITZ RE, RA CONOVER 1980 Somatic embryogenesis in cell cultures of Carica stipulata. HortScience 15: 733-735
89. LITZ RE, RA CONOVER 1981 In vitro polyembyony in Carica papaya L. ovules. Z Pflanzenphysiol 104: 285-288
90. LITZ RE, RA CONOVER 1982 In vitro somatic embryogenesis and plant regeneration from Carica papaya ovular callus. Plant Sci Lett 26: 153-158
91. LITZ RE, RJ KNIGHT, S GAZIT 1984 In vitro somatic embryogenesis from Mangifera indica L. callus. Sci Hortic 22: 233-240
92. LIU JR, KC SINK, FG DENNIS 1983 Adventive embryogenesis from leaf explants of apple seedlings. HortScience 18: 871-873
93. MEHTA U, IV RAMANIYA RAO, HY MOHAN RAM 1982 Somatic embryogenesis in bamboo. In A Fujiwara, ed, Plant Tissue Culture 1982. Proc 5th Int Cong Plant Tissue Cell Cult, Tokyo, pp 9-110
94. MEINS F 1983 Heritable variation in plant cell cultures. Annu Rev Plant Physiol 34: 327-346
95. MONACO LC, MR SONDAHL, A CARVALHO, OJ CROCOMO, WR SHARP 1976 Applications of tissue culture in the improvement of coffee. In J Reinert, YPS Bajaj, eds, Applied and Fundamental Aspects of Plant Cell Tissue and Organ Culture. Springer Verlag, Berlin, pp 109-129
96. MOTT RL 1981 Trees. In BV Conger, Cloning Agricultural Crops Via In Vitro Techniques. CRC Press, Boca Raton, pp 217-254
97. MULLINS MG, C SRINIVASAN 1976 Somatic embryos and plantlets from an ancient clone of the grapevine (cv Cabernet Sauvignon) by apomixis in vitro. J Exp Bot 27: 1022-1030
98. MURASHIGE T 1974 Plant propagation through tissue culture. Annu Rev Plant Physiol 25: 135-166

99. MURASHIGE T, F SKOOG 1962 A revised medium for rapid growth and bioassays with tobacco tissue cultures. Physiol Plant 15: 473-497
100. NAG KK, BM JOHRI 1969 Organogenesis and chromosomal constitution in embryo callus of Nuytsia floribunda. Phytomorphology 19:405-408
101. NAGMANI R, JM BONGA 1985 Embryogenesis in subcultured callus of Larix decidua. Can J For Res 15: 1088-1091
102. NAGMANI R, S VENKETESWARAN 1983 Morphogenetic responses of cultured hypocotyl and cotyledonary segments of Leucaena. In Vitro 19: 265
103. NARAYANASWAMI S 1977 Regeneration of plants from tissue cultures. In J Reinert, YPS Bajaj, eds, Plant Cell, Tissue and Organ Culture, Springer Verlag, Berlin, pp 179-250
104. NASSUTH A, TM WORMER, F BOUMAN, G STARITSKY 1980 The histogenesis of callus in Coffea canephora stem explants and the discovery of early embryoid initiation. Acta Bot Neerl 29: 49-54
105. NORSTOG K 1965 Induction of apogamy in megagametophytes of Zamia integrifolia. Am J Bot 52: 993-999
106. NORSTOG K, E RHAMSTINE 1967 Isolation and culture of haploid and diploid cycad tissues. Phytomorphology 17: 374-381
107. NWANKWO BA, AD KRIKORIAN 1983 Morphogenetic potential of embryo and seedling-derived callus of Elaeis guineensis Jacq. var pisifera Becc. Ann Bot 51: 65-76
108. ORTON TJ 1983 Experimental approaches to the study of somaclonal variation. Plant Mol Biol Rep 1: 67-76
109. OSBORNE DJ 1984 Concepts of target cells in plant differentiation. Cell Differ 14: 161-169
110. PAIVA M, J JANICK 1983 In vivo and in vitro production of alkaloids in Theobroma cacao L. Acta Hortic 131: 265-273
111. PANNETIER C, J BUFFARD-MOREL 1982 Production of somatic embyros from leaf tissues of coconut Cocos nucifera L. In A Fujiwara, ed, Plant Tissue Culture 1982. 5th Int Cong Plant Tissue Cult, Tokyo, pp 755-756
112. PARANJOTHY K, R OTHMAN 1978 Embryoid and plantlet development from cell cultures of Hevea. In TA Thorpe, ed, Abst 4th Int Cong Plant Tissue and Cell Cult. Calgary, pp 42
113. PENCE VC, PM HASEGAWA, J JANICK 1979 Asexual embryogenesis in Theobroma cacao L. J Am Soc Hortic Sci 104: 145-148
114. PENCE VC, PM HASEGAWA, J JANICK 1980 Initiation and development of asexual embryos of Theobroma cacao L. in vitro. Z Pflanzenphysiol 98: 1-14
115. PENCE VC, PM HASEGAWA, J JANICK 1981 In vitro cotyledonary development and anthocyanin synthesis in zygotic and asexual embryos of Theobroma cacao L. J Am Soc Hortic Sci 106: 381-385
116. PENCE VC, PM HASEGAWA, J JANICK 1981 Sucrose-mediated regulation of fatty acid composition in asexual embryos of Theobroma cacao. Physiol Plant 53: 378-384
117. PEREZ C, B FERNANDEZ, R RODRIGUEZ 1983 In vitro plantlet regeneration through asexual embryogenesis in cotyledonary segments of Corylus avellana L. Plant Cell Rep 2: 226-228
118. PHILIPS VJ 1984 In vitro organogenesis and plantlet formation in cashew (Anacardium occidentalis L.). Ann Bot 54: 149-152

119. PIERSON ES, AAM VAN LAMMEREN, JHN SCHEL, G STARITSKY 1983 In vitro development of embryoids from punched leaf discs of Coffea canephora. Protoplasma 115: 208-216
120. RABESCHAULT H, JP MARTIN, S CAS 1972 Recherches sur la culture des tissus de palmier à huile, Elaeis guineensis Jacq. Oléagineaux 27: 531-534
121. RADOJEVIC L 1979 Somatic embryogenesis and plantlets from callus cultures of Paulownia tomentosa Stued. Z Pflanzenphysiol 91: 57-62
122. RADOJEVIC L, R VAJICIC, M NESOVIC 1975 Embryogenesis in tissue culture of Corylus avellana L. Z Pflanzenphysiol 77: 33-41
123. RAGHAVEN V 1983 Biochemistry of somatic embryogenesis. In DA Evans, WR Sharp, PV Ammirato, Y Yamada, eds, Handbook of Plant Cell Culture. MacMillan, New York, pp 655-671
124. RAJASEKARAN K, J VINE, MG MULLINS 1982 Dormancy in somatic embryos and seeds of Vitis: changes in endogenous abscisic acid during embryogeny and germination. Planta 154: 139-144
125. RAMAWAT KG, HC ARYA 1976 Growth and morphogenesis in callus cultures of Ephedra gerardiana. Phytomorphology 26: 395-403
126. RANGAN TS 1984 Clonal propagation: somatic embryos of Citrus. In IK Vasil, ed, Cell Culture and Somatic Cell Genetics of Plants, Vol 1. Academic Press, New York, pp 68-73
127. RANGAN TS, T MURASHIGE, WP BITTERS 1968 In vitro initiation of nucellar embryos in monoembryonic Citrus. Hortic Sci 3: 126-227
128. RAO PS 1965 In vitro induction of embryonal proliferation in Santalum album L. Phytomorphology 15: 175-179
129. RAO PS, VA BAPAT, M MHATRE 1984 Regulatory factors for in vitro multiplication of sandalwood tree (Santalum album Linn). II. Plant regeneration in nodal and internodal stem explants and occurrence of somaclonal variations in tissue culture raised plants. Proc Indian Nat Acad Sci Biol 50: 196-202
130. RAO PS, P OZIAS-AKINS 1985 Plant regeneration through somatic embyrogenesis in protoplast cultures of sandal wood (Santalum album L.). Protoplasma 124: 80-86
131. REINERT J 1959 Uber die Kontrolle der Morphogenese und die Induktion von Adventivembryonen an Gewebekulturen aus Karotten. Planta 53: 318-333
132. RAO AN 1982 Tissue culture of economically important plants. In AN Rao, ed, Tissue Culture of Economically Important Plants, COSTED and ANBS, Singapore, pp 269-286
133. REISCH B 1983 Genetic variability in regenerated plants. In DA Evans, WR Sharp, PV Ammirato, Y Yamada, eds, Handbook of Plant Cell Culture, Vol 1. MacMillan, New York, pp 748-769
134. REYNOLDS JF, T MURASHIGE 1979 Asexual embryogenesis in callus cultures of palms. In Vitro 15: 383-387
135. SHARP WR, MR SONDAHL, LS CALDAS, SB MARAFFA 1980 The physiology of in vitro asexual embryogenesis. Hortic Rev 2: 268-310
136. SHARP WR, DA EVANS, PV AMMIRATO, Y YAMADA, eds, 1984 Handbook of Plant Cell Culture, Vol 2. MacMillan, New York
137. SOMMER HE, CL BROWN 1980 Embryogenesis in tissue cultures of sweetgum. For Sci 26: 257-260

138. SONDAHL MR, WR SHARP 1977 High frequency induction of somatic embryos in cultured leaf explants of Coffea arabica L. Z Pflanzenphysiol 81: 395-408
139. SONDAHL MR, TL SALISBURY, WR SHARP 1979 Characterization of embyrogenesis in coffee callus by scanning electron microscope. Z Pflanzenphysiol 94: 185-188
140. SONDAHL MR, WR SHARP 1979 Research in Coffea and applications of tissue culture methods. In WR Sharp, PO Larsen, EF Paddock, V Raghavan, eds, Plant Cell and Tissue Culture, Principles and Application. Ohio State Univ Press, Columbus, pp 527-584
141. SONDAHL MR, DA SPAHLINGER, WR SHARP 1979 A histological study of high frequency and low frequency induction of somatic embryos in cultured leaf explants of Coffea arabica L. Z Pflanzenphysiol 94: 101-108
142. SONDAHL MR, T NAKAMURA, HP MEDINA FILHO, A CARRVALHO, LC FAZUOLI, WM COSTA 1984 Coffee. In PV Ammirato, DA Evans, WR Sharp, Y Yamada, eds, Handbook of Plant Cell Culture, Vol 3. MacMillan, New York, pp 564-590
143. SPIEGEL-ROY P, J KOCHBA 1973 Stimulation of differentiation in orange (Citrus sinensis) ovular callus in relation to irradiation of the media. Rad Bot 13: 97-103
144. SPIEGEL-ROY P, J KOCHBA 1977 Application of tissue culture for plant improvement. In W Barz, E Reinhard, MH Zenk, eds, Plant Tissue Culture and its Biotechnological Applications. Springer Verlag, Berlin, pp 404-413
145. SPIEGEL-ROY P, J KOCHBA 1980 Embryogenesis in Citrus tissue cultures. In A Fiechter, ed, Advances in Biochemical Engineering, Vol 16. Springer Verlag, Berlin, pp 27-48
146. SPIEGEL-ROY P, J KOCHBA, S SAAD 1983 Selection for tolerance to 2,4-dichlorophenoxyacetic acid in ovular callus of orange (Citrus sinensis). Z Pflanzenphysiol 109: 41-48
147. SRINIVASAN C, MG MULLINS 1980 High frequency embryo production from unfertilized ovules of grapes. Sci Hortic 13: 245-252
148. STAMP JA, GG HENSHAW 1982 Somatic embryogenesis in cassava, Manihot esculenta. Z Pflanzenphysiol 105: 183-188
149. STAMP JA, GG HENSHAW 1984 Adventitious regeneration in cassava. In PG Alderson and LA Withers, eds, Plant Tissue Culture and its Agricultural Applications. Proc 41st Conf Easter School Ser, Agic Sci, Univ Nottingham (in press)
150. STARITSKY G 1970 Embryoid formation in callus of coffee. Acta Bot Neerl 19: 509-514
151. STEWARD FC, MO MAPES, K MEARS 1958 Growth and organized development of cultured cells. II. Organization in cultures grown from freely suspended cells. Am J Bot 45: 705-708
152. STREET HE 1976 Experimental embryogenesis – the totipotency of cultivated plant cells. In CF Graham, PF Wareing, eds, The Developmental Biology of Plants and Animals. Blackwell Sci Publ, Oxford, pp 73-90
153. STREET, HE, LA WITHERS 1974 The anatomy of embryogenesis in culture. In HE Street, ed, Tissue Culture and Plant Science. Academic Press, New York, pp 71-100
154. TAKENO K, M KOSHIOKA, RP PHARIS, K RAJASEKARAN, MG MULLINS 1983 Endogenous gibberellin-like substances in somatic embryos of grape (Vitis vinifera X Vitis rupestris) in

relation to embryogenesis and the chilling requirement for subsequent development of mature embryos. Plant Physiol 73: 803-808

155. TATENO Y, Y NEMOTO, H SUGAWARA 1982 Bibliographical base for plant tissue and cell culture (IRIS). *In* A Fujiwara, ed, Plant Tissue Culture 1982. Proc 5th Int Cong Plant Tissue Cell Cult, Tokyo 1982, pp 787-788
156. TISSERAT B 1984 Date palm. *In* WR Sharp, DA Evans, PV Ammirato, Y Yamada, eds, Handbook of Plant Cell Culture. MacMillan, New York, pp 505-545
157. TISSERAT B, DA DEMASON 1980 A histological study of development of adventive embryos in organ cultures of *Phoenix dactylifera* L. Ann Bot 46: 465-472
158. TISSERAT B. EB ESAN, T MURASHIGE 1979 Somatic embryogenesis in angiosperms. Hortic Rev 1: 1-78
159. TISSERAT B, T MURASHIGE 1977 Effects of ethephon, ethylene and 2,4-dichlorophenoxyacetic acid on asexual embryogenesis *in vitro*. Plant Physiol 60: 437-439
160. TISSERAT B, T MURASHIGE 1977 Probable identity of substances in *Citrus* that repress asexual embryogenesis. In Vitro 13: 785-789
161. TRAN THANH VAN, K 1981 Control of morphogenesis *in vitro*. Annu Rev Plant Physiol 32: 291-311
162. TULECKE W, G MCGRANAHAN 1985 Somatic embryogenesis and plant regeneration from cotyledon tissue of walnut, *Juglans regia* L. Plant Science 40: 53-67
163. TURNHAM E, DH NORTHCOTE 1982 Use of acetyl coenzyme A carboxylase EC 6.4.1.2 activity and changes in wall composition as measures of embryogenesis in tissue cultures of oil palm *Elaeis guineensis*. Biochem J 208: 323-332
164. VARDI A, P SPIEGEL-ROY, E GALUN 1975 *Citrus* cell cultures; isolation of protoplasts, plating densities, effect of mutagens and generation of embryos. Plant Sci Lett 4: 231-236
165. VARDI A, P SPIEGEL-ROY 1982 Plant regeneration from *Citrus* protoplasts: variability in methodological requirements among cultivars and species. Theor Appl Genet 62: 171-176
166. VASIL IK 1984 Cell Culture and Somatic Cell Genetics. Vol 1. Laboratory Procedures and their Applications. Academic Press, New York
167. VASIL IK, V VASIL 1980 Clonal propagation. Int Rev Cytology Suppl 11A: 146-173
168. WANG YC, J JANICK 1984 Inducing precocious germination in asexual embryos of cacao. Hortic Sci 19: 839-841
169. WRIGHT DC, J JANICK, PM HASEGAWA 1983 Temperature effects on *in vitro* lipid accumulation in asexual embryos of *Theobroma cacao* L. Lipids 18: 863-867
170. WRIGHT DC, AK KONONOWICZ, J JANICK 1984 Factors affecting *in vitro* acid content and composition in asexual embryos of *Theobroma cacao*. J Amer Soc Hortic Sci 109: 77-81
171. YIE ST, SI LIAW 1977 Plant regeneration from shoot tips and callus of papaya. In Vitro 13: 564-568
172. ZATYKO JM 1979 Complete red currant (*Ribes rubrum* L) plants from adventive embryos induced *in vitro*. Curr Sci 48: 456-457

173. ZATYKO JM, F KISS, F SZALAY 1981 Induction of adventive embryony in cultured ovules of black currant (Ribes nigrum L.). Hortic Res 21: 99-101
174. ZATYKO, JM, I SIMON, C SZABO 1975 Induction of polyembryony in cultivated ovules of red currant. Plant Sci Lett 4: 281-283
175. ZHAN XC 1983 Regulation of somatic embryogenesis and organogenesis by plant hormones. Acta Phytophysiol Sin 9: 317-325
176. ZILIS MR, MM MEYER 1977 Embryoid production in the woody plants Hamamelis virginiana and H. vernalis. Hortic Sci 12: 389

10. FIGURES

FIGURE 1. Cotyledonary somatic embryo derived from protoplasts of Santalum album, grown on modified M & S medium with IAA 1 mg/l and BAP 1 mg/l. Rao and Ozias-Akins (130).
FIGURE 2. Somatic embryos of Carica papaya 7 weeks after transfer from White's medium with sucrose 60 g/l, l-glutamine 400 mg/l and coconut water 20% v/v, to liquid basal medium. Litz and Conover (90).
FIGURE 3. Somatic embryo derived from the megagametophyte of Larix decidua. Nagmani and Bonga (102).
FIGURE 4. Somatic embryo of Seyval hybrid grape formed on low sucrose, hormone-and-vitamin-free medium after conditioning on M & S medium with 2,4-D 1 mg/l, BAP 0.1 mg/l and sucrose 30 g/l. Krul and Worley (76).
FIGURE 5. Secondary somatic embryos from the hypocotyl of a germinating embryo of Mangifera indica. Litz, Knight and Gazit (91).
FIGURE 6. Somatic embryos of Juglans regia cv. Scharsch-Franquette formed by indirect somatic embryogenesis from callus tissue. Tulecke and McGranahan (162).
FIGURE 7. Root of a germinating somatic embryo of Juglans regia cv. Early Ehrhart with two early stage somatic embryos growing from earlier formed root cap tissue. Tulecke and McGranahan (162).
FIGURE 8. The same embryo as above, ten days later, showing the formation of globular somatic embryos at the root tip, an example of repetitive embryogenesis. Tulecke and McGranahan (162).

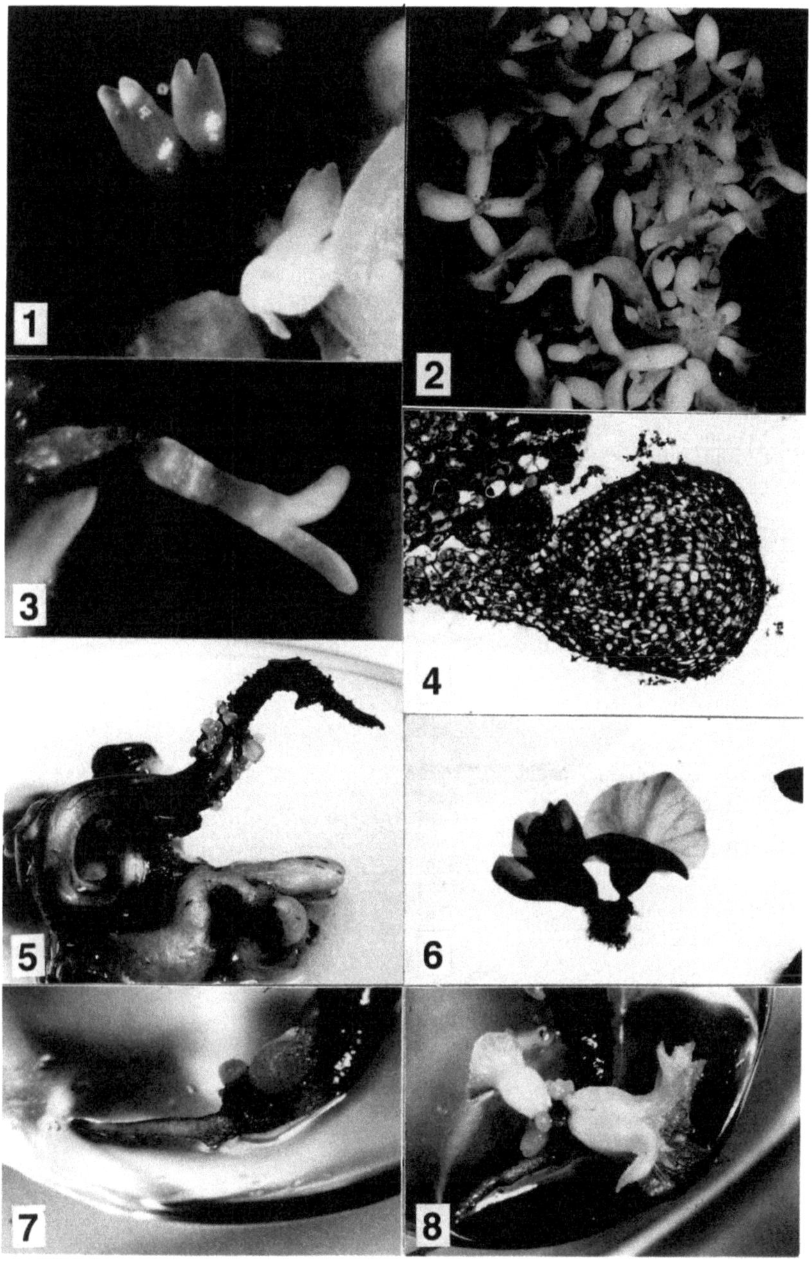
1
2
3
4
5
6
7
8

Table 1. Somatic embryogenesis in woody perennials, giving the plant source, tissue of origin, whether plants were grown in soil, and references.

Scientific name	Tissue of origin	Plants in soil	References
Cycacadaceae			
Cycas circinalis	embryo	-	106
Zamia integrifolia Ait.	embryo,	+	105
	female gametophyte	-	105
Zamia floridana L.	female gametophyte	-	82
Cupressaceae			
Biota orientalis Endl.	cotyledon	-	69
Pinaceae			
Abies balsamea L.	embryo	-	15
Larix decidua Mill	female gametophyte	-	101
Picea abies (L.) Karst.	cotyledons	-	75
	embryo	+	47
Picae glauca (Moench.) Voss	embryo	-	37
Pinus banksiana Lamb.	embryo	-	37
Pinus lambertiana Dougl.	embryo	-	45
Pseudotsuga menziesii (Mirb.) Franco	embryo	-	32
	stems	+	1
Ephedraceae			
Ephedra gerardiana Wall	callus	-	125
Arecaceae			
Chaemaedora costaricana Oerst	ovule	+	134
Cocos nucifera L.	leaf	-	111
	leaf, embryo	+	44
Elaeis guineensis Jacq.	embryo	-	120
	embryo, leaf	+	107
	leaf	-	49, 57
Howeia forsteriana Becc	ovule	-	134
Phoenix dactylifera L.	ovule	-	134
	embryo	+	156-160
Gramineae			
Bambusa arundinacea Retzius	embryo	+	93
Anacardiaceae			
Anacardium occidentale L.	cotyledon	+	118
Mangifera indica L.	nucellus	-	85, 87, 91
Aquifoliaceae			
Ilex aquifolium L.	cotyledon	+	50
	hypocotyl	+	51
Araliaceae			
Hedera helix L.	callus	+	9

Table 1. Continued.

Scientific name	Tissue of origin	Plants in soil	References
Bignoniaceae			
Paulownia tomentosa Steud.	ovules	+	121
Buxaceae			
Simmondsia chinensis (Link) C.K. Scneider	embryo	-	83
Corylaceae			
Corylus avellana L.	embryo	-	122
	cotyledon	+	117
Caricaceae			
Carica candamaricensis Hook.	hypocotyl	+	58
Carica papaya L.	callus	-	171
	ovules	+	86, 90
Carica stipulata Badillo	peduncle	+	88
Euphorbiaceae			
Hevea brasiliensis Muell.-Arg.	ovule	-	26, 112
Manihot esculenta Crantz	cotyledon	+	148
Hamamelidaceae			
Hamamelis vernalis Sarg.	embryo	-	176
Hamamelis virginiana L.	embryo	-	176
Liquidambar styraciflua L.	hypocotyl	+	137
Juglandaceae			
Juglans hindsii (Jeps.) Redh.	cotyledon	+	162
Juglans regia L.	cotyledon	+	162
Pterocarya sp.	cotyledon	+	162
Leguminosae			
Albizzia lebbeck L.	hypocotyl	-	43
Leucaena diversifolia (Schlect) Benth.	hypocotyl, cotyledon	+	102
Loranthaceae			
Arceuthobium pusillum Bieb.	embryo	-	14
Dendrophthoe falcata (Lf.) Ettings	embryo	-	56
Nuytsia floribunda (Labillardiere) R. Brown	embryo	-	100

Table 1. Continued.

Scientific name	Tissue of origin	Plants in soil	References
Myrtaceae			
Eucalyptus citriodora Hook.	callus	-	81
Eucalyptus grandis W. Hill	callus	-	81
Eugenia jambos Alston	ovules	+	84
Eugenia malaccensis L.	ovules	+	84
Myrciaria cauliflora Ber.	embryo, nucellus	-	87
Rosaceae			
Eriobotrya japonica Lindl.	ovule	-	87
Malus sp.	nucellus	-	38
	leaf	-	92
	seed	-	53
Prunus cerasus L.	petiole	-	34
Pyrus sp.	nucellus	-	54
Rubiaceae			
Cinchona ledgeriana Moens.	leaf	-	74
Coffea arabica L.	leaf	+	138
Coffea canephora Pierre	stem	+	150
	stem	-	105
	leaf disk	-	119
Coffea dewevrei Wildem	leaf	+	138
Rutaceae			
Citrus grandis L.	nucellus	-	61, 127
Citrus limonia Osbeck	nucellus	-	127
Citrus paradisi Macf.	stem, leaf	-	13
Citrus sinensis Osbeck	nucellus	+	164, 165
Santalaceae			
Santalum album L.	stem	+	79, 80 129, 130
Sterculiaceae			
Theobroma cacao L.	cotyledon	-	70
	embryo	-	73
	embryo	-	113, 114
Saxifragaceae			
Ribes nigrum L.	ovule	+	173
Ribes rubrum L.	ovule	+	172, 174
Vitaceae			
Vitis vinifera L.	ovule	-	97
	ovule	+	147
	ovule	+	76, 78

Table 2. General protocol for somatic embryogenesis and plant propagation.

Stage of the process	Factors involved
1. Selection of material	Genetic base, tissues used, stage of development
2. Conditions	Light, dark, temperature
3. Media	Conditioning, inductive, sequence, frequency of transfer
4. Somatic embryogenesis	Direct, adventive, repetitive Indirect, suspension of cells, protoplasts, screening
5. Somatic embryos	Maturation, selection, dormancy, cold treatment, abnormalities, growth requirements
6. Germination	Primary root, cotyledon expansion and greening, epicotyl growth, leaves
7. Propagation in soil	Transfer to non-sterile environment, dilute nutrient solution, day length, ambient humidity, fungicides
8. Acclimatization	Gradual leaf development in ambient temperature and humidity, hardiness to atmospheric fluctuations
9. Evaluation	Test for clonal or variant characters, selection and screening of plants

6. AMMONIA: ITS ANALOGUES, METABOLIC PRODUCTS AND SITE OF ACTION IN SOMATIC EMBRYOGENESIS

D.J. DURZAN

1. INTRODUCTION

Nearly 20 years have passed since ammonia was recognized as an absolute requirement for somatic embryogenesis (SE) (3, 136) and presumably for androgenesis and gynogenesis, i.e., gametophytic embryogenesis (GE). Today we still face a major question as to how ammonia is involved in such a transformation in somatic and gametophytic cells. The intervening years have indicated that the answer to this question is not simple.

Experience now indicates that several signals may be involved in the developmental transformations. In woody species the role of "signal networks" in SE is poorly understood, and conceptually new. For highly touted tissue culture media evoking such signals, Steward (119) has pointed out that "...nature easily overrides whatever limitations or virtues are implied by these numerous specific recipes...." He also stated that "...there is no universally applicable formula, although many guiding principles may be, and have been enunciated...". Furthermore, SE is often developmentally variant, aberrant and sometimes lethal (120). Some of these variations may be useful, but their importance remains to be established at least for forest trees (36, 47).

This review deals specifically with the role of ammonia, its early products, and the associated known enzyme systems that account for ammonia release and uptake as these relate to embryogenesis of all types (9).

Key to symbols: 2,4-D, 2,4-dichlorophenoxyacetic acid; GDH, glutamic acid dehydrogenase; GS, glutamine synthetase; GOGAT, glutamate synthetase; GS/GOGAT, glutamine (amide):2-oxoglutarate aminotransferase [NADP]; N, nitrogen; GE, gametophytic embryogenesis; SE, somatic embryogenesis; Bzl-^{6}ade, N^6-benzyladenine; Ado Met, S-adenosylmethionine; ACC, 1-aminocyclopropane carboxylic acid; NHCP, nonhistone chromosomal protein; ATP, adenosine triphosphate.

Since casein hydrolysate or L-amino acid supplements are often factors in promoting SE (60 and Appendix to this chapter), their relation to ammonia and senescing cells requires attention. Our discussion will include the transfer and allocation of reduced nitrogen (N) among metabolic paths and the catabolism of nitrogenous compounds back to ammonia. Model systems to demonstrate "process control" for these metabolic changes are few but intriguing.

Complexity emerges at the inorganic level with the formation of simple pivotal nitrogenous products that are spontaneous products of the interaction of ammonia and carbonic acid (e.g., carbamates, cyanate, etc.) or products of the catabolism of amino acids (e.g., cyanide, ammonia).

Cell wall fragments (oligosaccharins) and proteinase inhibitors are now viewed by some investigators as chemical messengers that activate genes in morphogenesis and during natural plant protection (67, 85, 130). Indeed complex carbohydrates can be trapped on filters under an apical meristem of conifers (109). Complex carbohydrates are also observed around cells in callus capable of somatic polyembryogenesis in several conifers (55). Furthermore, ions, such as calcium, may mediate many of the phytohormone-directed responses in plants (130). Since ammonia strongly interacts with all of these, there is again a need to draw together all of the relevant details for the role of ammonia in embryogenesis.

The expectation is that through some understanding of the role of ammonia and other simple forms of reduced N we will improve our control over the more morphogenetically recalcitrant cells and encourage systematic and persistent investigations on morphogenesis until useful conclusions are reached.

2. LOCALIZATION OF AMMONIA ASSIMILATION

A primary consideration in embryogenesis (zygotic, somatic or gametophytic) is the need for cells that are in physiological states that are responsive to ammonia and other reduced forms of N (Appendix). Selection of the responsive physiological state may be based on the developmental stage (e.g., just after fertilization) or dependent on pretreatments of the plant with auxins, cytokinins, and other factors to encourage the formation of embryogenic callus (69). Once embryogenic cells are obtained and the growth regulators are removed, the sources of ammonia and reduced N often become important. In some cases nitrogenous compounds may arise from senescent cells beneath or next to the embryogenic callus. This view is supported by the numerous reports of SE in explants of tropical fruit trees that become senescent as judged by their brown and black colorations (75).

During organogenesis in conifers, urea can be used as a sole source of N. Urea is also a major product of purine (cytokinin) catabolism. The epidermal cells of hypocotyls and cotyledons are the most active in urea assimilation, urease induction, and ammonia formation. The detection of ammonia production in these cells can be demonstrated histochemically (31, 32). The very same cells are most responsive to cytokinins and produce numerous adventitious buds (16, 17). This responsiveness to urea may have a selective advantage for the micropropagation of trees grown under intensive management where nitrogenous fertilizers are routinely employed.

While little is known about ammonia in embryogenesis, Miflin and Lea (90) and Lea et al. (71) have reviewed the literature on enzymes assimilating ammonia and their distribution in plants and isolated organelles. We will focus initially on the surface where ammonia or its analogues first contact the cell or protoplast and on reactions in the fertilized egg in zygotic embryogenesis. We assume that enzymes function at the ports of entry of ammonia so that a contiguity is seen between assimilation and subsequent metabolic events.

Animal eggs and embryos have provided model physiological systems for ammonia effects. Ammonium chloride and other weak bases, when added to unfertilized sea urchin eggs, result in the activation of events which normally follow fertilization such as an increase of protein synthesis and the initiation of DNA synthesis (138). Fertilization triggers an Na-H ion exchange that results in an increase in the intracellular pH and a decrease in the pH of the surrounding medium. The diffusion of a weak base, however, can bypass the Na-H exchange so that the amount of weak base entering the egg is proportional to the initial increase in rate of protein synthesis.

In alfalfa, SE is under the influence of an increase in the average intracellular pH (112). This work supports the view that amino acids provide an easily used source of nitrogen. Apparently proline "relieves the cell of the necessity of extensive nitrate reduction, and of balancing excess anion uptake under constant cation levels. The net result is a high intracellular pH and prolific regeneration." The question still remains as to how ammonia provides the metabolic basis for SE. Insight is needed into enzymatic reactions, the inorganic chemistry of ammonia and the pH changes associated with its assimilation (section 5.).

Enzymes accounting for the assimilation of ammonia isolated from the plant cultures include glutamine synthetase (GS) and glutamic acid dehydrogenase

(GDH). The general view is that ammonium rather than ammonia enters the cell by the GS pathway. Glutamine is a primary product of ammonium assimilation by the GS-GOGAT (glutamine(amide):2-oxoglutarate aminotransferase) pathway. Glutamic acid is a product of GDH.

Glutamine is synthesized in two compartments in leaf cells of most higher plants, the cytosol and chloroplasts. Associated with these compartments are four isoforms of the enzyme (84). Glutamine also seems to be involved in protoplast streaming of conifer cell suspensions (40). We do not know if these attributes make the leaf and shoot explants an advantageous source of competent cells for SE in some species. With conifers, explants from embryonic cells are currently more suitable for SE (see Appendix).

In zygotic embryogenesis, chloroplast development is dramatic. In jack pine embryos of mature seeds, amyloplasts are prominent. Upon germination the amyloplasts are transformed into chloroplasts bearing starch (Fig. 1).

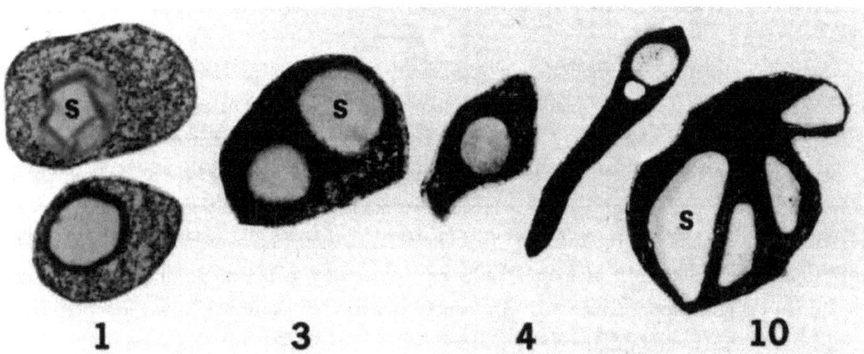

FIGURE 1. Stages of amyloplast-chloroplast development in germinating embryos of jack pine. Numbers refer to days of imbibition: S, starch. Amyloplasts containing starch change into chloroplasts by day 10 (44). Amyloplasts predominate in cells capable of SE (Figure 7).

For this reason, green somatic cell suspensions are usually dedifferentiated to starch laden cells by growing them in darkness and highly heterotrophic media. In darkness, protein degradation and gluconeogenesis predominate. Both processes are thought to lead to the release of ammonia or at least of N-rich compounds such as the amides. Where N is fed to cells as free amino acids and where shifts in N and hormonal levels are employed, SE appears more probable but not reliably so (cf. Appendix). The reproducibility, localization, and specificity of these effects of N must be improved.

In root cells, the enzymes for glutamine synthesis are affected by the adenylate energy charge (4, 135). In conifer explants from hypocotyls and cotyledons, the success of establishing callus is also related to the energy charge of cells at the time of explantation (43). Here the ability of cells to synthesize macromolecules from pivotal N compounds using adenosine triphosphate (ATP) appears to be an important factor in establishing callus and cell suspensions.

The conversion of ammonia to glutamic acid by GDH is influenced by calcium, especially in tobacco cells (51). While other nutrient elements may be involved in regulating ammonia assimilation, our experience with cell suspensions has shown that exogenous calcium levels are especially important in eliciting a useful auxin-cytokinin response from some but not all conifers (41, high calcium for cultures of white spruce from calcareous soils; 34, low calcium for Douglas-fir).

When cells leading to SE are examined, membranes produced internally and external to the cell wall are an important feature of cellular development (39) (Fig. 2). In our experience, membranes of uncertain origin are found attached to the surface of conifer cells undergoing morphogenesis (Fig. 3). How these relate to cutin and suberin (65), levels of amino acids and myo-inositol, or to prooxidation states of cells (20) is unknown. Since the membrane has a natural autofluorescence, this trait can be used to detect embryos in a callus (e.g., 55) or in cell suspensions. It is conceivable that surface reactions on membranes (in far-from-equilibrium physicochemical process) may be involved in the initial effects of ammonia. In chemical model systems, reactions of this type generate spatial structures which change patterns over time (5).

In Douglas-fir, the membrane and its adhering materials, when hydrolyzed with acid, releases high levels of glutamic acid, glycine, alanine, phenylalanine, glucose, galactose, mannose, xylose, and arabinose. Hydroxyproline, proline, and histidine were not detected but these results require reaffirmation (Durzan, unpublished).

Durzan and Steward (45) have attempted to place ammonia assimilation, totipotency, internal events in cells, and totipotency in the context of the life and cell cycle (Fig. 4). This depiction represents two aspects of the same concept.

Every one of the many nuclear-cytoplasmic interactions in cells (Fig. 4) is potentially a target for regulatory control (cf. Fig. 6). The control of gene expression involves replication, transcription, translation and enzyme action. These processes are subject to the intervention of ammonia and reduced N either directly or indirectly. For example in slugs, ammonia determines the choice of morphogenetic pathway (113). Recent evidence with bacteria reveals how N assimilation is regulated genetically (section 4.1.).

In embryogenesis, the assimilation of ammonia should relate where appropriate, to free nuclear divisions and culminate with internal cell division. The volume of the original cell is not greatly increased and the daughter cells are also surrounded by a membrane (Fig. 2B). Up to four free nuclei have been observed in somatic polyembryonic cell suspensions of sugar pine (Gupta and Durzan, unpublished).

At the physiochemical level, the internal divisions of embryonic cells imply that a product of ammonia assimilation or activity causes a change in surface tension at gas/liquid/solid interfaces. Where somatic embryos have been obtained from gymnosperms, a slight but significant drop in the temperature-corrected surface tension of the medium has been observed (Durzan, unpublished data). It is possible that the formation of an embryonic daughter cell is the result of a reorganized subcellular infrastructure or macromolecular scaffold, e.g., microtubles (57; Pescrata, volume 1).

Because of the variety of effects of reduced N on SE (Appendix), each plant species and cell line seems to have developed their own regulatory strategies. For this reason more holistic models of SE which involve metabolic networks, physiological states, and process control are needed.

3. PORTS OF ENTRY FOR REDUCED NITROGEN

Key early products in the biosynthesis of N compounds from ammonia are glutamine and glutamic acid. These are starting points for the synthesis of different families of protein amino acids (section 7.1.). Because of the requirement for ammonia in embryogenesis we assume that ammonia is available as NH_3 or NH_4+ and that these substrates do not interact to form carbamate with $C0_2$ or $HC0_3$--normally present in the culture medium. The requirement for ammonia can be substituted by other forms of organic N (Appendix).

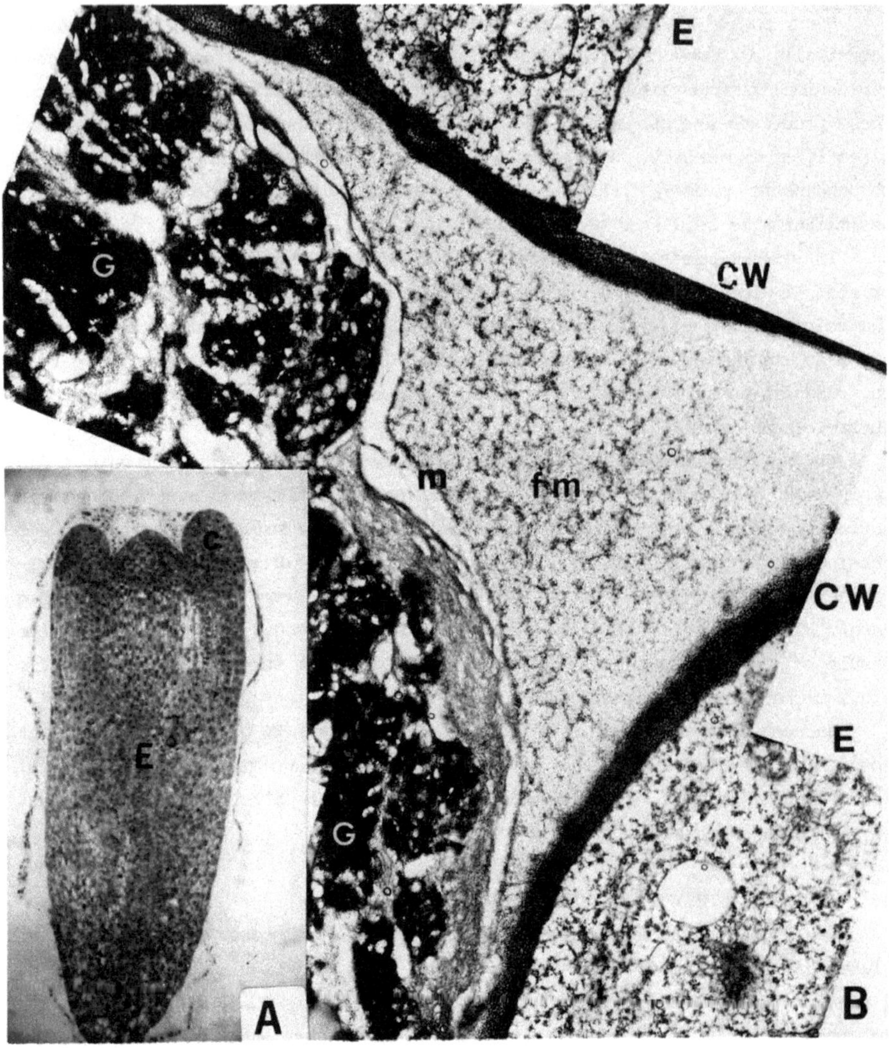

FIGURE 2. Naturally occurring autofluorescing membrane (m) of undefined origin external to the zygotic embryo (E) and attached to cell wall (CW) is a possible site of ammonia activity. A. Cross-section of a Douglas-fir zygotic embryo (ca. 1 mm length) (c) cotyledons. B. A membrane is attached to surface of cell walls of a Douglas-fir zygotic embryo (globular stage). The embryo interfaces with a nourishing female gametophyte (G) and a weakly electron-dense fibrillar material (fm) in a mucilaginous matrix. The matrix may contain oligosaccharins and is bounded by the membrane and the cell wall (x12,000).

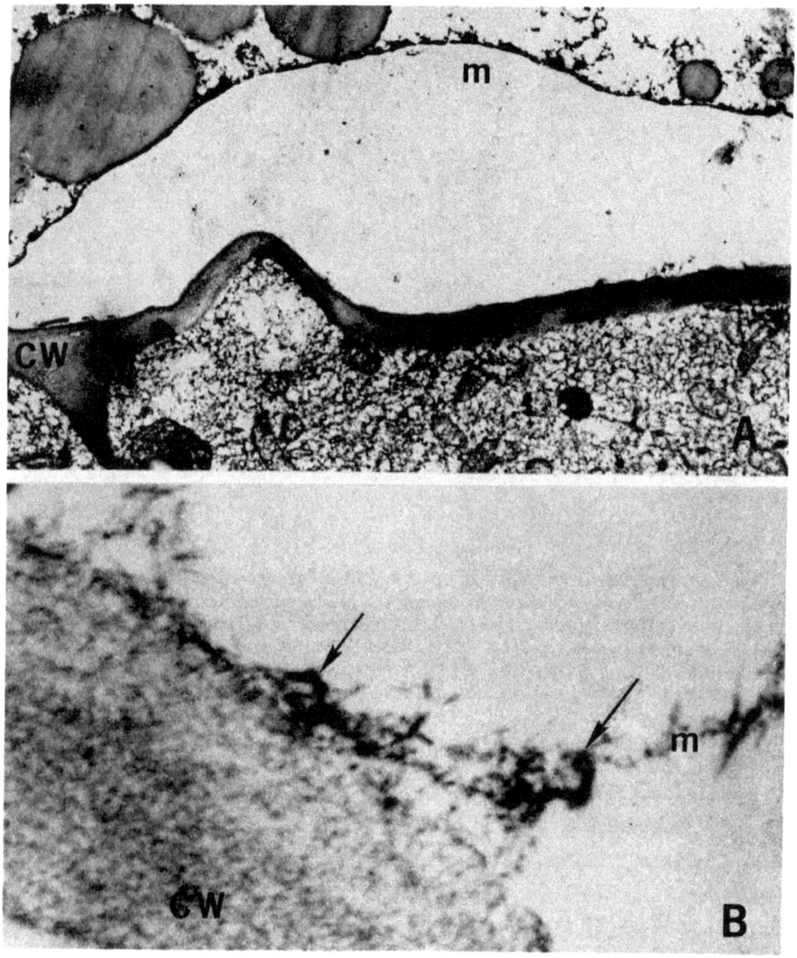

FIGURE 3. Membranes on the surface of somatic cells of conifers displaying morphogenesis in suspension culture. A. Electron micrograph of a Douglas-fir cell derived from a cotyledon callus and developing into a sphaeroblast (37, 38). The membrane (m), which is separated from the cell wall (CW) contains globular and fibrillar material (x11,000). B. Electron micrograph of section of a white spruce cell stained with $FeCl_3$ to reveal presence of phenolic and tannin materials (arrows) in the membrane (x14,000).

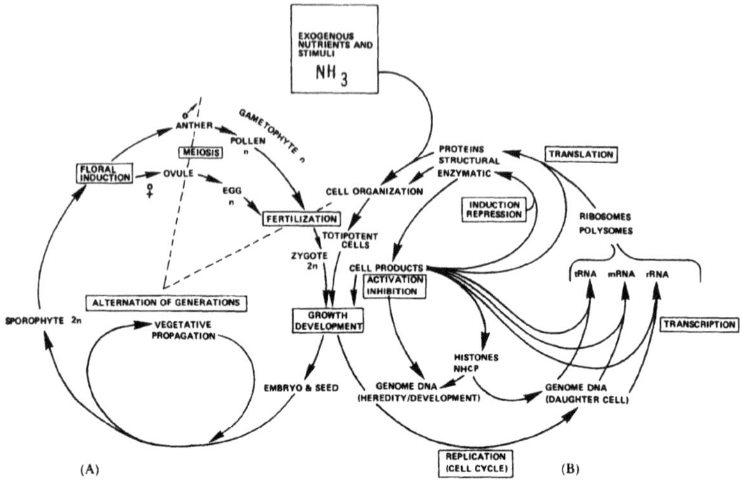

FIGURE 4. Life cycle diagram integrating cellular metabolism and heredity with development, and structure through the activities of totipotent cells. (A) Life cycle: sporophyte and gametophyte. (B) Nuclear-cytoplasmic cycle of cell determination (11). NHCP, nonhistone chromosomal protein; tRNA, transfer-RNA; mRNA, messenger-RNA; rRNA, ribosomal-RNA; n, haploid; 2n, diploid. Ammonia assimilation contributes to cell organization and cell products associated with totipotent cells.

3.1. GDH, GS/GOGAT

Glutamate dehydrogenase (GDH E.C. 1.4.1.2 and E.C. 1.4.1.4), which has a low affinity for ammonia, may account for ammonia assimilation at high levels (> 1 mM). In this reaction, glutamate is formed from α-ketoglutarate. Unfortunately, ammonia levels and comparisons of Km values among GDH (which range from 0.5 to 70 mM) are difficult to interpret. Miflin and Lea (90) emphasize that care must be taken in quoting such values because they are influenced by pH and concentration of substrates among other things.

At low ammonia levels (<0.1 mM) another mechanism is thought to account for ammonia uptake. GS coupled with glutamate synthetase (GOGAT) interact to produce two equivalents of glutamate. One of the two glutamatic acid residues accepts ammonia to yield glutamine. The net reaction is similar to GDH except for the added expenditure of energy as ATP to drive the reaction. In bacteria and Chlorella, GDH's which have a high affinity for ammonia have now been reported. This underscores the reluctance by some investigators to accept the GS/GOGAT system in ammonia assimilation and the difficulties in generalizing from these studies (45).

In animal cells, GDH has been considered as operating in reverse to the above with the production of ammonia especially in the mitochondrial matrix. In this system carbamoyl phosphate synthetase binds the ammonium ion to form carbamoyl phosphate without the need for a carrier (62). In plants, another port of entry for ammonia may be carbamoyl phosphate. Glutamine rather than glutamic acid is the postulated N donor for the activity of carbamoylphosphate synthetase. Considerably more research is needed to resolve the roles of ammonia and glutamine in N assimilation.

Most amino acids for protein synthesis originate from glutamate by transamination (90). Transamination reactions leading to the elaboration of amino acids for protein synthesis are thought to account for ammonia assimilation in growing and dividing cells in culture media at pH's 4 to 7.

Embryogenic cells of the carrot contain slowly migrating bands of GDH. Nonembryogenic lines revealed additional faster migrating bands (72). The significance of these differences for SE and ammonia assimilation is not clear. The production of these enzymes may eventually relate to genetic models of N control (cf. Figs. 5 and 6).

Specific proteins are involved in embryogenesis (98, 121). These may be derived from de novo synthesis of free amino acids or from residues arising by protein turnover. Their occurrence also varies according to the stage of embryonic development. Heat-shock proteins, which are a response to the way embryonic cells are cultivated or treated, are useful to study gene expression during development. Apparently some of the same proteins appear during temperature stress, N starvation, and anoxia (137). The heat-shock system is also of interest to SE because it offers a set of coordinately regulated genes, i.e., a common control over a group of genes, to study with developmental processes. As such, their coordination will have to tie in with classical views of ammonia nutrition, de novo amino acid synthesis and protein turnover (e.g., in conifers; 29, 93). In sugar pine, other unknown factors that stain with acetocarmine and block permeability to Evan's blue are involved in SE (55).

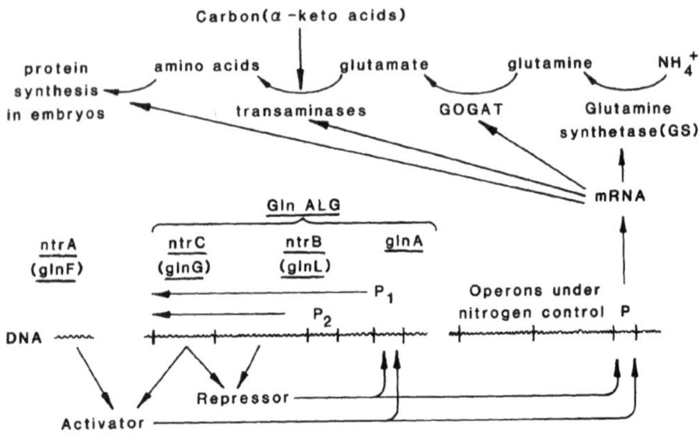

FIGURE 5. Genetic model for N control in plant cells. The model is based on work with bacterial cells (80). Operons under N control may be linked to embryogenesis and activated or repressed by nutrient conditions. Nitrogen control will also have to include processes that involve the assimilation of N and carbon. Unfortunately this model cannot be tested without the availability of mutants that reveal the control for each step.

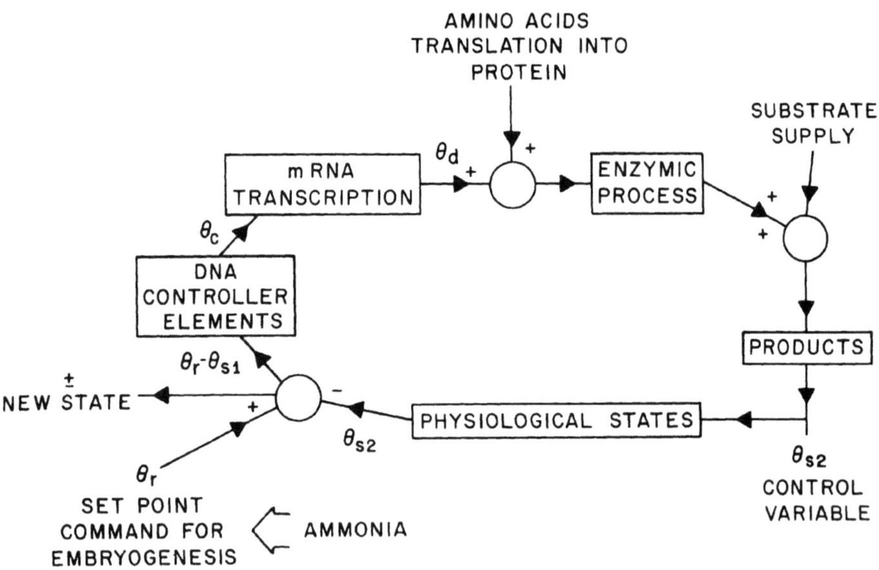

FIGURE 6. Signal-flow diagram for a simple "process control" system in SE. All signals on this diagram are designated by θ). The set point or command signal

θ_r is fed into the cell's network where it is compared with the current physiological state θ_{s1}. The difference between the set point and the signal θ_{s2} is called the deviation or error signal. Ammonia and 2,4-D are command signals. These comprise the forcing signals to elements in DNA that control development.

Responsive elements in turn generate controlling events which may depend on amplifying integrating or differentiating the initial signal. The signal θ_c represents an RNA-polymerase complex acting on DNA which evokes mRNA production required for embryogenesis. Another signal based on mRNA production θ_d evolves at a site or protein synthesis. This step is promoted by casein hydrolysate and occurs ahead of the enzymic processes. A change in the products produced by enzymic reactions and promoted by _myo_-inositol (oligosaccharins), represented as θ_{s2}, enters the system to provide process control over the next physiological state needed for embryo development. This is achieved by comparison with the previous set point θ_r. Physiological states and signals associated with the set point may be characterized by metabolic phenotypes (_cf_. Durzan, this volume).

3.2. Alternative ports of entry

Other modes of entry for ammonia into the primary amino acid pool are possible (45, 90). In many plants a wide range of α-keto acids or aldehydes is available for the synthesis of a corresponding amino acids, e.g., alanine from pyruvate and aspartate from oxaloacetate. Most of these reactions, however, are believed to occur after ammonia enters via GS/GOGAT or GDH (71).

4. GENETIC REGULATION

4.1. Glutamine synthetase

We do not yet know how GS or GS/GOGAT relate to genetic changes leading to embryogenesis. Some clues are emerging from the gene regulation of N uptake in bacteria (80). We must, however, acknowledge major differences that will continue to limit realistic comparisons of genetic regulation between prokaryotic and eukaryotic systems (82).

One major problem with eukaryotes is the lack of classical genetic studies of ammonia assimilation. Most studies are based on a single variable and steady states of cells grown on media that differ only in one component. The understanding of N control systems has recently progressed through genetic analysis using the powerful tools of recombinant DNA and by careful observations on mutant phenotypes (Fig. 5).

GS is responsible for the biosynthesis of glutamine, the assimilation of N into glutamate and for the activation of transcription of operons in amino acid utilization. In earlier views, activity of GS is regulated by the reversible adenylation of specific tyrosyl residues on each subunit of this dodecameric protein. Under N starvation, GS in bacterial cells is deadenylated and highly active.

GS alone is no longer considered a "regulatory protein." Instead, three specific genes ntr A, ntr B, and ntr C have been identified as contributors to N control. The model proposes that products of ntr B and ntr C products repress or activate the transcription of genes under N control including gln A (Fig. 5). In this way the enzymes capable of supplying cells with ammonia become subject to N regulation. The ntr and gln notations for genes are used by different laboratories. Figure 5 shows the correspondence and use of these notations in the N control model. The control over enzyme expression is demonstrated by mutations to produce characteristic phenotypes (80).

Mutation studies of the N control gene system (gln) have revealed a hierarchy of regulatory factors based on seven genes. Products of the gln genes and their interactions regulate the epistatic relationships among phenotypes. Three of the seven genes are organized as a complex operon called gln ALG. The operon has two promoters (gln Ap or P_1; gln Lp or P_2) and an autogenously regulated regulator. The latter is the product of the gln G gene. Autogenous control describes the action of a gene product that either inhibits or activates the expression of the gene coding for it.

In cells grown on a N-limited medium, autogenous regulation of gln G is evident. Transcription of gln G is initiated at the gln A promoter (P_1). By contrast, the gln G product is present in low concentration in cells grown on a poor carbon source with excess glutamine and ammonia. Here transcription is initiated at the gln L promoter (P_2). This low level of the gln G product is sufficient to completely repress the promoter gln Ap or P_1. When the product is converted to its active form it will fully activate transcription initiated at gln Ap. The understanding of GS is still rudimentary and the model quite complex.

The synthesis of enzymes and of membrane transport systems involving N are stimulated by N deficiency. This stimulation makes it possible for the cell to obtain ammonia or glutamate from other N compounds. In the model, N regulation has been divided into two subsystems a) those enzymes and permeases whose only function is N assimilation and b) those which can also supply the cell with energy and carbon. In an ammonia-poor medium, organic N compounds (urea, amides, amino and imino acids, ureides) can satisfy the requirement for reduced N.

Embryogenesis in woody perennials may also be viewed as a limited response to N status (starvation or assimilation). Embryogenesis may be regulated through the N sensing system as depicted by the bacterial model. Recently, tryptophan and ammonium citrate have been used to promote SE in rice and soybean (23, 116). Ammonia is released within cells through degradation of N compounds (cf. section 7.). A simple "process control" model involving physiological states and based on the regulation of SE by ammonia is postulated in Fig. 6. Plant cell growth patterns have been predicted from a feedback model, involving the control of nutrient uptake (59).

Active oxygen may also act as a modulator in gene expression (20). Oxygen is known to be important for the allocation of N during protein synthesis (93). Hence, while several models are available, none have been applied specifically to the study of SE in cells of woody perennials.

4.2. Transcription

The rate of transcription of DNA to produce specific nuclear or messenger RNA's is the most frequent way to control gene expression in eukaryotic cells (27).

During coordinate gene expression in carrot embryogenesis, the participation of nonhistone chromosomal proteins (NHCP) in transcription has been observed (81). Transcription is followed by the formation of new proteins in somatic embryonic cells (121).

Sengupta and Raghaven (114) found that embryonic cells synthesized poly(A)-containing RNA at a higher rate than nonembryonic cells (102). The addition of poly(A) residues to messenger-RNA's (mRNA) is a prerequisite in message processing. Authors postulate that the removal of auxin from the medium may control the biochemical events of embryogenesis by eliciting the synthesis of poly(A)-containing mRNA.

In bacteria, N starvation reduces the free amino acid content. This depletes the pool of charged transfer RNA (tRNA) species and creates ribosome stalling (66). Stalling contributes to the termination of transcription. This process is called attenuation (section 7.2.). It involves the regulation of operons associated with amino acid biosynthesis from ammonia. The role of ammonia in affecting the operation of these systems requires further study.

4.3. Stress and DNA repair

Organogenesis and embryogenesis are usually the outcome of stress upon the genomes of cells *in vitro*. How stress relates to the origin of embryonic somatic cells, genome shock (83) and to the transformation of cells to somatic embryos (11) remains unknown. In SE, senescing, highly oxidized, and pigmented dead cells often, but not always, appear before or concomitant with SE (Tulecke, this volume; 75).

Cellular prooxidant states arise in stressed cells where the intracellular levels of activated forms of oxygen is increased (20). Products of this activity may somehow encourage SE perhaps by the isolation of damaged cells and release of compounds promoting SE.

The major forms of active oxygen that create problems in prooxidant states are superoxide, its conjugate acid or the hydroperoxy radical, singlet oxygen, the hydroxyl radical, and hydrogen peroxide. It is curious that catalase and peroxidase are enzymes commonly observed in tissues before developmental events appear visible (131, 132). Unless a wide range of molecular repair mechanisms are available we can expect that SE arising from heterogenous populations of cells in such systems may be highly variable and aberrant for some genetic parameters (e.g., allozymes, transposable elements, unique base sequences, and restriction site variation).

Some plant cells under stress may develop unstable genetic elements and exhibit an unacceptably high mutation rate (89). In animal cells, the spontaneous decay of DNA is greater than generally recognized (73) and errors appear in DNA synthesis (76, 133). Given the reality of somaclonal and gametoclonal aberrations in plants, (36, 47, 56, 115), we can expect that conifers, which have the highest levels of DNA in the plant and animal kingdom (101), would be most susceptible to stress unless they have evolved efficient repair mechanisms. We do not know if this is true.

As for the genome-stress concept, the recent demonstration with tobacco cells that cell wall fragments (oligosaccharins) can control organogenesis is a significant development (85, 123). How these fragments evoke organogenesis and why embryogenesis is not yet evident in tobacco will require clarification. The chemical and structural basis for the action of oligosaccharins will have to be better understood. To further complicate matters, in tissues under stress, polyamines, guanidines, and amino acids react enzymatically and nonenzymatically with sugars and phenols to produce a range of brown and black conjugates having redox properties.

Trueness to type is fundamental in dealing with somatic embryos, clonal propagation, and the long life cycle of trees (63). Maintenance of trueness to type depends on repair pathways that excise segments of altered DNA. In cultured pea roots there is an excised extrachromosomal DNA in cells that differentiate from the G_2 phase of the cell cycle. This DNA is absent from cells that differentiate from the G_1 phase (68). Cells that differentiate from the G_2 phase tend to become highly elongated. Conceivably, extrachromosomal DNA may be responsible for the formation of suspensors in zygotic and somatic polyembryogenesis. Mitochondrial DNA amplification has been implicated in organogenesis in hazelnut embryos (7).

Many different enzymes initiating DNA repair have been discovered (73). For example, DNA glycosylases, which catalyze the cleavage of base-sugar bands in DNA is universally distributed. They act only on altered or damaged nucleotide residues. Fragmentation of DNA by oxidation or ionizing radiation occurs through ring-opening.

Urea and N-substituted urea derivatives are generated as remnants of pyrimidine breakdown. These modified residues remain attached to the deoxyribose in DNA. Among the types of repair enzymes known, the urea-DNA glycosylases seem unable to release N-substituted urea derivatives (formylpyruvylurea and formylurea) from fragmented pyrimidine residues in DNA. To further complicate matters, calmodulin, a calcium-containing protein, has been implicated in DNA repair (21). Some repair mechanisms show error-prone and inducible-repair functions (73).

In summary, DNA repair enzymes act on DNA in novel and unexpected ways. In some cases, repair involves urea derivatives that may produce ammonia. Analogues of urea, such as thiourea and guanidine (section 6.6.) have different effects on somatic cells. We do not know if these effects involve repair enzymes. We suspect that some of the above observations may help account for the variation and errors in somatic embryogenesis that arise in response to simple nitrogenous compounds.

5. EFFECTS OF AMMONIA

5.1. Acidification

In actively growing cultures, the pH of the medium falls rapidly as ammonia is assimilated. In some cells, a carrier-mediated transport of ammonia across the plasmalemma and tonoplast results in a rapid membrane depolarization and an unspecific diffusion of amines (8).

With rhizoid cells of liverworts, the addition of ammonium salts causes an initially rapid alkalinization of the cell interior (64). The pH of the vacuoles increases from 4.1 to 5.7. This is a result of two processes: 1) acidification of the exterior medium through deprotonation of the actively transported NH_4^+, and 2) alkalinization of the cell interior through protonation of NH_3 which is taken up to a significant extent from high external concentrations (8). Acidification of the medium is further accomplished by removal of cations, net accumulation of anions, deprotonation of weak acids (bicarbonate, organic acids) and by oxidation of substrates.

In somatic embryogenesis, where NH_4NO_3 is a primary N source, the initial acidification of the medium is often followed by an increase in pH. This occurs when nitrate-N is removed from the medium and converted by cells to glutamine and asparagine. These amides are then released to condition and raise the pH of the medium. With neutral pH's (6.8 to 7.0) ammonia is usually a better source of N than nitrate except at low pH (ca. 5.5). Recently microcomputer-controlled systems have been used to sample, assay and collect data related to the effect of ammonium on nitrate uptake in wheat (52).

The principal shortcoming of ammonium sulfates, phosphates, and chlorides is their strong acidity (see Minocha, volume 1). For this reason it is impossible to compare the effects of ammonia and nitrate without first understanding the effects of pH. When evaluating the acidification of the medium, acid- and base-neutralizing factors should also be recognized (129). Unfortunately the pH is usually given more attention than the acid-neutralizing capacity of the medium.

Ammonia buffers are rarely useful because of their toxicity to cells (97). At pH's 9 to 12 reactions with ammonia may create carbamates. By lowering the pH from 9.4 to 6.0 carbamate will decompose rapidly to ammonia and carbon dioxide (see 5.4.1.)

5.2. Effect of ammonia on nitrate assimilation

Very little is known about the effects of ammonia on nitrate assimilation. In some plants, the effect may vary from no effect to partial inhibition (10). Furthermore, the effect may be independent of nitrate reductase activity in roots. The uptake of N as NH_4NO_3 is optimal at pH 6.0 (100).

The addition of calcium to culture media is often beneficial for ammonium N uptake by cells. With NH_4NO_3, the ammonium is absorbed at a faster rate than nitrate but the rate of absorption of N changes under the influence of calcium and potassium (100).

In bacteria and algae, the effect of nitrate on ammonia is regulated by GS products of ammonium assimilation (section 4.1.). In higher plants, other factors generated during NH^+_4 assimilation may inhibit nitrate uptake. This process may be moderated by potassium (110). Trewavas (125) has speculated that nitrate may be considered as a plant hormone.

Effects of one nitrogenous compound on another has an important bearing in SE. Shifts in N levels involving nitrate and ammonia are seen in SE in Liquidambar (33, p. 291-292) and in Picea abies (69). Four aspects emerge when considering the energetic costs of shifts in N assimilation (127). These are 1) the active costs involved in converting nitrate to ammonia, 2) the establishment costs for structural components involved in N assimilation, 3) the energy required to convert fixed carbon into dry matter, and 4) maintenance costs manifest as carbon respired to maintain cells growing on each N source at any given growth rate.

5.3. Effect on sulfate assimilation

The effect of ammonium and nitrate on enzymes of assimilatory sulfate reduction has been examined in Lemna (12) and tobacco (105). If N is not growth-limiting, the assimilation of sulfur (sulfate) is positively controlled by the N pathways (105). Increased recovery of radioactive sulfur from protein was observed when ammonium was added to the nutrient solution, e.g., more sulfate was assimilated and used for protein synthesis. Brunold and Suter (12) postulate that the higher extractable activity of adenosine 5'-phosphosulfate sulfotransferase after ammonium addition to the medium regulates the formation of sulfur amino acids during increased protein synthesis.

In unpublished studies with Douglas-fir sphaeroblasts, sulfur amino acids do not appear in significant quantities in the medium or in cells until 2 to 3 weeks after the cultures are started when the pH rises (D.J. Durzan, unpublished data). This shows that the products of sulfur assimilation may appear in quantity after ammonia and nitrate are assimilated.

5.4. Effect of carbon dioxide assimilation

5.4.1. Role of carbamates and other spontaneous reaction products of ammonia

5.4.1.1. Carbamates. Spontaneous carbamate formation involves a reversible reaction of carbon dioxide with amines or ammonia. Carbamates are produced on the amino group of amino acid residues in proteins. This process affects the function of enzymes and proteins that react with gaseous environments. Major examples are hemoglobin and ribulose bisphosphate carboxylase-oxygenase (6, 78). The only known enzyme system in plants that has free carbamic acid as a postulated intermediate is urease.

In cell environments at pH >7 carbamates may affect the fate of ammonia in embryogenesis. Ammonia may spontaneously react with carbon dioxide to form simple, transient, unstable carbamate and cyanate intermediates (22, 61, 62). While the recorded pH of the culture medium may be <7, the pH in the vicinity of the ammonia and microenvironment of the cell surface may be >7. Indeed, the pH at locations on the plasmalemma or at the cell surface, as measured by several methods, can be well above pH 7 (79). At high pH's the spontaneously formed carbamate could react with or be transferred by carbamoyl transfer reactions. Unfortunately, these transient chemical events are difficult to study and involve the formation of numerous yet unidentified compounds (35).

Where urea-^{14}C is the main source of reduced N for a developing spruce seedling, carbamate formation tends to reduce the recovery of carbon-14 from proteins when compared with bicarbonate-^{14}C alone (31, 35). How these reactions shift the physiological state of cells or contribute to the formation of urea cytokinins (70) or to embryonic states induced by 2,4-D (55) remains unknown.

The chromatographic detection of compounds from cell extracts has indicated the presence of compounds having loosely bound ammonia. When ethanolic extracts of conifer cells are separated on two-directional paper chromatograms, which are then sprayed with the Sakaguchi reagent for monosubstituted guanidines, occasionally an anomalous blue reaction is seen rather than the typical red-orange reaction products. The blue coloration has been attributed to compounds that bind ammonia loosely. These are not always carbamates. The complex reacts with the hypobromite and phenols or oxines to form an indophenol product giving a blue coloration (F. Feigl and V. Anger, personal communication). Except for the few examples below, reactions and artifacts binding ammonia are not considered further.

5.4.1.2. Ribulose bisphosphate carboxylase-oxygenase. Carbon dioxide in the solid, liquid, and gaseous environment of the culture medium may affect the role of carbonic anhydrase in ammonia assimilation. Since natural embryogenesis occurs in darkness, the dark assimilation of CO_2 and bicarbonate should be examined. A pH 8.3, carbon dioxide tension controls the sexual differentiation of Hydra (77).

Photosynthetic assimilation of bicarbonate through involvement of carbamates in aquatic plants may prove informative (79). In photosynthesis, CO_2 reacts with ribulose-1,5-bisphosphate carboxylase-oxygenase (rubisco) and a metal ion to form a carbamate at an ε-amino group of lysine (92). Here CO_2 rather than carbonate or bicarbonate is involved. Enzyme-bound carbamates are labile. They do not withstand the rigorous handling of protein during isolation and study. Bound carbamates are distinct from free carbamates and hydantoins which also may form naturally or as artifacts during embryogenesis.

5.4.1.3. Carbonic anhydrase. Carbonic anhydrase reacts with carbon dioxide to provide bicarbonate at membranes. In animals the formation of carbamate in hemoglobin at high pH contributes to the physiological transport of carbon dioxide (6).

During the action of carbonic anhydrase (74) Zn^{+2} delivers OH^- to the carbon of CO_2 for the formation of bicarbonate. Zinc also forms complexes with amino acids and carbamino acids added to culture media (Table 2) Their role in controlling carbonic anhydrase and SE is unknown. In developing embryos of *Pistacia vera* cv. Kerman, zinc deficiency leads to the accumulation of soluble N rich in amides and arginine. If the deficiency is severe enough it leads to the abortion of zygotic embryos (46).

6. OTHER PIVOTAL PRODUCTS

6.1. Urea

Urea is sometimes a supplemental source of ammonia in cell and tissue cultures. The two equivalents of the strong base, ammonia, are neutralized by carbonic acid, a weak acid. Furthermore, urea can be generated internally through the enzymatic degradation of arginine, purines, ureides, and possibly monosubstituted guanidines. All of these are commonly occurring N compounds which may provide an internal source for ammonia during embryogenesis.

Table 1. Structure of some simple and pivotal, carbon-nitrogen compounds and carbamoyl derivatives of amino acids that are derived from ammonia. These compounds may be useful in controlling embryogenesis.

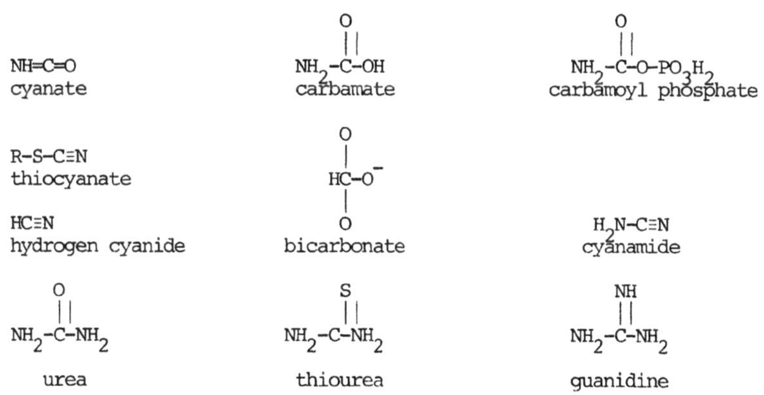

Table 2. Structure of amino acid analogues and reaction product of a carbamino acid and amino acid with zinc.

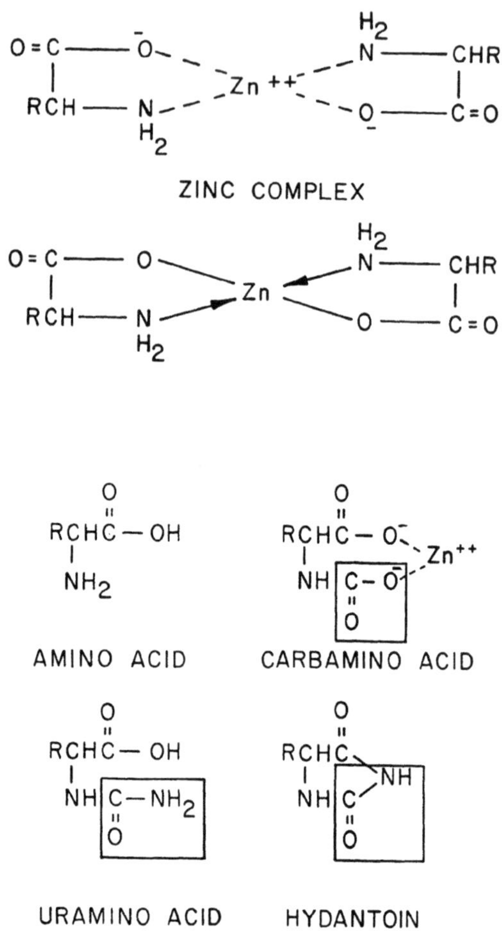

Table 3. Degradation of N^6-benzylamino[8-^{14}C]-purine by zygotic excised embryos, globular sphaeroblasts and 4-month-old cotyledon callus of Douglas-fir from which sphaeroblasts were derived. Experimental materials were exposed for 4 and 20 days to 10 µCi g^{-1} fresh weight per ml of medium. Tissues were washed and extracted with 80% ethanol to remove all alcohol-soluble compounds. Extracts were taken to dryness and redissolved in pH 2.2 Li citrate buffer. Compounds in extracts were separated on an automated amino acid analyzer. Radioactivity was determined by liquid scintillation counting. Unidentified compounds are represented by their elution times (min) from the columns of an amino acid analyzer programmed for physiological fluids. Mean of duplicate studies.

	Percent Distribution of Radioactivity					
	4 Days			20 Days		
Metabolite	Embryos	Sphaero-blasts	Callus	Embryos	Sphaero-blasts	Callus
Nucleotides & Anions	100	90.4	57.1	21.2	59	22.7
Urea (42 min)	--	--	6.8	8.1	23.2	5.3
60 min	--	--	23.5	3.2	--	15.1
92 min	--	--	--	0.7	--	--
200 min	--	--	3.2	64.6	--	52.6
230 min	--	9.6	9.5	0.9	14.1	4.3
290 min	--	--	--	0.9	--	--
Arginine (600 min)	--	--	--	0.2	--	--
Total cpm g fwt	25,000	19,000	38,000	136,000	116,000	143,000

The importance of cytokinins for SE has been suggested by the work of Krogstrup (69), Hakman et al. (58), and others. The catabolism of N^6-benzyladenine produces two equivalents of urea for each adenine degraded. Hence, one criterion for true-to-type SE may be a comparison of how zygotic and somatic embryos metabolize radiolabeled purines. This test has been applied to sphaeroblasts of Douglas-fir (Durzan, unpublished, cf. Table 3). Results are mixed but show that zygotic embryos did not degrade N^6-benzyladenine to urea at least during the first four-days of exposure. After 20 days, zygotic embryos could utilize the C-8 of benzyladenine more so than sphaeroblasts and callus.

6.2. Urease reversal

When water is limiting, ammonium carbonate becomes an nonenzymatic source for the formation of carbamate. At high pH, urea may be formed from carbamate through a reversal of the urease reaction (13). The addition of NH_4OH substantially increases the formation of urea while the addition of CO_2 drastically reduces it. These conditions are very toxic to cells. It seems unlikely that urea may be viewed as a natural product during the processing of ammonia through reversal of urease. However, these reactions may interfere with the interpretation of N^{15} and P^{31} NMR spectra of lyopholized cells capable of regenerating somatic embryos (112, 117).

6.3. Role of amines and carbamates

Polyamines have frequently been implicated in embryogenesis (Appendix; Bagni, volume 1). Amines and polyamines are strong bases with a tendency to react with carbon dioxide to form carbamates. Reactions are usually controlled by steric hindrance and basicity. Steric hindrance by a bulk substituent adjacent to the amino group lowers the stability of the carbamate formed by the CO_2-amine reaction (111). Polyamines are involved in a wide range of macromolecular activities (24).

The main interest in carbamic acid as a product of ammonia assimilation relates to the formation of carbamoyl phosphate via carbamoyl phosphate synthetase (62). Carbamoyl phosphate is unstable in aqueous solutions (1). However, carbamates and cyanates commonly occur in conifer cultures fed urea. Carbamic acid is an intermediate in reactions controlled by urease. Urease contains nickel as a cofactor and is among several proteins containing this element (126). Cyanate is the only other known compound formed spontaneously from carbamate (61).

In animal cells, ammonium ions increase the rate of *de novo* pyrimidine nucleotide synthesis (95). Mitochondrial carbamoyl phosphate synthesis from ammonium is a source of excess carbamoyl phosphate channeled into pyrimidines synthesis where the feedback regulation of this pathway by uridine triphosphate is bypassed.

In white spruce cell suspensions and in germinated seeds, ^{14}C-labeled pyrimidines can be recovered from nucleic acids in tissues fed ^{14}C-urea (31). More work is needed to show the relevance of carbamates in pyridine and nucleic acid synthesis in zygotic and somatic embryogenesis.

6.4. Cyanate -^{14}C

Urea decomposes slowly to cyanate and ammonium ions. This occurs in cell and tissue culture media where urea is used as a source of N. Freshly prepared urea is used to keep cyanate levels as low as possible. This prevents the carbamoylation of free amino groups of proteins at the cell surface (19).

Carbon-14 can be recovered from protein in white spruce cells fed cyanate-^{14}C. Levels of C-14 incorporated in these cells are far lower than those in cells fed bicarbonate-^{14}C (35).

Cyanate reacts with amino, sulfhydryl, phenolic, hydroxyl, carboxyl and imidazole groups of proteins to yield carbamoyl derivatives. Only the reaction product with the amino group is stable (19). These reactions may be of value in probing the role of ammonia in embryogenesis. Some unusual products are O-carbamoyltyrosine and urethanes.

6.5. Cyanide

Cyanide is useful to study cytokinin function by inhibiting cyanide-sensitive respiration (91). Cyanide is also a product of the conversion of 1-aminocyclopropane carboxylic acid (ACC) to ethylene and carbon dioxide. This is followed by the production of asparagine from the cyanide. When apple tissue was treated with Na ^{14}C-cyanide, or with ACC labeled on the C-1 of the cyclopropane ring, the label was recovered from the γ-carboxyl of asparagine (26, 140). Cyanide exposure is related to the senescence and possibly xylogenesis of cells (107). High levels of asparagine and xylogenesis are often seen when sphaeroblasts are formed (34, 38)

Bicarbonate and carbon dioxide markedly and directly enhance ethylene production from ACC (86). The effect is inhibited by light which depletes CO_2 pools. McRae et al. (86) postulate that CO_2 interacts directly with ACC. The authors suggest that this may be achieved by forming an ACC-carbamate complex (78).

6.6. Thiourea and guanidine

A pivotal, naturally occurring metabolite yielding ammonia and carbon dioxide is urea. Arginine and ureides are useful exogenous sources of urea. Analogues of urea such as thiourea and guanidine (Table 1) have strikingly different effects even though ammonia may be a common degradation product.

Thiourea is a useful supplement in media for cell suspensions of fruit tissues (94, 96, cf. Romani, this volume). Thiourea has an invigorating effect on cell growth and encourages the formation of uniform cell suspensions. Thiourea and its analogues bind iron (53) and in this way may somehow affect prooxidant states of cells (see 4.3.). In some instances, dithiocarbamate prevents oxidation reactions in cells (20). Diethyldithiocarbamate is used in micrografting coniferous tissues (cf. Jelaska, volume 3) and other thiocarbamates are herbicides.

Guanidine is highly toxic to cells. ^{14}C-guanidine fed to spruce callus leads to the production of a range of yet unidentified products (Durzan, unpublished). Guanidine is a strong organic base, equivalent in strength to the alkali hydroxides. Guanidines, like the polyamines, trap carbon dioxide from the air.

The substitution of oxygen in urea by sulfur or N yields very different responses in cells. Growth is increased with the sulfur analogue, thiourea, but inhibited by guanidine. If thiocyanates are formed in the medium, these could precipitate amines and other basic inhibitory compounds including the alkaloids (50). The role of these substances in the control of embryogenesis remains to be studied. Recently nitroguanidines have been recognized as a new class of cytokinins (108). We do not know if nitroguanidines form as products of non-enzymatic or cellular activity.

6.7. Cyanamide

Cyanamide is used at low concentrations as a nitrogenous fertilizer and to break the dormancy of grapevines (2). Cyanamide may be useful to invigorate cells of woody plants particularly from mature explants and resting buds. Calcium cyanamide is commonly used as a fertilizer in soil. The compound breaks down into urea, ammonium, and nitrate. Under some soil conditions dicyandiamide, guanylurea, and guanidine are formed (103).

Cyanamide increases levels of free arginine in a number of plants. It inhibits arginase and catalase activity (2). Cyanamide and thiourea promote mitochondrial respiration. (+)-α-(N-1-phenethyl) urea and the homologous monoguanidinium compounds stereospecifically inhibit calcium but not ADP-stimulated mitochondrial respiration (28). These compounds lead to an increased content of peroxides and glutathione. The latter are connected with a redox system that promotes germination of seeds (122). These reactions may be useful in the "conversion" or germination of somatic embryos. Cyanamide, however, has not yet found widespread use in culture systems. Its role in invigorating cells from mature tissues for embryogenesis remains uninvestigated.

7. N TRANSFERS AND RELEASE IN EMBRYOGENESIS: INTERNAL SOURCES OF AMMONIA

7.1. Nitrogen transfers and release

Reinert *et al.* (104) observed that the quantity rather than the quality of the supply of exogenous N was important for embryogenesis. Many sources of reduced N are effective (Appendix). An example of the use of casein hydrolysate (500 ppm) to stimulate somatic polyembryogenesis in sugar pine callus is shown in Figure 7.

Exogenously applied reduced nitrogenous sources may contribute to the release of ammonia from nitrogenous metabolites in cells (Fig. 8). Physiological processes that release ammonia include: protein turnover, gluconeogenesis, photorespiration, and the activities of enzyme systems, e.g., arginase and urease, ammonia lyases, and oxidases as in the formation of ethylene. Ammonia can regulate carbon metabolism in photosynthesis in leaf discs (99)

These reactions will not be discussed except to point out that the suitability of a variety of N sources in the Appendix must relate to the versatility of N and its ability to reallocate through a variety of metabolic pathways represented in Figure 8.

Historically, the dominant controls in N metabolism were thought to affect protein synthesis and turnover rates, nucleic acid synthesis, and oxygen availability (45). However, while the processing of amino acid supplements in the culture medium through an amino moiety or through to ammonia seem clear enough, many other aspects of amino acid metabolism related to embryogenesis are not understood. More sophisticated methods are needed to sort out the role of different amino acids, particularly how they contribute to conditioning of the medium and senescence or transformation of cells. Since we are dealing with populations of nitrogenous compounds we also cannot ignore their physicochemical properties and role as osmolytes (139).

7.2. Attenuation of amino acid biosynthesis

When microorganisms are deprived of a particular substrate such as ammonia or related amino acids, they respond by increasing the rate of synthesis of mRNA encoding the enzymes required to produce the missing ingredients (66). Cells are also able to sense the reduction of the intracellular level of a crucial amino acid and reduce levels of their cognate tRNA's. Each pool is influenced differently by rates of synthesis, transport, and use of the relevant amino acid. Similar influences may be expected for the synthesis of the corresponding tRNAs and aminoacyl-tRNA synthetases.

The extent of tRNA charging is believed to accurately reflect the cell's ability to provide amino acids for protein synthesis (cf. Figs. 6 and 8). A deficiency for a given charged tRNA would lead to ribosome stalling. Stalling occurs on each mRNA transcript during the attempted translation of the corresponding codons into amino acid residues in protein. These consequences of amino acid starvation are called "attenuation." They are believed to control the termination of transcription and regulate operons of amino acid biosynthesis.

While the above is based on results mainly with microorganisms, the growing list of amino acid additions to culture media and shifts in growth regulators (e.g., NAA, BAP, 2,4-D) during subculture suggests that attenuation is worthy of study in plant cells. This may be especially true where cells produce and condition the medium with amino acids (amides, arginine) commonly found in the translocation stream (42). The addition of amino acids to media would have a direct feedback effect while shifts in growth regulators may indirectly affect the attenuation of biosynthetic operons for the amino acids.

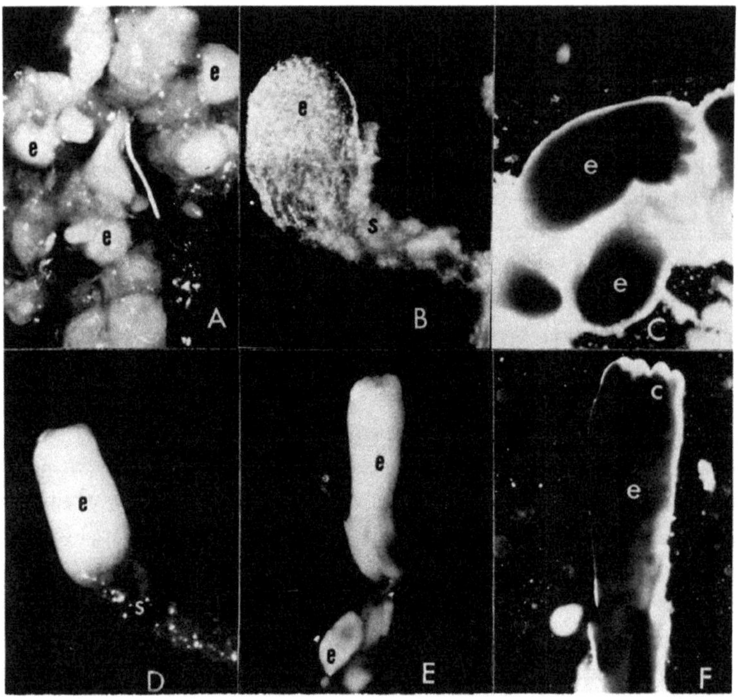

FIGURE 7. Somatic polyembryogenesis in sugar pine. White somatic embryos with suspensors and cotyledons are produced from a mucilaginous white callus containing cells from the root tip zone of 5-year-old seeds of sugar pine. A) mucilaginous callus with multiple somatic embryos (e) (x25); B) globular somatic embryo with suspensor cells (s) teased away from callus (x74); C) somatic embryos at various stages of development in callus (x48); D) elongated somatic embryo with suspensor and developing cotyledon primordia (x30); E) elongated somatic embryo having multiple embryos associated with suspensor (x28); F) somatic embryo with cotyledons (c) which are starting to produce chlorophyll (x32). Embryonic structures (A-F) are produced in response to 2,4-D and casein hydrolysate and myo-inositol. Embryos appear after 2 to 3 weeks of callus establishment and removal of 2,4-D (55).

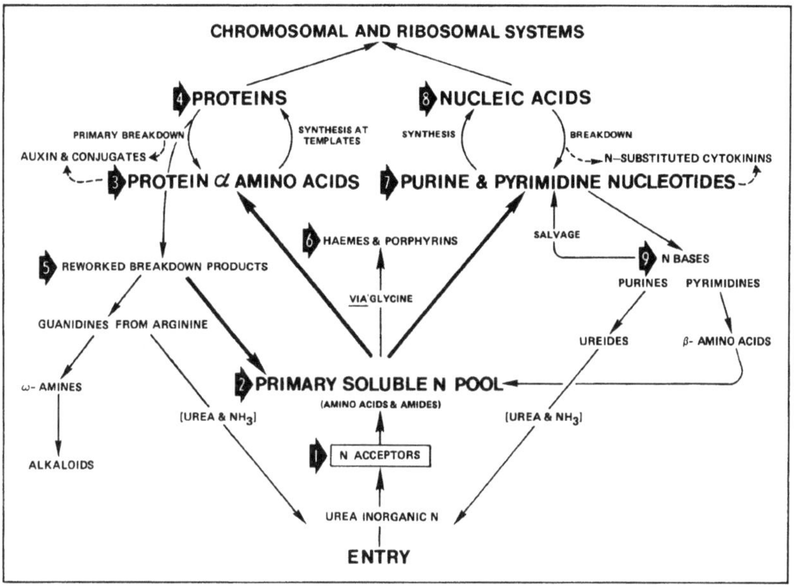

FIGURE 8. General relationships of ammonia and urea to soluble N compounds and chromosomal and ribosomal systems (45).

Ammonia and reduced N are viewed as starting and ending points of the sequential flow of events in N metabolism. Some of these events have been studied during embryogenesis in a variety of plants. The chart does not deal with the mechanism by which RNA, DNA, and ribosomes are involved in protein and nucleic acid synthesis. It does, however, recognize the incorporation of proteins and nucleic acids into the structure of chromosomes and ribosomes.

Key to the numbered arrows 1-9 on the chart.

1. Denotes acceptors for ammonia or reduced N from carbohydrate metabolism with 1 to 5 carbon atoms (α-keto acids).

2. The pool of soluble N compounds, which includes free amino acids and amides, provides the primary products for ammonia assimilation and some end-products of degradative metabolism arising from the turnover of macromolecules.

3. The 20 protein amino acids, which are products resulting from N transfers from compounds in the primary insoluble N pool, represent the principal amino acid families that contribute to protein synthesis and that arise from its breakdown.

4. Proteins are involved in reversible cycles of synthesis and breakdown or turnover; the N from breakdown products contributes to resynthesis or may be reworked as shown below. In cell and tissue cultures, breakdown usually occurs in darkness and synthesis in light.

5. The reworked breakdown products contribute much N in the form of amides (asparagine and glutamine) and soluble free amino acids (protein and nonprotein) to the reusable soluble N pool. N is also directed to other end-products, including alkaloids from amines and other major secondary products of metabolism (e.g., guanidines, γ-glutamyl peptides, etc. not specifically shown on the charts). Whereas the N of protein breakdown is returned as N-rich storage products, much of the carbon is respired away via the citric acid cycle, which also replenishes the carbon and energy needed when these N-rich reserves are reworked in subsequent synthesis.

6. Carbon as succinate from the citric acid cycle with N from amino acids, e.g., glutamate and glycine, leads to a wide range of haemes and porphyrins.

7. N as carbamoyl phosphate, glycine, aspartate, and glutamine with carbon from ribose produces the purine and pyrimidine ribonucleotides for nucleic acid synthesis.

8. In the synthesis of nucleic acids some ribonucleotides are reduced and modified to deoxyribonucleotides required for DNA in the nucleus or other organelles. Some of the bases in nucleic acids are further modified *in situ* during the synthesis of mRNA that serves as the template for protein and enzyme synthesis. Protein synthesis occurs at surfaces (e.g., templates) that provide information (via mRNA) and receive amino acids (via tRNA) for the condensation at a ribosomal surface which itself is a ribosomal protein. Nucleic acids are also involved with nucleoproteins which, in the form of chromatin with its genes, provides the physical basis of inheritance (*cf*. Fig. 6).

9. The primary products of nucleic acid breakdown (the nucleotides) yield by further breakdown their respective N-bases. These have several possible fates. The pyrimidines yield β-amino acids, the N of which may be donated to the reusable soluble pool. The purines and stored ureides, mainly in leguminous trees, may contribute N to the soluble pool via urea and ammonia as transient intermediates. Some of the bases may be salvaged for resynthesis of nucleic acids.

8. NEED FOR DIAGNOSTIC SPECIFICITY

Automated methods are now available for the separation and conductometric determination of ammonia and dissolved carbon dioxide (18). Much understanding would be gained by following the fate of labeled ammonia under conditions inductive for embryogenesis and repeating studies based on isotopic competition (30).

The role of reduced nitrogen compounds in the formation of somatic embryos of alfalfa has been studied using N^{15} nuclear magnetic resonance (NMR) (117).

In this system, the production of somatic embryos is increased 5- to 10-fold by alanine and proline. An increased level of alanine and proline was the only difference between the regenerating and nonregenerating states in alfalfa. When the same system is examined by solid state ^{31}P NMR, regeneration of somatic embryos occurs "without" the production of a low average intracellular pH (112).

Recently ^{15}N has been replaced by ^{13}N as novel methods are being developed to study nitrate and ammonium assimilation by plants (106, 124). For short-term studies (<2 hr) ^{13}N is easier to quantitate and is fed at greater specific activities than ^{15}N. At high specific activities, the steady state levels of a metabolite, e.g., ammonia or amino acids, are not altered (25). Examples are the assimilation of $^{13}NH_4^+$ by symbiotic and free-living cyanobacteria and the labeling of unique polypeptides of arginine and aspartate (87, 88). Unfortunately limitations due to isotopic effects, costs, and access to equipment still impose barriers to the widespread use of ^{13}N.

Where CO_2 impacts on N assimilation, the rate of CO_2 production can be measured using doubly labeled water, typically $H^2H^{18}O$ (14). In animals the ^{18}O label is lost as H_2O and CO_2. The ^2H label is lost only as water. The CO_2 production can be calculated from the turnover rates of the isotopes. Recently the method has become more attractive with the substitution of easily measured tritiated water (H^3HO) for deuterated water, and by the analysis of ^{18}O using cyclotron-generated proton activation, rather than by mass spectrometry (15). Unfortunately analytical problems still remain with the detection of carbamate intermediates (78).

Initial reactions of ammonia may involve membranes and other cell surfaces. Novel artificial membrane mimetic systems are available as polymerized surfactant vesicles that could serve as models (49). Together with a greater understanding of channel-forming membrane proteins (128) and new technologies involving electrodes (48), and the dynamic study of membranes by deuterium NMR (118), we have every hope that greater diagnostic specificity will someday be applied to embryogenesis in forest trees.

We must recognize that in some instances of SE, no nitrogenous factor has been implicated or identified. As such, the requirement for ammonia may not be ubiquitously absolute. However, the conditions may be such that ammonia is somehow released internally by cells initiating SE. Whether this is actually the case is unproven. In this context, the characterization of the physiological state of cells by maps of metabolic phenotypes (cf. Durzan, this volume) and by sensitive immunoassay methods (134) for growth regulators would be of great assistance in clarifying how imposed treatments modify the end-products of ammonia assimilation.

9. REFERENCES

1. ALLEN CM, ME JONES 1964 Decomposition of carbamyl phosphate in aqueous solutions. Biochemistry 3: 1238-1237
2. AMBERGER A 1985 Uptake and metabolism of hydrogen cyanamide in plants. In RJ Weaver, JO Johnson, AS Wicks, eds, Grape Bud Dormancy and H_2CN_2 in Grapes, Proc Int Seminar, Aug. 20, 1984, Dept Viticulture and Enology, Proc Conf Univ California, Davis, pp 5-19
3. AMMIRATO PV 1983 Embryogenesis. In DA Evans, WR Sharp, PV Ammirato, Y Yamada, eds, Handbook of Plant Cell Culture, Vol. 1, MacMillan Publ, New York, pp 82-123
4. ATKINSON DE 1977 Cellular energy metabolism and its regulation. Academic Press, New York
5. AVNIR D, M KAGAN 1984 Spatial structures generated by chemical reactions at interfaces. Nature 307: 717-720
6. BAGGOTT J 1978 The contribution of carbamate to physiological carbon dioxide transport. Trends in Biochem Sci Sept, p 207-208
7. BERNARD-DURIX F, A KOVOOR 1985 Differential replication of mitochondrial DNA in hazel (Corylus avellana) callus cultures during phytohormone-dependent callogenic and organogenic development. In Genetic Manipulations of Crops. Proc Int Meet, Beijing, China (in press)
8. BERTL A, H FELLE, FW BENTRUP 1984 Amine transport in Riccia fulitans. Plant Physiol 76: 75-78
9. BORNMAN CH 1984 Application of in vitro culture technology in clonal forestry. In Proc Int Symp Rec Adv For Biotechnol, July 10-13, 1984, Michigan Biotechnol Inst, Traverse City, Michigan, pp 178-194
10. BRETELER H, M SIEGERIST 1984 Effect of ammonium on nitrate utilization by roots of dwarf bean. Plant Physiol 75: 1099-1103
11. BRINK RA 1962 Phase change in higher plants and somatic cell heredity. Q Rev Biol 37: 1-22
12. BRUNOLD C, M SUTER 1984 Regulation of sulfate assimilation by nitrogen nutrition in the duckweed Lemna minor L. Plant Physiol 76: 579-583
13. BUTLER LG, FJ REITHEL 1977 Urea-catalyzed urea synthesis. Arch Biochem Biophys 178: 43-50
14. BUTTEMER WA, KA NAGY, WW WEATHERS 1982 Comparison of doubly labeled water and time-activity estimates of avian energetics. Physiologist 25: 339
15. CAHILL TA 1983 Crocker nuclear laboratory. Annu Rep 1982-83, Univ California, Davis, 85 pp
16. CAMPBELL RA, DJ DURZAN 1975 Induction of multiple buds and needles in tissue cultures of Picea glauca. Can J Bot 53: 1652-1657
17. CAMPBELL RA, DJ DURZAN 1976 Vegetative propagation of Picea glauca by tissue culture. Can J For Res 6: 240-243
18. CARLSON RM 1978 Automated separation and conductometric determination of ammonia and dissolved carbon dioxide. Anal Chem 50: 1528-1531
19. CARRERAS JA, A CHABAS, D DIEDERICH 1976 Physiological and clinical applications of protein carbamylation. In S Grisolia, R Baguena, F Mayor, eds, Urea Cycle, J Wiley and Sons, New York, pp 501-548
20. CERUTTI PA 1985 Prooxidant states and tumor promotion. Science 227: 375-381.
21. CHAFOULEAS JG, WE BOLTON, AR MEANS 1984 Potentiation of bleomycin lethality by anticalmodulin drugs: A role for calmodulin in DNA repair. Science 224: 1346-1348

22. CHRISTENSSON F, HCS KOEFOED, AC PETERSEN, K RASMUSSEN 1978 Equilibrium constants in the ammonium carbonate-carbaminate system. The acid dissociation constant of carbamic acid. Acta Chem Scand A 32: 15-17
23. CHRISTIANSSON ML, DA WARWICK, PS CARLSON 1983 A morphogenetically competent soybean suspension culture. Science 222: 623-634
24. COHEN SS 1978 What do the polyamines do? Nature 274: 209-210
25. COOPER AJL 1984 Radioactive isotopes of nitrogen. Chem Eng News, Nov 26, p 5
26. DAGANI R 1984 Mechanism of ethylene synthesis in plants clarified. Chem Eng News, Feb 13: 21-23
27. DARNELL JE Jr 1982 Variety in the level of gene control in eukaryotic cells. Nature 297: 365-371
28. DAVIDOFF F, D HASS, D BERTOLINI 1976 (+)-α-(N-1-phenethyl) urea stereospecifically inhibits Ca^{2+}- but not ADP-stimulated mitochondrial respiration. Science 193:66-67
29. DAVIES DD 1982 Physiological aspects of protein turnover. In D Boulter, B Parthier, eds, Nucleic Acids and Proteins in Plants. I. Structure, Biochemistry and Physiology of Proteins. Springer-Verlag, New York, pp 189-228
30. DOUGALL DK 1971 Isotope competition experiments using mixtures of protein amino acids. Coll Int CNRS 193: 367-371
31. DURZAN DJ 1973a The metabolism of ^{14}C-urea by white spruce seedlings in light and darkness. Can J Bot 51: 1197-1211.
32. DURZAN DJ 1973b The incorporation of tritiated water into amino acids in the presence of urea by white spruce seedlings in light and darkness. Can J Bot 51: 351-358
33. DURZAN DJ 1982a Nitrogen metabolism and vegetative propagation of forest trees. In JM Bonga, DJ Durzan, eds, Tissue Culture in Forestry, Martinus Nijhoff/Dr W. Junk, The Hague, pp 256-324
34. DURZAN DJ 1982b Somatic embryogenesis and sphaeroblasts in conifer cell suspension. In A Fujiwara, ed, Plant Tissue Culture, Proc 5th Int Cong Plant Cell Tissue Cult, Tokyo, Japan, 1982, pp 113-114
35. DURZAN DJ 1983 The fate of carbon during the assimilation of carbamoyl phosphate in white spruce seedlings as revealed by [^{14}C]-carbamoyl phosphate, [^{14}C]-cyanate, and [^{14}C]-bicarbonate labeling patterns. Physiol Plant 59: 233-241
36. DURZAN DJ 1984a Potential for genetic manipulation of forest trees: Totipotency, somaclonal aberration, and trueness to type. In Proc Int Symp Rec Adv For Biotechnol, July 10-13, 1984, Michigan Biotechnol Inst, Traverse City, Michigan, pp 104-125
37. DURZAN DJ 1984b Special problems: Adult vs. juvenile explants. In WR Sharp, DA Evans, PV Ammirato, Y Yamada, eds, Handbook of Plant Cell Culture, Vol 2, MacMillan Publ Co, New York, pp 471-503
38. DURZAN DJ 1984c Initial pH and early changes in nitrogen metabolism associated with sphaeroblast development in cell suspension cultures of white spruce. In FJ Novak, L Havel, J Dolezel, eds, Proc Int Symp Plant Tissue Cell Cult Appl Crop Improv September 24-29, 1984, Olomouc, Czechoslovakia, pp 193-194
39. DURZAN DJ 1986 Nitrogenous extractives extracellular to the lignocellulosic cell wall. In JW Rowe, ed, Natural Products Extraneous to the Lignocellulosic Cell Wall of Woody Plants, Chap 4, Springer-Verlag, New York, (in press)
40. DURZAN DJ, G BOURGON 1976 Growth and metabolism of cells and tissue of jack pine (*Pinus banksiana*). 7. Observations on cytoplasmic streaming and effects of L-glutamine and its analogues on subcellular activities. Can J Bot 54: 507-517

41. DURZAN DJ, SC CHAFE, SM LOPUSHANSKI 1973 Effects of environmental changes on sugars, tannins, and organized growth in cell suspension cultures of white spruce. Planta (Berl) 113: 241-249

42. DURZAN DJ, V CHALUPA 1976 Growth and metabolism of cells and tissue of jack pine (Pinus banksiana). 6. Free nitrogenous compounds in cell suspension cultures of jack pine as affected by light and darkness. Can J Bot 54: 496-506

43. DURZAN DJ, V CHALUPA, AJ MIA 1976 Growth and metabolism of cells and tissues of jack pine (Pinus banksiana). 1. The establishment and some characteristics of a proliferated callus from jack pine seedlings. Can J Bot 54: 437-445

44. DURZAN DJ, AJ MIA, BSP WANG 1971 Effects of tritiated water on the metabolism and germination of jack pine seeds. Can J Bot 49: 2139-2149

45. DURZAN DJ, FC STEWARD 1983 Nitrogen metabolism. In FC Steward, RGS Bidwell, eds, Plant Physiology: An Advanced Treatise, Vol 8, Academic Press, New York, pp 55-265

46. DURZAN DJ, K URIU 1986 Metabolic networks in developing pistachio embryos (Pistachia vera cv. Kerman). New Discoveries and Technologies. Alan R. Liss, Inc , NY (In press)

47. EVANS DA, WR SHARP, HP MEDINA-FILHO 1984 Somaclonal and gametoclonal variation. Am J Bot 71: 759-774

48. FAULKNER LR 1984 Chemical microstructures on electrodes. Chem Eng News 62: 28-45

49. FENDLER JH 1984 Polymerized surfactant vesicles: novel membrane mimetic systems. Science 223: 888-894

50. FULTON CC 1940 New precipitating agents for alkaloids and amines. Am J Pharm 112:1-35

51. FURUHASHI K, Y TAKAHASHI 1982 Glutamate dehydrogenase of green tobacco callus tissue. In A Fujiwara, ed, Plant Tissue Culture, 1982, Proc 5th Int Cong Plant Tissue Cell Culture, Tokyo, pp 241-242

52. GOYAL SS, RC HUFFAKER 1985 Induction of NO_3^- transport system in wheat seedlings: Effect of NH_4^+ and NO_2^-. Plant Physiol 77(Suppl) Abst 169

53. GRANIFO J, R LATORRE, J COSTAMAGNA 1981 Characterizacion de algunos nuevos complejos de Fe(II) en alto spin, con tioureas substituidas compo ligantes. Contr Cient y Technol No. 51, Quimica XVIII 67-78

54. GUPTA PK, DJ DURZAN 1985 Shoot multiplication from mature trees of Douglas-fir (Pseudotsuga menziesii) and sugar pine (Pinus lambertiana). Plant Cell Reports 4: 177-179

55. GUPTA PK, DJ DURZAN 1986 Somatic polyembryogenesis from callus of mature sugar pine embryos. Bio/Technology 4:643-645

56. HACKETT WP 1983 Phase change and intra-clonal variability. HortScience 18: 12-16

57. HAHNE G, F HOFFMANN 1984 Dimethyl sulfoxide can initiate cell divisions of arrested callus protoplasts by promoting cortical microtubule assembly. Proc Nat Acad Sci USA 81: 5449-5453

58. HAKMAN I, LC FOWKE, S VON ARNOLD, T ERIKSSON 1985 The development of somatic embryos of Picea abies (Norway Spruce). Plant Sci Lett 38:53-59

59. HO SV, ML SHULER 1977 Predictions of cellular growth patterns by a feedback model. J Theor Biol 68: 415-435

60. JOHRI BM (ed) 1982 Experimental Embryology of Vascular Plants. Springer-Verlag, New York

61. JONES ME 1963 Carbamyl phosphate. Science 140: 1373-1379

62. JONES ME 1976 Partial reactions of carbamyl-phosphate synthetase: A review and an inquiry into the role of carbamate. In S Grisolia, R Baguena, F Mayor, eds, Urea Cycle, J Wiley and Sons, New York, pp 107-122

63. KESTER DE 1983 The clone in horticulture. HortScience 18: 831-837
64. KLEINER D 1981 The transport of NH_3 and NH_4+ across biological membranes. Biochem Biophys Acta 639: 41-52
65. KOLATTUKUDY PE 1980 Biopolyester membranes of plants: Cutin and suberin. Science 208: 990-1000
66. KOLTER R, C YANOFSKY 1982 Attenuation in amino acid biosynthetic operons. Annu Rev Genet 16: 113-134
67. KORNFIELD R, S KORNFIELD 1985 Assembly of asparagine-linked oligosaccharides. Annu Rev Biochem 54: 631-634
68. KRIMER DB, J VAN'T HOF 1983 Extrachromosomal DNA of pea (Pisum sativum) root-tip cells replicates by strand displacement. Proc Nat Acad Sci USA 80: 1933-1937
69. KROGSTRUP P 1984 Micropropagation of conifers. PhD Thesis. Royal Vet Agric Univ, Copenhagen 161 pp (in Danish)
70. KUROSAKI F, S TAKAHASHI, K SHUDO, T OKAMOTO 1981 Structural and biological links between urea- and purine cytokinins. Chem Pharm Bull 29: 3751-3753
71. LEA PJ, RM WALLSGROVE, BJ MIFLIN 1985 The biosynthesis of amino acids in plants. In GC Barrett, ed, Chemistry and Biochemistry of the Amino Acids. Chapman and Hall, London, pp 197-226
72. LEE DW, DK DOUGALL 1973 Electrophoretic variation in glutamate dehydrogenase and other isozymes in wild carrot cells cultured in the presence and absence of 2,4-dichlorophenoxyacetic acid. In Vitro 8: 347-352
73. LINDAHL T 1983 DNA repair enzymes. Annu Rev Biochem 51: 61-87
74. LIPSCOMB WN 1983 Structure and catalysis of enzymes. Annu Rev Biochem 52: 17-34
75. LITZ R 1985 Somatic embryogenesis in tropical fruit trees. In RH Henke, KW Hughes, MJ Constatin, A Hollaender, eds, Tissue Culture in Forestry and Agriculture. Basic Life Sciences, Vol 32, Plenum Press, New York, pp 179-193
76. LOEB LA, TA KUNKEL 1983 Fidelity of DNA synthesis. Annu Rev Biochem 51: 429-457
77. LOOMIS WF 1957 Sexual differentiations in Hydra. Science 126: 735-739
78. LORIMER GH 1983 Carbon dioxide and carbamate formation: The makings of a biochemical control system. Trends Biochem Sci 8: 65-68
79. LUCAS WJ 1983 Photosynthetic assimilation of exogenous HCO_3^- by aquatic plants. Annu Rev Plant Physiol 34: 71-104
80. MAGASANIK B 1982 Genetic control of nitrogen assimilation in bacteria. Annu Rev Genetics 16: 135-168
81. MATSUMOTO H, D GREGOR, J REINERT 1975 Changes in chromatin of Daucus carota cells during embryogenesis. Phytochemistry 14: 41-47
82. MCCLINTOCK B 1967 The role of the nucleus. Genetic systems regulating gene expression during development. Dev Biol Suppl 1: 84-112
83. MCCLINTOCK B 1984 The significance of responses of the genome to challenge. Science 226: 792-801
84. MCNALLY SF, B HIREL, P GADEL, AF MANN, GR STEWART 1983 Glutamine synthetases of higher plants. Evidence for a specific isoform content related to their possible physiological role and their compartmentation within the leaf. Plant Physiol 72: 22-25
85. MCNEIL M, AG DARVILL, SC FRY, P. ALBERSHEIM 1984 Structure and function of the primary cell walls of plants. Annu Rev Biochem 53: 625-663
86. MCRAE DG, JA COKER, RL LEGGE, JE THOMPSON 1983 Bicarbonate/CO_2-facilitated conversion of 1-aminocyclopropane-1-carboxylic acid to ethylene in model systems and intact tissues. Plant Physiol 73: 784-790

87. MEEKS JC, CS ENDERLIN, KL WYCOFF, JS CHAPMAN, CM JOSPEH 1983a Assimilation of $^{13}NH_4^+$ by *Anthoceros* grown with and without symbiotic *Nostoc*. Planta 158: 384-391
88. MEEKS JC, KL WYCOFF, JS CHAPMAN, CS ENDERLIN 1983b Regulation of expression of nitrate and dinitrogen assimilation by *Anabaena* species. Appl Env Microbiol 45: 1351-1359
89. MEINS FJ Jr 1983 Heritable variation in plant cell culture. Annu Rev Plant Physiol 34: 377-346
90. MIFLIN DJ, PJ LEA 1982 Ammonia assimilation and amino acid metabolism. *In* D Boulter, B Parthier, eds, Nucleic Acids and Proteins in Plants. 1. Structure, Biochemistry and Physiology of Proteins, Springer-Verlag, New York, pp 5-64
91. MILLER CO 1978 Cytokinin modification of metabolism of p-coumaric acid by a cell suspension of soybean (*Glycine max* L. Merrill). Planta 140: 193-199
92. MIZIORKO HM, GH LORIMER 1983 Ribulose-1,5-bisphosphate carboxylase-oxygenase. Annu Rev Biochem 52: 507-535
93. MOTHES K 1958 Ammoniakentgiftung und Aminogruppenvorrat. *In* W Ruhland, ed, Handbuch der Pflanzenphysiologie. Springer-Verlag Berlin, VIII:716-762
94. NITSCH JP, T ASHIRA, LME ROSSINI, C NITSCH 1970 Bases physiologique de la production de chair de pomme et de poire *in vitro*. Bull Soc Bot Fr 117: 479-492
95. PAUSCH JG, DO KEPPLER, W GEROK 1977 Increased *de novo* pyrimidine synthesis in liver induced by ammonium ions in amounts surpassing the urea cycle. Eur J Biochem 76: 157-163
96. PECH JC, RJ ROMANI 1979 Senescence of pear fruit cells cultured in a continuously renewed, auxin-deprived medium. Plant Physiol 63: 814-817
97. PERRIN DD, B DEMPSEY 1974 Buffers for pH and Metal Ion Control. Chapman and Hall, London, 176 pp
98. PITTO L, F SOSCHIAVO, G GINLIANO, M TERZI 1983 Analysis of heat-shock protein pattern during somatic embryogenesis of carrot. Plant Mol Biol 2: 231-237
99. PLATT SG, Z PLATET, JA BASSHAM 1977 Ammonia regulation of carbon metabolism in photosynthesizing leaf discs. Plant Physiol 60: 739-742
100. PRIANISHNIKOV DN 1951 Nitrogen in the life of plants. Kramer Business Service, Madison, Wisconsin
101. PRICE HJ, AH SPARROW, AF NAUMAN 1973 Evolutionary and developmental considerations of the variability of nuclear parameters in higher plants. I. Genome volume, interphase chromosome volume, and estimated DNA content of 236 gymnosperms. *In* Basic Mechanisms in Plant Morphogenesis, Brookhaven Nat Lab Symp No 25, June 4-6, 1973, Office Techn Serv, Dept Commerce, Washington, DC, pp 390-421
102. RAGHAVEN V 1983 Biochemistry of somatic embryogenesis. *In* DA Evans, WR Sharp, PV Ammirato, Y Yamada, eds, Handbook of Plant Cell Culture, Vol 1, MacMillan Publ, New York, pp 655-671
103. RATHSACK K 1955 Über Umsetzungsprodukte des Cyanamids in Boden. Landw Forsch 6: 116-123
104. REINERT J, M TAZAWA, S SEMENOFF 1967 Nitrogen compounds as factors of embryogenesis *in vitro*. Nature 216: 1215-1216
105. REUVENY Z, DK DOUGALL, PM TRINITY 1980 Regulatory coupling of nitrate and sulfate assimilation pathways in cultured tobacco cells. Proc Nat Acad Sci USA 77: 6670-6672
106. RHODES D, AC MYERS, G JAMIESON 1981 Gas chromatography-mass spectrometry of N-heptafluorobutyrl isobutyl esters of amino acids in the analysis of the kinetics of $[^{15}N]H_4^+$ assimilation in *Lemna minor* L. Plant Physiol 68: 1197-1205

107. ROBERTS LW, S BABA, T SHIRAISHI, AR MILLER 1982 Progress in cytodifferentiation under in vitro conditions. In A Fujiwara, ed, Plant Tissue Culture 1982, Proc 5th Int Cong Plant Tissue Cell Cult, Tokyo, pp 87-90
108. RODAWAY S, AW LUTZ 1985 Nitroguanidines: A new class of synthetic cytokinins. Plant Physiol 77(Suppl) Abst 109
109. ROMBERGER JA, CA TABOR 1975 The Picea abies shoot apical meristem in culture. II. Deposition of polysaccharides and lignin-like substances beneath cultures. Am J Bot 62: 610-617
110. RUFTY TW Jr, WA JACKSON, CD RAPER Jr 1982 Inhibition of nitrate assimilation in roots in presence of ammonium: The moderating influence of potassium. J Exp Bot 33: 1122-1137
111. SARTORI G, DW SAVAGE 1983 Sterically hindered amines for CO_2 removal from gases. I and EC Fundamentals 22: 239-249
112. SCHAEFFER J 1985 Regeneration in alfalfa tissue culture. Characterization of intracellular pH during somatic embryo production by solid-state P-31 NMR. Plant Physiol 79:584-589
113. SCHINDLER J, M SUSSMAN 1977 Ammonia determines the choice of morphogenetic pathways in Dictostelium discoideum. J Mol Biol 116: 161-169
114. SENGUPTA C, V RAGHAVEN 1980 Somatic embryogenesis in carrot cell suspension. II. Synthesis of ribosomal RNA and poly(A) + RNA. J Exp Bot 31: 259-268
115. SINDEN SL, JF SHEPARD 1983 Variation at the cellular level. HortScience 18: 837-840
116. SIRIWARDANA S, MW NABORS 1983 Tryptophan enhancement of somatic embryogenesis in rice. Plant Physiol 73: 142-146
117. SKOKUT TA, J MANCHESTER, J SHAEFER 1985 Regeneration of alfalfa tissue culture. Stimulation of somatic embryo production by amino acids and N-15 NMR determination of nitrogen utilization. Plant Physiol 85:579-583
118. SMITH RL, E OLDFIELD 1984 Dynamic structure of membranes by deuterium NMR. Science 225: 280-288
119. STEWARD FC 1983 Reflections on aseptic culture. In DA Evans, WR Sharp, PV Ammirato, Y Yamada, eds, Handbook of Plant Cell Culture, Vol 1, MacMillan Publ, New York, pp 1-10
120. STEWARD FC, HW ISRAEL, RL MOTT, HJ WILSON, AD KRIKORIAN 1975 Observations on the growth and morphogenesis in cultured cells of carrot (Daucus carota L.). Proc Roy Soc London Ser B 273: 33-53
121. SUNG ZR, R OKIMOTO 1983 Coordinate gene expression during somatic embryogenesis in carrots (Daucus carota L.). Proc Nat Acad Sci USA 80: 2661-2665
122. TAYLORSON RB, SB HENDRICKS 1977 Dormancy in seeds. Annu Rev Plant Physiol 28: 331-354
123. TRAN THANH VAN K, P TOUBART, A COUSSON, AG DARVILL, DJ GOLLIN, P CHELF, P ALBERSHEIM 1985 Oligosaccharins can control morphogenesis in tobacco explants. Nature 314: 615-617
124. THAYER JR, RC HUFFAKER 1982 Kinetic evaluation, using N-13 reveals two assimilatory transport systems in Klebsiella pneumoniae. J Bact 149: 198-202
125. TREWAVAS AJ 1983 Nitrate as a plant hormone. In MB Jackson, ed, Interactions Between Nitrogen and Growth Regulators in the Control of Plant Development, Brit Plant Growth Regulat Group Monogr 9: 97-110
126. THOMSON AJ 1980 Proteins containing nickel. Nature 298: 602-603
127. TURPIN DH, DB LAYZELL, IR ELRIFI 1985 Modeling the C economy of Anabaena flos-aquae. Plant Physiol 78: 746-752
128. UNWIN PNT, PD ENNIS 1984 Two configurations of a channel-forming membrane protein. Nature 307: 609-614

129. VAN BREEMEN N, CT DRISCOLL, J MULDER 1984 Acid deposition and internal proton sources in acidification of soils and waters. Nature 307: 599-604
130. VANDERHOEF L, T KOSUGE 1984 The molecular biology of plant hormone action: Research directions for the future. Am Soc Plant Physiol, Rockville, Maryland, 40 pp
131. VAN FLEET DS 1952 Histochemical localization of enzymes in vascular plants. Bot Rev 18: 354-398
132. VAN FLEET DS 1959 Analysis of the histochemical localization of peroxidase related to the differentiation of plants. Can J Bot 37: 449-458
133. VON HIPPEL PH, DG BEER, WD MORGAN, JA MCSWIGGEN 1984 Protein-nucleic acid interactions in transcriptions: A molecular analysis. Annu Rev Biochem 53: 389-446
134. WEILER EW 1984 Immunoassay of plant growth regulators. Annu Rev Plant Physiol 35:85-95
135. WEISSMAN GS 1976 Glutamine synthethase regulation by energy charge in sunflower roots. Plant Physiol 57: 339-343
136. WETHERELL DF 1978 In vitro embryoid formation in cells derived from somatic plant tissues. In KW Hughes, R Henke, M Constantin, eds, Propagation of Higher Plants Through Tissue Culture. Proc Symp Univ Tenn, April 16-19, 1978, Tech Info Centre US Dept Energy 102: 124
137. WILLIAMS B 1984 Anaerobic Stress and induction of thermotolerance. Proc Bay Area Plant Molecular Biol Minisymp on Plant Stress. Dec. 19, 1984, Stanford, Univ, California
138. WINKLER MM, JL GRAINGER 1978 Mechanism of action of NH_4Cl and other weak bases in the activation of sea urchin eggs. Nature 273: 536-538
139. YANCEY PH, ME CLARK, SC HAND, RD BOWLUS, GN SOMERO 1982 Living with water stress: Evolution of osmolyte stystems. Science 217: 1214-1222
140. YANG S-F, NE HOFFMAN 1984 Ethylene biosynthesis and its regulation in higher plants. Annu Rev Plant Physiol 35: 155-189

KEY TO SYMBOLS IN APPENDIX

Basal Media[1]

B, Blayde's medium
MS, Murashige-Skoog
MT, Murashige-Tucker
B-5, Gamborg *et al*
LS, Linsmaeir and Skoog
W, White
WC, Wild Carrot

Chemicals and Ingredients

BA, or BAP, N^6-benzyladenine
CW, coconut water
2,4-D, 2,4-dichlorophenoxyacetic acid
GA, gibberellic acid
K, kinetin
ME, malt extract
NAA, α-naphthalene acetic acid
NOAA, β-naphthoxyacetic acid
IAA, β-indoleacetic acid
PVP, polyvinylpyrrolidone

[1]See article cited in Tables (References) for citations and modifications to these basal media.

Appendix

Selected alphabetical and chronological listing of examples of somatic embryogenesis with special reference to nitrogen sources. A. Daucus carota. B. Woody perennial species. Concentrations in µM unless otherwise noted.

Species and explant source	N source and factors	Medium, observations	Reference
A. Daucus carota[1]			
phloem explants	Coconut milk, casein hydrolysate, sequential	W, Cell suspension cultures	1958 Steward FC et al. Am J Bot 45: 705
	treatments in cell suspension	summary of research	1975 ibid Proc Roy Soc (Lond) 273:33
		Callus on solid media	1958 Reinert J Ber Deutsch, Bot Ges 71:15
	Reduced N	Absolute requirement for SE	1965 Halperin W, WF Wetherell Nature 205:519 1966 Halperin W Am J Bot 53:443
	Nitrate, NH_4Cl, glycine asparagine, glutamine arginine, yeast extract	Effective sources of reduced N	1966 Kato H, M Takeuchi. Sci Papers Coll Gen Education (Japan) 16:245
	Quantity and not quality of N is important		1967 Reinert J et al., Nature 216:1215
	Kinetin	Restoration of declining morphogenetic capacity in long-term culture.	1972 Wochock Z, DF Wetherell Experientia 28:104

[1]A recent volume on "Somatic Embryogenesis" 1985 M Terzi, L Pitto, and ZR Sung, Proc. Workshop San Miniato, Italy, May 28-31, summarizes much of the knowledge on carrot embryogenesis.

Species and explant source	N source and factors	Medium, observations	Reference
petiole	0.1 mM NH_4Cl promotes embryogenesis in a NO_3^- medium. Few embryos per culture with 5-55 mM nitrate	Reaffirmation of role of NH_4, gln and ala also good N sources	1976 Wetherell DF, D K Dougall Physiol Plant 37:97
	NH_4^+ as sole N source	WC, 10 mM NH_4Cl & 20 mM KCl controlled at pH 5.4 with KOH and $KHCO_3$	1978 Dougall, D K and D C Verma In Vitro 14:180
	Amino acids nitrate, DCMU kinetin, IAA	W, heterotrophic nutrition is followed by a mixotrophic and autotrophic period	1978 Neumann K H et al., Z Pflanzenernaehr Bodenkd 141:299
	Polyamines		1978 Montague, M et al. Plant Physiol 62:430
	Arginase decarboxylase activity		1979 ibid 63:341
	KNO_3, NH_4NO_3 amino acids	NH_4SO_4 not useful, gln most effective of amino acids as sole N	1979 Kamada, H and H Harada. Z Pflanzenphysiol 91:453
hypocotyl cells (31 to 47 μ dia)	DNA, RNA, protein content, template activity using synchronous cells	Changes in macromolecular levels are important in media with and without zeatin and 2,4-D (medium of Lin and Staba)	1980 Fujimara, T et al., Physiol Plant 49:255

Species and explant source	N source and factors	Medium, observations	Reference
hypocotyl	Endogenous amino acids, nitrate reductase aminotransferase activity	MS, gln, ala and gln stimulate embryogenesis	1982 Kamada H, H Harada Proc 5th Intl Cong Plant Tiss Cell Cult 115
hypocotyl	Control at the transcriptional level	Reduction in replicon size of DNA in medium with zeatin vs. zeatin and 2,4-D	1982 Fujimura T, A Komamine Proc 5th Intl Congr Plant Tiss Cell Cult 105-106
temperature-sensitive mutant cell lines	Proline and 2-amino-5-ethoxycarbonyl pyrimidine-4-(3H)-one	B5, N-compounds reduced morphogenesis	1982 Nuti Ronchi V et al., Proc 5th Intl Congr Plant Tissue Cell Cult 117
	Heat-shock proteins in all stages after globular stage	α-Amanitin does not alter protein pattern	1983 Pitto L et al., Plant Mol Biol 2:231
petiole explant	Polyamines	WC, Arginine decarboxylase and polyamines required, inhibition of embryogenesis by α-difluoromethylarginine but not α-difluoromethylornithine	1984 Feirer R et al., Science 223:1433
B. Woody Perennial Species			
Abies balsamea shoots		Axillary bud formation and embryogenesis	1977 Bonga JM *In Vitro* 13:41
Albizzia lebbeck East Indian walnut	not identified	B5	1981 Ghaiyal PK, SC Maheshwari Naturwiss 68:379

Species and explant source	N source and factors	Medium, observations	Reference
Bambusa arundinacea Bamboo seeds (free cells and callus)	not identified	N_6 (medium of Chu Chih-ching) 2,4-D, BA, PVP	1982 Mehta U *et al.*, Proc 5th Intl Cong Plant Tissue Cell Cult 109.
Carica papaya petiole		W with 1 NAA, 10 2-IP to 0.1 NAA and 0.01 BA	1974 deBruijne *et al.*, Meded Fac Landbou Rijksuniv Ghent 39:637
Ovules	Glutamine embryogenic callus	W with 20% CW	1982 Litz RE, RA Cornover Plant Sci Lett 26:153
Citrus sinensis citrus "Shamouti" orange (Rutaceae) ovule	none identified	MS, 0.5 K to 1 gL^{-1} ME, adenine 20 mg L^{-1}	1973 Kochba J, P Spiegel-Roy Z Pfanzenzuchtg 69:156; 1977, Z Pfanzen-physiol 81:283
	habituation to auxin	auxin antagonists	1974 Button J *et al.*, J Expt Bot 25:446
Citrus endosperm	none identified	High levels of GA for triploid plantlets	1981 Wang T-Y, J Chang *In* Plant Tissue Culture, p 463 Pitman Intl Series in Applied Biology, Boston
protoplasts	not identified	MT medium with no growth regulators	1975 Vardi A *et al.*, Plant Sci Lett 4:231
Cocos nucifera stem, leaf, and rachilla of mature trees	Coconut milk	Sequential treatments on 2 media & several auxins & cytokinins	1984 Gupta PK *et al.* Plant Cell Repts 3:222

Species and explant source	N source and factors	Medium, observations	Reference
Coffea arabica cv. Bourbon	none identified, callus culture, cysteine	MS with 2 K, 2 2,4-D to 2.3 K & 0.5 NAA	1977 Sondahl MR, WR Sharp, Z Pflanzenphysiol. 81:395
	morphological mutants, cysteine	MS rooting of developed buds, BA, IAA, charcoal	1982 Sondahl MR, Proc 5th Intl Cong Plant Tissue Cell Cult 417
Coffea canephora young stem, shoot	none identified	embryoids on modified LS with 0.5 K, 0.5 2,4-D	1970 Staritsky G 1970 Acta Bot Neerl 19:509
Coffea dewevrei leaf	none identified	MS with 10 K, 6 2,4-D to 2.5 K and 0.5 NAA	1979 Sondahl MR, WR Sharp, In: Plant Cell and Tissue Culture. WR Sharp et al. (eds), Ohio State Univ Press, p 527
Corylus avellana L. immature embryos (hazel) Betulacea	casein hydrolysate callus culture	MS with removal of of 2,4-D	1975 Radojevic LZ Pflanzen-physiol 77:33; 1979, 91:51
Cycas circinalis megagametophyte	none identified callus culture	2,4-D removal, LS or W	1967 Norstog K and E Rhamstine Phytomorph 17: 1, 374
Ephedra gerardiana	none identified callus culture	MS, K and NAA	1976 Ramawat KG HC Arya Phyto-morph 26:395
Gnetum ula proembryos	Casein hydro-lysate, yeast extract	Proembryo development	1963 Vasil V In: Plant Tissue and organ culture. Symp Intl Soc Plant Morphol, Univ Delhi, p 178-280

Species and explant source	N source and factors	Medium, observations	Reference
Ilex aquifolium L. immature embryo	none identified, embryo culture	LS	1971 Hu CY, IM Sussex Phytomorph 21:103
Larix decidua female gametophyte complex from immature seed	casein hydrolysate	Embryos of undefined origin. Stage of development (2 wks after fertilization) is important. (0.5 Litvay medium)	1985 Nagmani R, J Bonga Can J For Res 15:1088
Liquidambar styraciflua seedlings	callus, initially no NH_4^+ and KNO_3 followed by NH_4NO_3 with NAA and BA	Blayde's and modified MS with 5.3 NAA and and 8.8 BA	1980 Sommer HE, CL Brown For Sci 26:257
Paulownia tomentosa ovules	casein hydrolysate, callus culture	MS, 4.5 2,4-D	1979 Radojevic LZ Pflanzenphysiol 91:57
embryo		MS, 5.7 IAA and 0.5 to 5 K followed by none	1979 Ibid
Picea abies L. (Karst) 7-9 day cotyledons, embryos	MS modified with 3.4 mM NH_4OH & 23.1 mM K NO_3	BAP, K, 2,4-D to induce moist white callus on epidermal cells of cotyledon s grown on reduced Ca & N sources on 1/2 MS. Transfer to hormone-free medium.	1984 Krogstrup P Ph D. Thesis Royal Ag Vet Univ, Copenhagen Denmark (in Danish)
	KNO_3 & NH_4NO_3 glycine and other free amino acids	BAP, 2,4-D on a medium of Von Arnold and Eriksson. Can J Bot 59:870, 1981.	1985. I Hakman et al Plant Sci Lett 38:53
Pinus lambertiana immature and mature embryos (callus suspensor cells)	Casein hydrolysate	Somatic polyembryogenesis from white callus on 5 ppm 2,4-D & 500 ppm casein hydrolysate followed by hormone-free medium	1985 Gupta PK, DJ Durzan Bio/Technology 4:in press; cf. Fig 7

Species and explant source	N source and factors	Medium, observations	Reference
Pinus resinosa proembryos	Ginkgo extract, embryo culture	Knop's medium without N	1969 Banerjee SN, NW Radforth Bot Mag 82:329
Pinus taeda and Pseudotsuga menziessi	none identified, cell suspension cultures	Embryoids in MS	1977 Winton L, S Verhagen Tappi p 21
Pinus taeda immature seeds	See Pinus lambertiana		1986 Gupta PK, DJ Durzan (unpublished)
Pseudotsuga menziessi, Douglas-fir 1 to 4-yr-old saplings cell suspensions	NH_4NO_3 and shift to urea and amino acids under low calcium conditions	Early stages of SE with some polarization, 2.5 NAA 0.1 BA to 0.01 NAA and 0.1 BA	1982 Durzan DJ Proc 5th Intl Cong Plant Tissue Cell Culture 113-114
germinating seeds (cotyledons)	none identified	MS, and a range of growth regulators	1980 Abo El-Nil M US Patent No 4,217,730
immature seeds	See Pinus lambertiana		1986 Gupta PK, DJ Durzan (unpublished)
Santalum album L. embryo	none identified, callus culture	W,9.1 2,4-D and 23.2 K to no growth regulators	1965 Rao PS Phytomorph 15:175
Protoplasts from cell suspension (20-yr-old tree shoots)	IAA, BA & casamino acids or yeast extract (decreased yield)	sequence of rapid transfers on MS 1/2 and W	1985 Rao PS, Ozias-Atkins P Protoplasma (in press)
		medium description	1978. Rao PS, VA Bapat Can J Bot 56:1153
20-yr-old stem	none identified, callus culture	MS, W with 2,4-D and K to 1.5 to 5.8 GA_3	1979 Lakshmi Sita SG et al. Plant Sci Lett 15:265

Species and explant source	N source and factors	Medium, observations	Reference
Mature trees (20-25 yrs)	(glycine, W) and (NH_4^+, MS)	W, MS, 2,4-D	1982 Lakshmi Sita SG et al. Curr Sci 51:88
Theobroma cocao immature embryos	casein hydrolysate, embryo culture	MS, 10% CW, 8 NAA, IAA or 2,4-D to no growth regulators	1979 Pence V et al. J Am Soc Hort Sci 104:145
Vitis sp. 'Seyval'		MS, 4.5 2,4-D, 0.4 BA to 10.7 NAA and 0.4 BA	1977 Krul WR JF Worley J Am Soc Hort Sci 102:360
Vitis vinifera nucellar callus	none identified	Nitsch's basal, NOAA, BA, sequential treatments	1976 Mullins MG, C Srinivasan J Expt Bot 27:1022
Zamia floridana megagametophyte	none identified		1948 LaRue CD Bull Torrey Bot Club 75:597
Zamia integrifolia megagametophyte, embryo	casein hydrolysate, L-alanine, amides	W and modifications, IAA, 2,4-D, K, sequential treatments	1965 Norstog K Am J Bot 52:993 1967 Norstog K, E Rhamstine Phytomorph 17:374

7. EMBRYO CULTURE

RONG H. HO

1. INTRODUCTION

The search for ideal trees with desirable characteristics has been on for many decades with the hope of using those trees to generate progenies or rooted cuttings. These propagules are needed to reforest the vast logged areas and to increase the fiber production. Selection of plus trees has been in progress and the trees have been screened for elite characteristics through progeny testing. Crosses of different ecotypes have been carried out firstly to capture the vigor from heterozygosity of wide crosses and secondly to increase the adaptability of the new genotypes. Inbreeding to obtain homozygous lines for subsequent outcrossing has been attempted to procure total expression from dominant or over-dominant genes. Hybridization of different species has been explored to obtain hybrid vigor and to enlarge the gene pool of a given species.

Seeds from crosses sometimes fail to germinate. This may be due to inhibitors, immature embryos, or degenerated endosperm. Some crosses such as those resulting from inbreeding or from interspecific hybridization, often produce inviable seeds because of prezygotic and postzygotic incompatibility. Prezygotic incompatibility includes gametophytic and sporophytic incompatibility in the pollen-pistil interaction. Postzygotic incompatibility is attributed to genic incompatibility, chromosome alteration, chromosome elimination, ploidy changes and incompatible cytoplasm. Postzygotic incompatibility has been a greater hindrance to inbreeding and interspecific hybridization than prezygotic incompatibility.

Problems caused by seed dormancy, embryos remaining immature, embryos with degenerated endosperms, and prezygotic and postzygotic incompatibility have, in many species and hybrids, been overcome by *in vitro* culture of embryos, embryo rescue, *in ovulo* embryo culture, and *in vitro* pollination and fertilization. The present review will deal with most aspects of embryo culture. Since little work has been done in this area, species other than forest trees will be included in the discussion when appropriate.

2. HISTORICAL BACKGROUND

Culture of plant embryos *in vitro* was first initiated by Hannig (48). He concluded that embryos required salts and carbohydrates for growth. He also found that immature embryos germinated precociously. Stingl (133) reported that embryos of one species could grow well on endosperms of others. Buckner and Kastle (28) obtained seedlings by culturing embryos without cotyledons.

Knudson (61) succeeded in culturing orchid (Orchidaceae) embryos on nutrient agar and obtained plantlets in the absence of symbiotic fungus. Two years later, Dietrich (39) observed that immature embryos embedded in a medium continued their embryonic growth to maturity and those incubated on the surface of the medium germinated prematurely.

The first attempt to culture embryos of forest tree species *in vitro* was made by Schmidt (121). He observed that the megagametophyte of one pine (*Pinus*) species was effective in promoting chlorophyll formation in the embryo of another species (Table 1). Culture of excised pine embryos and raising to seedlings was carried out some 8 years later by Brunner (27).

Refinements in culture technique shifted the emphasis from culturing mature to immature embryos. White (142) succeeded in culturing heart-shaped, but not globular, embryos to maturity. Radforth (103) supplemented Crone solution with yeast extract in cultures of ginkgo (*Ginkgo biloba*) proembryos which grew to 22 cells. Embryos at the globular stage were not successfully cultured to maturity until Overbeek *et al.* (100) supplemented a medium with unautoclaved coconut milk. Norstog (97) reported successful culture of the smallest embryos (less than 100 cells) on a medium supplemented with coconut milk and glutamine or a mixture of amino acids.

Embryo culture has been utilized to overcome stratification problems in forest tree species. Li (80) used the technique to hasten germination of ginkgo embryos. Embryos isolated from dormant seeds of pine and yew (Taxus) showed no dormancy in culture (73, 74). This technique also shortens the breeding cycle of deciduous trees (69).

Laibach (68) first demonstrated the use of in vitro technique to rescue embryos of interspecific hybrids. Tukey (139) used the same technique to rescue the aborting embryos from degenerative seeds of sweet cherry (Prunus avium). A similar technique was applied to grow sugar pine (P. lambertiana) hybrids to maturity (134). Since then it has been widely used to rescue various hybrids. To date, over 80 interspecific and intergeneric hybrids have been obtained by using this method (35).

Culture of flowers and ovaries was introduced by LaRue (72). In the same year, Withner (144) succeeded in culturing excised orchid ovules. Immature tomato (Lycopersicon esculentum) flowers were cultured to produce fruits containing no seeds (95). Two years later, Nitsch (96) reported production of tomato fruits with viable seeds. A significant event in ovule culture was made by Maheshwari (85). He successfully cultured ovules which had embryos at the 2-celled stage. An attempt to culture ovules of forest tree species was first made by Muzik (93) followed by Banerjee (11) who cultured pine ovules.

Introduction of in vitro pollination and fertilization is more recent. Kanta (57) used the technique of intraovarian pollination to bypass the style and effect pollination. In 1962, he succeeded in producing normal fruits with viable seeds (58). Since then a number of species and hybrids have been obtained through this technique (114, 130). In vitro pollination and fertilization has not been attempted with forest tree species. Cotton (Gossypium) hybrids were the only woody plants to be produced by this technique (130).

There are several review articles on embryo culture and related problems for plant species. The reader should look into the papers by Rappaport (116), Narayanaswami and Norstog (94), Raghavan (107, 108, 109, 110), Rangaswamy (112), Beasley (15), and

Collins, Taylor, and DeVerna (36). For embryo culture in forest tree species, the reader is referred to two brief reviews by Bonga (23) and Mott (91), and an atlas by Brown and Sommer (26).

3. EMBRYO CULTURE

Embryo culture has been defined as "*in vitro* development or maintenance of isolated mature or immature embryos" by Shaeffer (125). Two types of culture of immature or mature plant embryos have been distinguished, pregerminal and postgerminal (119). Pregerminal culture is the maintenance of immature embryos and their development into mature embryos; postgerminal culture is the germination of immature or mature embryos into normal seedlings, and later into viable plants.

Embryo development proceeds from the postzygotic stage to full establishment of polar meristems and cotyledons. It does not include the fertilized egg or the brief zygote stage. From the postzygotic stage to the initiation of cotyledons, the immature embryos are heterotrophic and rely on either megagametophyte, endosperm or other accessary cells for the nutritional resources for development. After the initiation of cotyledons, the embryos become autotrophic and grow easily to maturity on a nutrient medium (111).

The medium needed to fulfill the requirements of autotrophic embryos for growth is less complex in composition than that of the heterotrophic embryos which require a more complex composition of nutrients to mimic the *in vivo* source of nutrients for development. The culture of heterotrophic embryos, i.e. pregerminal culture, aims at understanding the controls of development and nutrient requirements. The culture of autotrophic embryos, i.e. postgerminal culture, is used to study the embryonic growth to seedling and the biochemical aspects of dormancy.

3.1. Embryo culture of gymnosperm tree species

Embryogenesis in gymnosperms has been divided into 3 phases; 1) the proembryo phase which includes all the stages after the zygote and before the elongation of the suspensor, 2) the early embryo phase which includes all the stages after the elongation of suspensor and before the initiation of cotyledons, and 3) the late embryo phase which includes all the stages from the initiation of cotyledons to full establishment of polar meristems and cotyledons (126). The immature embryos are globular-shaped until they reach the late early embryo phase. The establishment of embryo polarities (shoot and root) occurs in the late early embryo phase and from then on to maturity. These embryos are torpedo-shaped. The mature embryos with many cotyledons are embedded in a haploid megagametophyte which functions as an endosperm during germination.

From a historical perspective, the 3 phases of embryogenesis can be divided into 2 categories of embryo culture: 1) culture of nearly mature or mature embryos, and 2) culture of proembryos. The term proembryo is used to refer to all the developmental stages of the embryo preceding cotyledon initiation (111), and this will include the first 2 phases of embryogenesis. The culture of nearly mature or mature embryos will refer to culture of embryos at the stages in the late embryo phase.

3.1.1. Culture of nearly mature or mature embryos. Mineral salts are an important ingredient in media for embryo culture. In the early embryo culture studies, Tollen solution with a very simple salt composition was used for medium preparation (27). This medium was soon replaced by Knop (139), Crone (103), and Pfeffer solution (84). After the nutritional requirements of embryos became better known, more elaborate media, such as White (143), Nitsch (96), Heller (49), and Murashige and Skoog (MS) (92) were used for embryo culture. These media included carbohydrates, macronutrients, micronutrients, chelating agents, vitamins, amino acids, plant growth regulators, and miscellaneous additions, such as coconut milk and extracts of organic matter.

Brunner (27) cultured excised pine embryos on a medium with or without macerated megagametophyte. He concluded that a growth substance or substances from the gametophyte was essential for normal embryo growth and development. The promoting effect of water extracts from megagametophytes on growth of embryos to normal seedlings was reported by Berlyn (17) and Gjonnes (41). Mapes and Zaerr (86) observed enhancement of embryo growth at the gametophyte extract concentrations of 0.1 to 1% but at high concentrations (10%) there was an inhibitory effect in Douglas-fir (*Pseudotsuga menziesii*). However, Li and Shen (81), Hyun (55),

Sacher (120), and Asakawa (6) reported that gametophyte extracts provided little benefit in growth promotion of embryos. Extracts from wheat and rice bran were also found to be inhibitory on embryo growth (81) while yeast extract promoted callus formation in the cultured embryos (62).

Coconut milk has also been reported to play a promotive or an inhibitory role in embryo culture. Ball (10) and Konar and Oberoi (62) found that coconut milk promoted growth of embryos to normal seedlings. However, no significant effect of coconut milk on embryo growth was reported by Hyun (55) and Sacher (120). Inhibitory effects on growth were observed at a concentration of 5% coconut milk in loblolly pine (P. taeda) by Brown (24).

Carbohydrates in the medium greatly enhance embryo survival and embryo growth (6, 25, 80). Carbohydrate, most commonly sucrose, is used not only as an energy source but also as an osmotic agent. Ball (10) compared the growth of ginkgo embryos on 5 kinds of sugar and found that sucrose was the best for shoot growth and glucose for root growth, while mannitol was inhibitory to embryo growth. A wide range of sucrose concentrations from 0.2% to 16% has been used in embryo culture with the most often used range being from 2% to 4% (10, 13, 30, 80, 124).

The addition of vitamins and organic nitrogen to the culture medium affect embryo growth and development in some species. Bartels (14) defined the optimal concentration of thiamine for pine embryos as between 0.00000003 ng and 0.003 µg. Ball (10) reported a striking stimulation of shoot growth in ginkgo embryos when glutamine was added to the culture medium.

Plant growth regulators have a profound effect on embryo growth in culture. The effect is either inhibitory or stimulatory depending on the concentration of the regulator, the kind of regulator, and the species under study. Bartels (14) reported that the concentration of indoleacetic acid (IAA) at not more than 0.002 mg/l was optimal for pine embryo growth while Bonga (22) and Brown (24) found an inhibitory effect when the concentration was higher than 1 mg/l. Ball (10) observed that IAA caused severe inhibition of growth of both shoot and root in ginkgo embryos when the concentration was at 0.1 mg/l or higher. Whereas, in a medium containing 0.1 mg/l of IAA and 1 mg/l of 2,4-dichlorophenoxyacetic acid (2,4-D), arbor-vitae (Biota orientalis) embryos developed into normal seedlings (62). Pine embryos exhibited callus formation when 6 mg/l of 2,4-D was added to the medium (55). Sacher (120) obtained short root formation when 0.5 mg/l of naphthaleneacetic acid (NAA) was used in pine embryo culture. A similar promotive effect in root growth was observed in ginkgo embryos from gibberellic acid 3, 4 and 7 (GA_3, GA_4, GA_7) at any concentration between 0.0001 and 0.1 mg/l (30, 31). However, each of the 3 GAs inhibited epicotyl growth at any concentration higher than 0.001 mg/l. Howe (50) reported that the addition of GA_3 or indolepropionic acid (IPA) at 0.1 mg/l produced no significant stimulation to the growth and development of the pine embryos to seedlings.

Abscisic acid (ABA) is probably the only known natural plant growth regulator playing an inhibitory role in plant embryo growth. LePage-Degivry (74, 78) reported dormancy in immature embryos of mature yew (T. baccata) seeds caused by ABA or an ABA-like compound in the embryos. Leaching and/or chilling of the embryos in a liquid medium, or inclusion of GA_3 in the medium, were necessary to break dormany (76). The leached and chilled embryos would not grow if the medium contained ABA (77). LePage-Degivry and Garello (79) also found that yew embryos would germinate on an agar medium after culture in a liquid medium for 8 days. The agar medium contained sucrose, and calcium and potassium ions. In yew, the presence of inhibitors in embryos was the cause of dormancy while in ginkgo and eastern hemlock (Tsuga canadensis), inhibitors were reported to be present in the megagametophytes (80, 98). Later, Lee and Chang (73) germinated ginkgo seeds without seed coats on a medium and concluded differently from Li (80) that the presence of seed coats induced dormancy.

The megagametophyte plays a role in the formation of chlorophyll in the cultured embryos (21). Pine and spruce (Picea) embryos left embedded in gametophytes of their own species or placed in gametophytes of other species would form chlorophyll when cultured in darkness (121). Bogorad (21) found that pine embryos isolated from germinating seeds, i.e. embryos without megagametophytes, continued to form chlorophyll in culture in darkness. The rate of chlorophyll formation was 4 times greater when the embryos were isolated from seeds germinated in light and cultured in darkness than when isolated from seeds germinated in darkness for 6 days and cultured under light. Chlorophyll formation failed to occur when the embryos were cultured on a medium containing gametophyte extract, autoclaved gametophyte, or mineral nutrient with 2% sucrose. On the contrary, greening of pine embryos on an agar medium containing inorganic nutrients and 5% sucrose in darkness was reported by Schou (124). Engvild (40) made a similar observation and concluded that there appeared to be no basis for assuming that an active substance, specific to the chlorophyll synthesis, was transferred from the megagametophyte to the embryo to initate the greening of cotyledons.

Cotyledons have a haustorial function by providing a large surface area for uptake of nutrients or important metabolites for growth of the embryo axis. Excellent embryo growth was observed when cotyledons were embedded in the medium and the culture tube was inverted so that the root radicle grew downwards (10, 13, 25, 73). Li (80) and Bulard (29) found that good growth occurred when both cotyledons and radicle of an embryo were in contact with the medium. However, Berlyn (17) obtained abnormal root growth and epicotyl development when only the cotyledons were in contact with the medium. Later, Berlyn and Miksche (20) found that embryos without cotyledons elongated to the same extent as those with intact cotyledons when epicotyl or cotyledons were planted in a medium. Embryos cultured _in vitro_ did not require a haustorial function of cotyledons.

3.1.2. _Culture of proembryos_. Culture of proembryos has been attempted mainly to determine the extent of phylogenetic control of the early growth and development in plants. The role of the suspensor in embryo growth and development and the phylogenetic significance of cleavage and simple polyembryony have been the focus of these studies. Cleavage polyembryony occurs in the genera _Pinus_, _Cedrus_, _Tsuga_, _Pseudolarix_, and _Larix_, and simple polyembryony in _Picea_ and _Pseudotsuga_ (34). _Larix_ shows a delayed form of cleavage polyembryony (122) while simple polyembryony, rarely cleavage polyembryony, is the rule in _Abies_ (34).

Physiological and physical factors influence embryonic growth and development. Radforth (103) cultured ginkgo proembryos in the archegonia and obtained a symmetrical diameter growth while *in vivo* the proembryos developed asymmetrically. He also demonstrated that the formation of the conical type of suspensor could be deferred and concluded that factors external to the proembryos were able to influence its form of development. Radforth and Pegoraro (105) found that the cultured pine embryos had a radial or 3-dimensional growth pattern rather than the filamentous organization common *in vivo*. They indicated that the development of proembryos was not controlled entirely by phylogenetic factors.

In the culture of European larch (*L. decidua*) proembryos, Sterling (129) concluded that cleavage polyembryony was less an invariable characteristic of the embryo than a function of nutrient control. Woods (145) confirmed this. He stated that polyembryony in pine was a physiological problem and had no phylogenetic significance. Similarly, the suspensor had little phylogenetic significance (145) and the potential for differentiation *in vitro* surpassed that for differentiation *in vivo* (22). Suspensor cells could also continue to elongate independently of the association with embryonic cells but were closely associated with the growth and morphogenesis of embryos (138).

Radforth (103), Woods (145) and Radforth and Pegoraro (105) used Crone solution supplemented with dextrose, sucrose and yeast extract to culture pine proembryos and obtained some growth. Banerjee and Radforth (12) supplemented Crone solution with extract from ginkgo megagametophyte and sucrose to culture pine proembryos. They induced the formation of secondary embryos from both excised segments of the suspensor cells and the suspensor cells associated with the proembryos. The suspensor cells formed pits in the cell walls when pine proembryos were cultured in a phthalic buffer solution supplemented with IAA (22, 104). IAA added to Pfeffer solution with 2% sucrose promoted an embryonic cell (right after cleavage) in pine to form a mass of several hundred cells (84). The addition of thiamine to Pfeffer solution was necessary to induce the formation of multicell proembryos from small proembryos of *Keteleeria davidiana in vitro* (84). Thomas (135, 136) emphasized the importance of vitamins for normal development of pine proembryos. For pine proembryos at the cleavage stage, a nurse culture technique with calli was required to obtain growth comparable to that *in vivo* (137, 138).

3.2. Embryo culture of angiosperm tree species

Embryogenesis in angiosperms is very similar to that in gymnosperm species except for the absence of free nuclear division in the zygote (38). Angiosperm embryos show a distinct configuration in shape at different stages of embryogenesis. For the ease of distinguishing among developmental stages, embryogenesis has been divided into 3 phases: 1) the globular phase including all stages after zygote and before the initiation of cotyledons, 2) the heart-shaped phase including all the stages after the initiation of cotyledons and before the torpedo-shaped phase, and 3) the torpedo-shaped phase including all events to full establishment of cotyledons and polar meristems (109).

The mature embryo is usually in a torpedo shape and embedded in a triploid endosperm. In some species, the seed has little or no endosperm (perisperm in some species), and has a small embryo axis with a pair of large cotyledons (dicotyledon). Sometimes the maturing embryo develops further from a torpedo to a spatulate shape (51, 93) or from a torpedo to a walking-stick shape and finally to a full grown shape of either an inverted U (*Morus* species), a circular figure (*Atriplex* species), or a coil form (*Koelreuteria* species) (123).

Based on the heterotrophic and autotrophic nature of the embryos, the 3 phases will be discussed in 2 headings: 1) the culture of proembryos including the first 2 phases of embryogenesis, and 2) the culture of nearly mature or mature embryos including the torpedo phase and all the stages to maturity. Culture of palm embryos will not be included in the present discussion since Reynolds (118) has given a detailed description in vegetative propagation of palm trees.

3.2.1. Culture of nearly mature or mature embryos.

Most embryos are embedded in the endosperm and rely on it for supply of nutrients in germination. Through embryo culture it will be possible to determine what exogenous nutrients would conform to the composition of the endosperm. Culture media used in the early stages of study were simple in composition and consisted of macronutrients and carbohydrate (69, 139, 140). Later, more complex media were used for culture as more data became available about the nutrient requirements of embryos. The evolution in the use of media for angiosperm embryo culture is very similar to that for gymnosperm embryo culture. In fact, the same medium is often used for culture of both angiosperm and gymnosperm embryos.

Growth and survival of cultured embryos is greatly enhanced by the addition of carbohydrates to the medium. Glucose and sucrose are the two main carbon energy sources in culture. Addition of sucrose was essential for germination of early ripening varieties of peach (P. persica) embryos (60) and for growth of mature hybrid bamboo (Bambusa x Saccharum) embryos (4). In these experiments, glucose was used empirically in concentrations ranging from 0.5 to 3% and sucrose from 0.5 to 5%. Tukey (139, 140) used 0.5 to 2% of glucose to culture embryos of a number of Prunus species and found that 2% was most favorable for rapid embryo development. Bulard and Monin (32) supplemented a medium with 3% glucose to culture ash (Fraxinus excelsior) embryos to plantlets. Lammerts (69) found that the addition of sucrose satisfied the carbohydrate requirements of peach and apricot (P. armeniaca) embryos. He used 2% sucrose to culture immature embryos and 0.5% to culture mature ones. Kester (59) observed that shoot growth of peach embryos showed little response to a change in sucrose concentration and root growth was best at 4 or 5% of sucrose.

Although inorganic nitrogen is supplied in the macronutrients, addition of organic nitrogen has an influence on embryo growth and development. Root growth of decotyledonized peach embryos was enhanced by the addition of 1 mg/l of asparagine and 3 mg/l of glycine. The promotion was nullified and inhibition resulted when asparagine was raised to 10 mg/l and glutamic acid was added at 3 mg/l (59). Srivastava and Steinhauer (128) reported germination of oak (Quercus lebani) embryos and their development into normal seedlings on MS medium. A fragile callus was obtained when casein hydrolysate and some plant growth regulators were added to the medium. Deletion of casein hydrolysate was necessary to induce organogenesis on the callus.

Nearly mature or mature embryos are autotrophic. Cellular biosynthesis in the embryos should be able to meet their vitamin requirements. Myo-inositol was not required to stimulate the growth of cherry, peach and plum (*P. myrobolana*) embryos. However, the absence of myo-inositol would impede growth and development of the embryos without cotyledons (1). Shoot and root growth in peach embryos was significantly increased by the addition of thiamine at 0.2 mg/l, ascorbic acid at 20 mg/l, and riboflavin at 0.5 mg/l (59).

Failure of embryo culture on a medium containing macronutrients, micronutrients, organic nitrogens, vitamins and carbohydrates indicates that some embryos require additional nutrients for growth and development. Mature embryos embedded in endosperms or perisperms may not be as autotrophic as those without endosperms or perisperms. Mature embryos of rubber plant (*Hevea brasiliensis*) showed no apparent reaction to the addition of coconut milk while embryos at the torpedo-shaped stage failed to grow to plantlets (93). Kester (59) reported that yeast extract significantly increased shoot and root growth of partially decotyledonized peach embryos. Abou-Zeid *et al.* (3) found that extract from after-ripening cherry cotyledons stimulated growth in immature embryos or embryo axes of dormant embryos. The promotive effect was lost when the extract was autoclaved (2). Evidently the cotyledonary factor was heat sensitive.

Plant growth regulators induce growth and morphogenesis in plants. In peach embryo culture, morphological response to IAA varied with concentration; at 0.001 mg/l, no stimulatory effect was present and at 0.1 mg/l or higher, it resulted in callus formation and limited shoot and root growth (59). Callus was also formed in paulownia (*Paulownia tomentosa*) embryos when they were incubated on a medium supplemented with either IAA, NAA, or 2,4-D at 1 mg/l (106), and in oak when 2 mg/l of IAA with 5 mg/l of 2-isopentenyladenine (2iP), 30 mg/l of adenine sulphate, and 1 g/l of casein hydrolysate was added to MS medium (128). Bellarosa (16) obtained callus formation when oak (*Q. suber*) embryos were cultured on a medium with 2 mg/l each of NAA and benzylaminopurine (BAP). Deletion of NAA from the medium was necessary to produce normal plantlets. However, addition of BAP to a medium for *in vitro* germination of walnut (*Juglans regia*) resulted in an unfavorable effect on the growth of embryos (37). Yuce (146) obtained normal germination of olive (*Olea europaea*) embryos from seeds without any stratification when they were cultured on a medium with 2 or 4 mg/l of IAA and 1 or 3 mg/l of GA_3. Dormancy of ash embryos could also be broken by the addition of 1 mg/l of GA_3 to the medium (32).

Reduction of the size of cotyledons seemed to have some stratification-like effect in inducing after-ripening in dormant seeds. Abou-Zeid et al. (3), and Abou-Zeid and Neumann (2) found that up to 50% reduction in cotyledon size promoted shoot and root growth in cherry embryos and complete removal of cotyledons resulted in shoot development only. When cherry seeds were collected before initiation of dormancy, embryos with intact cotyledons had better growth and development than those without cotyledons (56, 102, 127). Kester (59) discovered that in peach embryos in vitro growth decreased as cotyledons were reduced in size from 0 to 90%; no germination was obtained for decotyledonized embryos. However, Muzik (93) found that as much as two thirds of cotyledons on rubber plant embryos could be removed without any effect on plant growth. Complete removal of cotyledons on ash embryos was no impediment for normal germination although contact of the cotyledons with the culture medium induced dormany (32).

3.2.2. Culture of proembryos. In the development of embryos in vitro to plantlets the growth pattern is apparently related to the age of each embryo when excised (141). Proembryos of some species do not enter embryonic development and have precocious germination while others proceed to develop into normal mature embryos before germination.

In holly (Ilex) trees, the mature seed has a heart-shaped embryo which requires a minimum of one year of dormancy to complete its development and to begin germination (51, 53). When cultured in vitro, rudimentary embryos were sufficiently autotrophic to support themselves if 4% glucose or non-reducing sugar (sucrose, raffinose, stachyose) was added to Linsmaier and Skoog (LS) medium as the only organic compound (83). In vitro culture for only 6 to 8 weeks was necessary to complete embryonic development. The development was best when proembryos were cultured in darkness, with 75% of the embryos reaching maturity (52, 54). As the heart-shaped embryos began to grow in darkness, they became progressively more insensitive to subsequent light inhibition; a

6-day initial dark incubation was sufficient to prevent the inhibitory effect of light. However, cotyledon development in about two thirds of the plantlets was abnormal. These cotyledons showed swelling, thickening and partial chlorosis, and clusters of embryoids that appeared on the upper surface of the cotyledons (54).

Normal embryonic development was absent and precocious germination resulted when proembryos of Prunus species were cultured on a medium containing macronutrients and 2% glucose (141). When excised as proembryos at the heart-shaped stage they produced green cotyledons and white leaves. Those excised at later stages grew into normal plantlets. Abnormalities in plantlets were also observed when rubber plant proembryos at the heart-shaped stage were cultured on MS medium supplemented with amino acids and plant growth regulators (33). Carron (33) and Muzik (93) were not able to germinate proembryos at the globular stage in rubber plant even with the addition of 33% of coconut milk to the culture medium. No germination was obtained in magnolia (Magnolia soulangeana and M. grandiflora) when embryos at the heart- or torpedo-shaped stage were cultured on Heller medium with 2% sucrose (75).

Although Anagnostakis (5) succeeded in germinating elm (Ulmus americana) embryos at the torpedo-shaped stage (based on the size of embryos reported), none survived because of a poor root system. Multiple shoots were regenerated from immature embryos of eastern cottonwood (Populus deltoides) isolated 10 or 21 days after pollination and cultured on MS medium supplemented with 150 mg/l of polyvinylpyrrolidone, Staba vitamins, 250 mg/l of glutamine and 0.5 mg/l of BAP (64). In older embryos (28 to 35 days after pollination), a single shoot was produced from most embryos but some developed multiple shoots later. The shoots were readily rooted in MS medium supplemented with 0.5 mg/l of IAA.

4. EMBRYO RESCUE

In successful crosses in forest tree breeding, fertilization occurs and a zygote is formed followed by normal proembryo formation and subsequent embryogenesis. However, empty seed is produced in some crosses resulting from embryo abortion which occurs in the process of embryogenesis. Embryo breakdown could occur in the proembryo or early embryo stage; the endosperm could degenerate at any stage of embryogenesis (117). Most of these abortions are attributed to postzygotic incompatibility.

Selfing to produce homozygous lines has been met with embryo abortion at the early embryo stage in both Douglas-fir (99) and white spruce (P. glauca) (88). In hybridization of P. pungens x P. engelmanii, the abortion occurred earlier at the proembryo stage (63). Proembryos degenerated at the 4-nucleate stage (47) or at the cleavage stage with 8 cells in the embryo initial in pine hybridization (66). Poplar (Populus) hybridization resulted in early abortion at the stage after zygote formation and the beginning stage of embryo development (87). These observations indicated that the megagametophyte or endosperm was either unable to nourish the developing embryos or antagonistic to the developing embryos.

In some postzygotic incompatibility, embryo abortion does not occur and a full seed is produced. However, the seed fails to germinate and produce a seedling. This has been attributed to either 1) embryo immaturity, 2) an embryo with degenerated megagametophyte or endosperm, 3) a degenerating embryo with megagametophyte or endosperm, or 4) dormancy.

Embryo immaturity was found in fully grown seeds of English holly (Ilex aquifolium). The embryos at the heart stage were cultured on LS medium with 4% sucrose. They developed into mature embryos, germinated,and formed normal seedlings (53). Berlyn (17) cultured pine embryos isolated from a seed with 4 embryos on White medium supplemented with 10 mg/l of ferric citrate, 2% sucrose, and green coconut milk or autoclaved megagametophyte. He succeeded in growing the largest embryo to a plantlet but not the other 3 smaller embryos although 2 of them had prominent cotyledon initials.

Cherry seeds are sometimes shrivelled because the embryo and endosperm ceased development prior to fruit ripening. Tukey (139) produced plantlets from such cherry seeds by culturing the embryos on Crone solution supplemented with potassium nitrate, 1% glucose and 0.2% asparagine. Stone and Duffield (134) obtained normal hybrid seeds of P. lambertiana x P. amandi and P. lambertiana x P. koraiensis but failed to germinate them. They removed the seed coats and cultured the embryos with megagametophytes on 1.5% plain agar to normal seedlings. The presence of the gametophyte was not inhibitory to germination.

5. *IN OVULO* EMBRYO CULTURE

In ovulo embryo culture is most useful when the proembryo or early embryo cannot be rescued by embryo culture and the sporophytic tissue is vital to assist in rearing the embryo to maturity. The ease of culture without the drawback of embryo isolation is another important factor in using this technique. However, compatibility between embryo and megagametophyte or endosperm tissue is essential for successful culture. A thorough understanding of the reproductive process is also necessary to ensure good timing of ovule isolation for culture, i.e. before embryo degeneration has occurred.

Plant embryos have been raised to maturity *in vitro* by culturing intact flowers, ovaries, placenta attached ovules, or individual ovules (15). The type of explants for culture depends on the genotype, species and culture conditions although isolated ovules appear to be the preferred explant. The embryo stage at the time of explant isolation is also a limiting factor for successful culture. Evidently embryos can be cultured at a much earlier stage in *in ovulo* embryo culture than in embryo culture. Stewart and Hsu (131) cultured cotton ovules to maturity which were isolated at the zygote stage.

In forest tree species, the globular embryo stage was the earliest stage in which the young embryo in an ovule was cultured. Globular embryos of rubber plant were cultured *in ovulo* on MS medium supplemented with 1 mg/l each of 2,4-D and BAP, 2 mg/l of NAA, 5% of coconut milk, and 7% sucrose and formed viable plantlets (42). However, Muzik (93) was not able to grow the immature ovules of rubber plant to maturity on either LaRue or White medium supplemented with 33% coconut milk, 0.2 mg/l of IAA, and 2% sucrose. He successfully cultured the mature ovules which eventually germinated in a fashion similar to that of seed germination *in vivo*. Successful ovule culture was also carried out by Li *et al*. (82) to rescue the embryos of a hybrid between *P. simonii* and *P. pyramidalis* due to the endosperm starting to degenerate at the free nucleate stage. The hybrid poplar embryos were at either the heart-, torpedo-, or the late torpedo-shaped stage when ovules were cultured on MS medium supplemented with 0.01 mg/l of IAA, 0.1 mg/l of BAP, and 2% sucrose. Stewart (130) reported a successful culture of one-month-old peach ovules to the mature embryo stage.

Radforth and Bonga (104) reported that IAA was an important growth regulator in inducing the formation of a meristematic layer and the development of secondary thickenings with pits on all the parenchyma cell walls surrounding the archegonial jackets in cultured ovules with fertilized eggs in black pine (P. nigra). Radojevic (106) emphasized the need to have the placenta present in the initial culture of fertilized ovules of paulownia. When 2,4-D was added to MS medium, calli on these cultured ovules produced embryoids. However, neither report mentioned development of the fertilized eggs in the cultured megagametophytes or ovules.

Banerjee (11) irradiated ovules with embryos at the different stages of development in red pine (P. resinosa) with gamma cell before culturing them on White medium with 15% coconut milk. After 18 days, he found that the zygotes in the ovules differentiated beyond the proembryo stage, the early embryos developed root initials and cotyledon primordia, and the embryos grew to stunted seedlings. Kriebel and Shafer (67) cultured the entire second-year megastrobili in 2 pine species and 4 hybrids on Nitsch medium with 2% sucrose and obtained limited development of embryos and megagametophytes (66).

6. IN VITRO POLLINATION AND FERTILIZATION

In vitro pollination and fertilization is a potential technique to bypass prezygotic incompatibility and thus achieves completion of sexual reproduction. The technique involves the continuation of culture of ovules with fertilized eggs to obtain mature embryos. Therefore, this technique could be considered as an extension of in ovulo embryo culture.

Prezygotic incompatibility includes gametophytic and sporophytic incompatibility. Gametophytic incompatibility is mostly caused by arrest of pollen tube growth in the style resulting in the failure of gamete delivery for syngamy. Sporophytic incompatibility is the failure of pollen germination on the stigma or pollen tube penetration into the stigma. Sporophytic incompatibility has been reported in some poplar crosses (43) while gametophytic incompatibility occurs in oak (101), birch (Betula) and alder (Alnus) (46). Failure in hybridization of hard pine species (subgenus Pinus) resulted from the inability of pollen tubes to function normally in the nucellus tissue of foreign species (65). The arrest of pollen tube growth in the nucellus was the reason for spruce hybrid failure (89).

The barriers in prezygotic incompatibility can sometimes be overcome by using mentor pollen to facilitate pollen germination and tube penetration, and to accomplish fertilization (132). Modification of the floral structure and intraovarian pollination can be an effective way to eliminate the barriers (57). In some cases, premature floral abscission, not incompatibility, is the reason, for the failure to form filled seeds. Application of plant growth regulators or other chemicals may prevent abscission and prolong the life of the flowers.

In many plant species, the barriers cannot be overcome easily or effectively by the above in vivo techniques. The use of in vitro methods will be necessary to circumvent the barriers. Rao and Rangaswamy (115) cultured whole pistils of tobacco (Nicotiana) to circumvent the fruit abscission, and Rangaswamy and Shivanna (113) cultured ovaries of petunia (Petunia) to overcome self-incompatibility. Tobacco placenta segments with attached ovules were pollinated in vitro by placing pollen on the placenta, ovule, or both (7). Stewart (130) succeeded in fertilizing the cultured free ovules of cotton species by placing pollen on the micropylar end to promote pollen germination and tube penetration.

There is no report on in vitro pollination and fertilization in forest tree species. There are reports dealing with prezygotic incompatibility. Application of the technique may produce some hybrids which otherwise cannot be obtained by in vivo techniques. However, care should be taken in selecting the explant source, the type of explant, the developmental stage of the maternal tissue, the genotype of pollen and pistil parent, pollen viability, pollen density in pollination, the position and timing of pollination, culture medium and environment.

7. CONCLUDING REMARKS

The main reason for embryo culture has been to overcome *in vivo* inviability of proembryos and immature and mature embryos. Other objectives often cited include studies of embryogenesis and embryo physiology, completion of embryo maturation, hastening of embryo maturation, and bypassing dormancy requirements. *In ovulo* embryo culture holds distinct advantages when maternal tissue is vital to embryo survival. All these techniques - embryo culture, embryo rescue, and *in ovulo* embryo culture, are designed to overcome postzygotic incompatibility. Prezygotic incompatibility has to be overcome by *in vitro* pollination and fertilization followed by culture of the fertilized ovules to obtain mature embryos.

In vitro pollination and fertilization can provide an excellent technique for studying pollen-pistil interaction, pollen germination on the stigma, pollen tube penetration in the style, pollen competition, and fertilization. Details of embryogenesis can be effectively furnished by data from *in ovulo* embryo culture. This technique can also supply us with a tool to analyze the nature of competition in simple and cleavage polyembryony in conifers. Individual embryos in an ovule with polyembryos resulting from multiple fertilizations or cleavage of an embryo can be reared to maturity.

Advancement in culture techniques will enable us to overcome the barriers in sexual reproduction. Desirable genes may be introduced into the newly fertilized eggs in the cultured ovules. Pollen may be modified to carry desirable genes for *in vitro* pollination and fertilization. However, the techniques cannot be expected to achieve crosses when the gametes fail to fuse or when lethal genes occur in the combinations.

8. REFERENCES

1. ABOU-ZEID A 1973 Vielseitige Beeinflussung des Wachstums und der Morphogenese bei der Kultur von *Prunus*-Embryoachsen *in vitro*. 1. Myo-inosit als kritischer Wachstumsfaktor beim radikularen Meristem der Embryoachsen dormanter Kirsch-, Pfirsich-, und Pflaumenembryonen. Angew Bot 47: 227-239
2. ABOU-ZEID A, KH NEUMANN 1973 Preliminary investigations on the influence of cotyledons on the development of cherry embryos (*Prunus avium* L.). Z Pflanzenphysiol 69: 299-305

3. ABOU-ZEID A, W GRUPPE, KH NEUMANN 1972 Eine Methode zur Anzuch normalwachsender Kirschsamlinge aus Embryoachsen. Gartenbauwiss 37: 399-407
4. ALEXANDER MP, TC RAMANA RAO 1968 In vitro culture of bamboo embryos. Curr Sci 14: 415
5. ANAGNOSTAKIS SL 1977 In vitro culture of immature embryos of American elm. HortScience 12: 44
6. ASAKAWA S 1959 Physiology of embryo dormancy. J Jap For Soc 41: 389-395
7. BALATKOVA V, J TUPY 1972 Some factors affecting the seed set after in vitro pollination of excised placenta of Nicotiana tabaccum L. Biol Plant 14: 82-88
8. BALL E 1956a Growth of the embryo of Ginkgo biloba under experimental conditions. I. Origin of the first root of the seedling in vitro. Am J Bot 43: 488-495
9. BALL E 1956b Growth of the embryo of Ginkgo biloba under experimental conditions. II. Effects of a longitudinal slit in the tip of the hypocotyl. Am J Bot 43: 802-810
10. BALL E 1959 Growth of the embryo of Ginkgo biloba under experimental conditions. III. Growth rate of the root and shoot upon media absorbed through the cotyledons. Am J Bot 46: 130-139
11. BANERJEE SN 1968 Effect of ionizing radiation on developing embryo-gametophyte complex of Pinus resinosa Ait. Bot Mag Tokyo 81: 11-21
12. BANERJEE SN, NW RADFORTH 1969 In vitro studies on the developing embryos of Pinus resinosa. Bot Mag Tokyo 82: 329-340
13. BARTELS H 1957a Kultur isolierter Koniferenembryonen. Naturwissenschaften 44: 290-291
14. BARTELS H 1957b Kultur isolierter Koniferenembryonen II. Naturwissenschaften 44: 595-596
15. BEASLEY CA 1977 Ovule culture: fundamental and pragmatic research for the cotton industry. In J Reinert, YPS Bajaj, eds, Applied and Fundamental Aspects of Plant, Tissue, and Organ Culture. Springer-Verlag, Berlin, pp 160-178
16. BELLAROSA R 1981 In vitro culture of Quercus suber L. embryos. In Proc IUFRO Sect S2 01 5. Internat sur la Culture 'in vitro' des Essences Foretieres, Fontainebleau, France, pp 119-125
17. BERLYN GP 1962 Developmental pattern in pine polyembryony. Am J Bot 49: 327-333
18. BERLYN GP 1967 The structure of germination in Pinus lambertiana Dougl. Yale Univ School For Bull No 71, 36 pp
19. BERLYN GP, RC BECK 1980 Tissue culture as a technique for studying meristematic activity. In CHA Little, ed, Control of Shoot Growth in Trees. Proc Joint Workshop IUFRO Working Party on Xylem Physiology and Shoot Growth Physiology. July 20-24, 1980, Fredericton, NB, Canada, pp 305-324
20. BERLYN GP, JP MIKSCHE 1965 Growth of excised pine embryos and the role of cotyledons during germination in vitro. Am J Bot 52: 730-736
21. BOGORAD L 1950 Factors associated with synthesis of chlorophyll in the dark in seedlings of Pinus jeffreyi. Bot Gaz 111: 221-241

22. BONGA JM 1960 Potentiality for growth and differentiation in the embryo-gametophyte complex of Pinus nigra var. austriaca as season advances. PhD Thesis, McMaster Univ, Canada, pp 109
23. BONGA JM 1977 Applications of tissue culture in forestry. In J Reinert, YPS Bajaj, eds, Applied and Fundamental Aspects of Plant, Tissue, and Organ Culture. Springer-Verlag, Berlin, pp 93-108
24. BROWN CL 1964 The seedling habit of longleaf pine. Rep No 10. Georgia For Res Council, Macon, Georgia, pp 68
25. BROWN CL, EM GIFFORD JR 1958 The relation of the cotyledons to root development of pine embryos grown in vitro. Plant Physiol 33: 57-64
26. BROWN CL, HE SOMMER 1975 An atlas of gynmosperm cultured in vitro: 1924-1974. Georgia For Res Council, Macon, Georgia, pp 271
27. BRUNNER G 1932 Beitrage zur Entwicklungsphysiologie der Kiefernkeimlinge. Jahrb Wiss Bot 76: 407-440
28. BUCKNER JD, JH KASTLE 1917 The growth of isolated embryos. J Biol Chem 29: 209-213
29. BULARD C 1952 Culture aseptique d'embryons de Ginkgo biloba: Role des cotyledons dans l'absorption du sucre et la croissance de la tige. CR Acad Sci 235: 739-741
30. BULARD C 1967a Modifications provoquees par divers regulateurs de croissance sur des plantules de Ginkgo biloba L. cultivees in vitro. Soc Bot Fr, Colloq Morphol Exp 1966: 119-130
31. BULARD C 1967b Un cas d'inhibition de croissance de l'epicotyl chez Ginkgo biloba L. obtenu sous l'influence de gibberellines. CR Acad Sci 265: 1301-1304
32. BULARD C, J MONIN 1963 Etude de comportement d'embryons de Fraxinus excelsior L. preleves dans des graines dormantes et cultives in vitro. Phyton 20: 115-125
33. CARRON MP 1981 Germination in vitro d'embryos immatures d'Hevea. Caoutchoues et Plastiques No 612: 93
34. CHAMBERLAIN CJ 1935 Gymnosperms; Structure and Evolution. Univ of Chicago Press, Chicago, pp 484
35. COLLINS GB, JW GROSSER 1984 Embryo culture. In IK Vasil, ed, Cell Culture and Somatic Cell Genetics of Plants. Vol 1 Laboratory Techniques. Acad Press, New York, pp 240-257
36. COLLINS, GB, NL TAYLOR, JW DEVERNA 1984 In vitro approaches to interspecific hybridization and chromosome manipulation in crop plants. In JP Gustafson, ed, Gene Manipulation in Plant Improvement. Plenum Press, New York, pp 323-383
37. COSSIO F, G MINOTTA 1983 Preliminary trials on the in vitro culture of embryos isolated from walnuts, and comparison of different mineral salt combinations. Rivista della Ortofloroflutticol Ital 67: 287-298
38. COULTER JM, CJ CHAMBERLAIN 1903 Morphology of Angiosperms. D Appleton and Co, New York, pp 348
39. DIETRICH K 1924 Uber Kultur von Embryonen ausserhalb des Samens. Flora 117: 379-417
40. ENGVILD KC 1964 Growth and chlorophyll formation of dark-grown pine embryos on different media. Physiol Plant 17: 866-874

41. GJONNES BS 1962 Vitrokulturer av gran [Picea excelsa (Lam.) Link.] Noen forsok med kimplanter. Vitro cultures of spruce, some experiments on seedlings. Medd Norske Skogsforsoksv 18: 85-105
42. GUO GF, XJ JIA, LX CHEN 1982 Induction of plantlet from cultured globular embryo in vitro of Hevea brasiliensis. Hereditas (Beijing) 4: 27-28
43. GURIES RP, RF STETTLER 1976 Pre-fertilization barriers to hybridization in the poplars. Silvae Genet 25: 37-44
44. HADDOCK PG 1954 Sapling sugar pines grown from excised mature embryos. J For 52: 434-437
45. HADDOCK PG 1981 Growth to early maturity of sugar pine trees derived from excised mature embryos germinated in vitro. For Chron 57: 116-119
46. HAGMAN M 1975 Incompatibility in forest trees. Proc Roy Soc Lond B 188: 313-326
47. HAGMAN M, L MIKKOLA 1963 Observations on cross-, self-, and interspecific pollinations in Pinus peuce Griseb. Silvae Genet 12: 73-79
48. HANNIG E 1904 Zur Physiologie pflanzlicher Embryonen. I. Uber die Kultur von Cruciferen Embryonen ausserhalb des Embryosacks. Bot Zeitg 62: 45-80
49. HELLER R 1953 Recherches sur la nutrition minerale des tissue vegetaux cultivees in vitro. Ann Sci Nat Bot Biol Veg 14: 1-223
50. HOWE VK 1962 Culture of excised embryos of Pinus ponderosa. Proc Iowa Acad Sci 69: 137-142
51. HU CY 1975 In vitro culture of rudimentary embryos of eleven Ilex species. J Amer Soc Hortic Sci 100: 221-225
52. HU CY 1976 Light-mediated inhibition of in vitro development of rudimentary embryos of Ilex opaca. Am J Bot 63: 651-656
53. HU CY, IM SUSSEX 1971 In vitro development of embryoids on cotyledons of Ilex aquifolium. Phytomorphology 21: 103-107
54. HU CY, JD OCHS, FM MANCINI 1978 Further observations on Ilex embryoid production. Z Pflanzenphysiol 89: 41-49
55. HYUN SK 1954 Induction of polyploidy in pine by means of colchicine treatment. Z Forstgenetik 3: 25-31
56. IVANICKA J, A MOKRA 1982 Development and cultivation of early-ripening cherry. Biologia 37: 3-12
57. KANTA K 1960 Intraovarian pollination in Papaver rhoeas L. Nature 188: 683-684
58. KANTA K, NS RANGASWAMY, P MAHESHWARI 1962 Test-tube fertilization in a flowering plant. Nature 194: 1214-1217
59. KESTER DE 1953 Factors affecting the aseptic culture of lovell peach seedlings. Hilgardia 22: 335-365
60. KESTER DE, CO HESSE 1955 Embryo culture of peach varieties in relation to season of ripening. Am Soc Hortic Sci 65: 265-273
61. KNUDSON L 1922 Nonsymbiotic germination of orchid seeds. Bot Gaz 73: 1-15
62. KONAR RN, YP OBEROI 1965 In vitro development of embryoids on the cotyledons of Biota orientalis. Phytomorphology 15: 137-140
63. KOSSUTH SV, GH FECHNER 1973 Incompatibility between Picea pungens Engelm. and Picea engelmanii Parry. For Sci 19: 50-60

64. KOUIDER M, RM SKIRVIN, KP SALADIN, JO DAWSON, JJ JOKELA 1984 A method to culture immature embryos of Populus deltoides in vitro. Can J For Res 14: 956-958
65. KRIEBEL HB 1972 Embryo development and hybridity barriers in white pines. Silvae Genet 21: 39-44
66. KRIEBEL HB 1975 Interspecific incompatibility and inviability problems in forest trees. Proc 14th Can Tree Improv Assoc, Part 2, Dept Environ, Ottawa, pp 67-79
67. KRIEBEL HB, TH SHAFER 1971 In vitro culture of second year cones of white pines - preliminary results. Proc Work Group Reprod For Trees, XV IUFRO Congr, Gainesville, Fla
68. LAIBACH F 1925 Das Taubwerden der Bastardsamen und die kunstliche Aufzucht fruh absterbender Bastardembryonen. Z Bot 17: 417-459
69. LAMMERTS WE 1942 Embryo culture an effective technique for shortening the breeding cycle of deciduous trees and increasing germination of hybrid seed. Am J Bot 29: 166-171
70. LARUE CD 1935 Cultures of spermatophyte tissue. Am J Bot 22: 914
71. LARUE CD 1936 The growth of plant embryos in culture. Bull Torrey Bot Club 63: 365-382
72. LARUE CD 1942 The rooting of flowers in sterile culture. Bull Torrey Bot Club 69: 332-341
73. LEE CL, H CHANG 1963 Comparative studies on the mature embryo of Pinus koraiensis grown in vitro. Sci Sinica 12: 1149-1159
74. LEPAGE-DEGIVRY MT 1968 Mise en evidence d'une dormance associee a une immaturite de l'embryon chez Taxus baccata L. CR Acad Sci 266: 1028-1030
75. LEPAGE-DEGIVRY MT 1970 Acid abcissique et dormance chez les embryons de Taxus baccata L. CR Acad Sci 271: 482-484
76. LEPAGE-DEGIVRY MT 1973a Etude en culture in vitro de la dormance embryonnaire chez Taxus baccata L. Biol Plant 15: 264-269
77. LEPAGE-DEGIVRY MT 1973b Influence de l'acide abscissique sur le developpement des embryos de Taxus baccata L. cultives in vitro. Z Pflanzenphysiol 70: 406-413
78. LEPAGE-DEGIVRY MT 1973c Intervention d'un inhibiteur lie dans la dormance embryonnaire de Taxus baccata L. CR Acad Sci 277: 177-180
79. LEPAGE-DEGIVRY MT, G GARELLO 1973 La dormance embryonnaire chez Taxus baccata: Influence de la composition du milieu liquide sur l'induction de la germination. Physiol Plant 29: 204-207
80. LI TT 1934 The development of ginkgo embryo in vitro. Nat Tsing Hua Univ (Beijing) 2B: 42-52
81. LI TT, T SHEN 1934 The effect of "pantothenic acid" on the growth of the yeast and on the growth of the radicle of Ginkgo embryo in artificial media. Nat Tsing Hua Univ (Beijing) 2B: 53-60
82. LI WD, RL WANG, XY ZHU 1983 On the embryonic development and ovule culture of interspecific hybrids between Populus simonii Carr. and P. pyramidalis Borkh. Acta Bot Sin 25: 409-416
83. LINSMAIER EM, F SKOOG 1965 Organic growth factor requirements of tobacco tissue cultures. Physiol Plant 18: 100-127

84. LOO, SW, FH WANG 1943 The culture of young conifer embryos *in vitro*. Science 98: 544
85. MAHESHWARI N 1985 *In vitro* culture of excised ovules of *Papaver somniferum*. Science 127: 342
86. MAPES MO, JB ZAERR 1981 The effect of the female gametophyte on the growth of cultured Douglas-fir embryos. Ann Bot 48: 577-582
87. MELCHIOR GH, FW SEITZ 1968 Interspezifische Kreuzungssterilitat innerhalb der Pappelsektion *Aigeiros*. Silvae Genet 17: 88-93
88. MERGEN F, J BURLEY, GM FURNIVAL 1965 Embryo and seedling development in *Picea glauca* (Moench) Voss after self-, cross- and wind-pollination. Silvae Genet 14: 188-194
89. MIKKOLA L 1969 Observations on interspecific sterility in *Picea*. Ann Bot Fenn 6: 285-339
90. MOORE MB 1976 Early development of ponderosa pine (*Pinus ponderosa* Laws.) embryos on a defined culture medium. Silvae Genet 25: 23-25
91. MOTT RL 1981 Trees. *In* BV Conger, ed, Cloning Agricultural Plants via *In Vitro* Techniques. CRC Press, Inc, Boca Raton, Florida, pp 217-254
92. MURASHIGE T, F SKOOG 1962 A revised medium for rapid growth and bioassays with tobacco tissue culture. Physiol Plant 15: 473-497
93. MUZIK TJ 1956 Studies on the development of the embryo and seed of *Hevea brasiliensis* in culture. Lloydia 19: 86-91
94. NARAYANASWAMI S, K NORSTOG 1964 Plant embryo culture. Bot Rev 30: 587-628
95. NITSCH JP 1949 Culture of fruits *in vitro*. Science 110: 499
96. NITSCH JP 1951 Growth and development *in vitro* of excised ovaries. Am J Bot 38: 566-577
97. NORSTOG K 1961 The growth and differentiation of cultured barley embryos. Am J Bot 48: 876-884
98. OLSEN JS, FW STEARNS, H NIENSTAEDT 1959 Eastern hemlock seeds and seedlings, response to photoperiod and temperature. Connecticut Agric Exp Sta Bull No 620, pp 42-43
99. ORR-EWING AL 1957 A cytological study of the effects of self-pollination on *Pseudotsuga menziesii* (Mirb.) Franco. Silvae Genet 6: 179-185
100. OVERBEEK J VAN, ME CONKLIN, AF BLAKESLEE 1941 Factors in coconut milk essential for growth and development of very young *Datura* embryos. Science 94: 350-351
101. PIATNITSKY SS 1947 On pollination in oaks and the germination of the pollen on the stigma. Dokl Akad Nauk USSR 56: 545-547
102. POULSEN GB 1983 Development of the embryo of the sour cherry cultivar Stevnsbaer. Tidsskrift Planteavl 87: 33-38
103. RADFORTH NW 1936 The development *in vitro* of the proembryo of *Ginkgo*. Trans Roy Can Inst Toronto 21: 87-94
104. RADFORTH NW, JM BONGA 1960 Differentiation induced as season advances in embryo-gametophyte complex of *Pinus nigra* var. *austriaca*, using indoleacetic acid. Nature 185: 332
105. RADFORTH NW, LC PEGORARO 1955 Assessment of early differentiation in *Pinus* proembryo transplanted to *in vitro* conditions. Trans Roy Soc Canada 49: 69-82

106. RADOJEVIC L 1979 Somatic embryos and plantlets from callus cultures of Paulownia tomentosa Steud. Z Pflanzenphysiol 91: 57-62
107. RAGHAVAN V 1966 Nutrition, growth and morphogenesis of plant embryos. Biol Rev 41: 1-58
108. RAGHAVAN V 1967 Plant embryo culture. In FH Wilt, NK Wessells, eds, Methods in Developmental Biology. TY Crowell Co, New York, pp 413-424
109. RAGHAVAN V 1976 Experimental embryogenesis in vascular plants. Acad Press, New York, 567 pp
110. RAGHAVAN V 1977 Applied aspects of embryo culture. In J. Reinert, YPS Bajaj, eds, Applied and Fundamental Aspects of Plant Cell, Tissue, and Organ Culture. Springer-Verlag, Berlin, pp 375-397
111. RAGHAVAN V 1980 Embryo culture. In IK Vasil, ed, Perspectives in Plant Cell and Tissue Culture. Acad Press, New York, pp 209-240
112. RANGASWAMY NS 1977 Applications of in vitro pollination and in vitro fertilization. In J Reinert, YPS Bajaj, eds, Applied and Fundamental Aspects of Plant Cell, Tissue, and Organ Culture. Springer-Verlag, Berlin, pp 412-425
113. RANGASWAMY NS, KR SHIVANNA 1972 Overcoming self-incompatibility in Petunia axillaris III. Two-site pollinations in vitro. Phytomorphology 21: 284-289
114. RAO PS 1965 The in vitro fertilization and seed formation in Nicotiana rustica L. Phyton 22: 165-167
115. RAO PS, NS RANGASWAMY 1972 In vitro development of the pollinated pistils of Nicotiana rustica L. Bot Gaz 133: 350-355
116. RAPPAPORT J 1954 In vitro culture of plant embryos and factors controlling their growth. Bot Rev 20: 201-225
117. REED SM, GB COLLINS 1980 Histological evaluation of seed failure in three Nicotiana interspecific hybrids. Tob Sci 24: 154-156
118. REYNOLDS JF 1982 Vegetative propagation of palm trees. In JM Bonga, DJ Durzan, eds, Tissue Culture in Forestry. Martinus Nijhoff/Dr W Junk Publ, The Hague, pp 183-207
119. RIJVEN AHGC 1952 In vitro studies on the embryo of Capsella bursapastoris. Acta Bot Neerl 1: 158-200
120. SACHER JA 1956 Observations on pine embryos grown in vitro. Bot Gaz 117: 206-214
121. SCHMIDT A 1924 Ueber die Chlorophyllbildung im Koniferenembryo. Bot Arch 5: 260-282
122. SCHOPF JM 1943 The embryology of Larix. Ill Biol Monogr 19: 1-97
123. SCHOPMEYER CS 1974 Seeds of woody plants in the United States. USDA For Serv, Wash DC, Agric Handbook No 450, pp 883
124. SCHOU L 1951 On chlorophyll formation in the dark in excised embryos of Pinus jeffreyi. Physiol Plant 4: 617-620
125. SHAEFFER WI 1984 Usage of vertebrate, invertebrate and plant cell, tissue, and organ culture terminology. In Vitro 20: 19-24
126. SINGH H 1978 Embryology of Gymnosperms. Gebruder Borntraeger, Berlin, pp 297
127. SPITSYN IP 1972 Embryo culture of sour and sweet cherries and of their hybrids in an artificial nutrient medium. Biol Sbornik 1972: 71-76

128. SRIVASTAVA PS, A STEINHAUER 1982 In vitro culture of embryo segments of Quercus lebani: Organogenesis and callus growth as a differential response to experimental conditions. Z Pflanzenphysiol 106: 93-96
129. STERLING C 1949 Preliminary attempts in larch embryo culture. Bot Gaz 111: 90-94
130. STEWART JM 1981 In vitro fertilization and embryo rescue. Environ Exp Bot 21: 301-315
131. STEWART JM, CL HSU 1977 In ovulo embryo culture and seedling development of cotton (Gossypium hirsutum L.). Planta 137: 113-117
132. STETTLER RF, RP GURIES 1976 The mentor pollen phenomenon in black cottonwood. Can J Bot 54: 820-830
133. STINGL G 1907 Experimentelle Studie uber die Ernahrung von pflanzlichen Embryonen. Flora 97: 308-331
134. STONE EC, JW DUFFIELD 1950 Hybrids of sugar pine by embryo culture. J For 48: 200-201
135. THOMAS MJ 1970a Premieres recherches sur les besoins nutritifs des embryons isoles du Pinus silvestris L. Embryons indifferencies et en voie de differentiation. CR Acad Sci 270: 1120-1123
136. THOMAS MJ 1970b Premieres recherches sur les besoins nutritifs des embryons isoles du Pinus silvestris L. Embryons differencies. CR Acad Sci 270: 2648-2651
137. THOMAS MJ 1972 Comportement des embryons de trois especes de Pins (Pinus mugo Turra., Pinus silvestris L. et Pinus nigra Arn.), isoles au moment de leur clivage et cultives in vitro, en presence de cultures-nourrices. CR Acad Sci 274: 2655-2658
138. THOMAS MJ 1973 Etude comparee des developpements in situ and in vitro des embryons de Pins: influence des cultures-nourrices sur les embryons isoles durant les premieres phases de leur developpement. Soc Bot Fr, Mem 1973 Coll Morphol 147-148
139. TUKEY HB 1933 Artificial culture of sweet cherry embryos. J Hered 24: 7-12
140. TUKEY HB 1934 Artificial culture methods for isolated embryos of deciduous fruits. Am Soc Hortic Sci 32: 313-322
141. TUKEY HB 1938 Growth patterns of plants developed from immature embryos in artificial culture. Bot Gaz 99: 630-665
142. WHITE PR 1932 Plant tissue cultures. A preliminary report of results obtained in the culturing of certain plant meristems. Arch Exp Zellf 12: 602-620
143. WHITE PR 1943 A handbook of Plant Tissue Culture. J Cattell Press, Lancaster, Pa, pp 277
144. WITHNER CL 1942 Nutrition experiments with orchid seedlings. Am Orchid Soc Bull 11: 261-263
145. WOODS AB 1953 The development in vitro of the proembryo of Pinus. MSc Thesis, McMaster Univ, Canada, pp 57
146. YUCE B 1979 Studies on the germination of olive seed. Tarimsal Arastirma Dergisi 1: 189-203

Table 1. Chronicle of embryo and ovule culture in forest tree species in conifers.

Name	Species	Summary
Schmidt, 1924	pines, spruce, arbor-vitae	Medium included protein, amino acid, carbohydrate and potassium nitrate. It was the first attempt to culture embryos, and growth and development of embryos were observed.
Brunner, 1932	pine	Tollen medium. Abnormal growth occurred in the medium and normal growth in the medium with macerated gametophyte.
Li, 1934	ginkgo	Medium had macronutrients and glucose. Delayed germination was not due to embryo immaturity.
Li and Shen, 1934	ginkgo	Medium contained macronutrients and glucose. Growth was inhibited by methanol extracts of rice and wheat bran, and ginkgo gametophyte.
LaRue, 1935	Douglas-fir, pine, spruce, arbor-vitae	White medium included heteroauxin. Embryos were grown into normal seedlings.
LaRue, 1936	Douglas-fir, pine, spruces, hemlock, arbor-vitae	White medium. Embryos were grown into normal seedlings.
Radforth, 1936	ginkgo	Crone solution with yeast extract. Proembryos grew up to about 22 cells.
Loo and Huang, 1943	pine, *Keteleeria*	Pfeffer solution. Pine proembryos grew to more than 100 cells and *Keteleeria* to a lesser degree.
Sterling, 1949	larch	Pfeffer solution. Larch proembryos had abnormal growth and cleavage of the first embryonal tubes.
Bogorad, 1950	pine	Pfeffer solution with autoclaved megagametophyte. Chlorophyll was formed in embryos from germinated seeds but not from ungerminated seeds.
Stone and Duffield, 1950	pine hybrids	Embryos in megagametophyte germinated on agar medium.
Schou, 1951	pine	Medium had salts and carbohydrates. Carbohydrate was essential for chlorophyll formation in cultured embryos.
Bulard, 1952	ginkgo	Knop solution. Cotyledons in the medium were necessary for embryo development and sugar was an essential ingredient.

Name	Species	Summary
Woods, 1953	pine	Crone solution. Physiological factors were influential on proembryo growth.
Haddock, 1954	pine	Excised embryos grew in an agar medium and seedlings established in soil produced normal trees.
Hyun, 1954	pine	Knop solution supplemented with IBA, 2,4-D, coconut milk and megagametophyte extract. Abnormal seedlings produced from colchicine treatment.
Radforth and Pegoraro, 1955	pine	Crone solution. Proembryos grew to various sizes and groups of embryonic cells.
Ball, 1956a	ginkgo	Tukey solution. Differentiation appeared normal in the early stage of development in culture.
Ball, 1956b	ginkgo	The base of hypocotyl was split into two equal parts and each grew into a normal root.
Sacher, 1956	pine	White medium. Embryos grew into normal seedlings with initial rapid growth.
Bartels, 1957a	pines, Douglas-fir	White medium. Cotyledons in contact with medium promoted hypocotyl and root growth.
Bartels, 1957b	pine	White medium. Embryos grew well on 4% glucose or 7% sucrose.
Brown and Gifford, 1958	pine	Gautheret macronutrients and Nitsch micronutrients. Cotyledons in the medium were much more beneficial for root growth.
Asakawa, 1959	pine	Gautheret macronutrients and Nitsch micronutrients. Embryos from stratified seeds grew better in the presence of sucrose than those from unstratified seeds.
Ball, 1959	ginkgo	Knop solution. Cotyledons in the medium promoted root growth at 4% sugar, but not shoot growth. Shoot growth obtained upon addition of glutamine or coconut milk.
Olson, Stearns and Nienstaedt, 1959	hemlock	Knop solution. It was suggested that megagametophyte had inhibitors responsible for delayed germination in unstratified seeds.
Bonga, 1960	pine	Phthalic buffer solution and IAA. Nearly mature or mature embryos grew to seedlings.

Name	Species	Summary
Radforth and Bonga, 1960	pine	IAA influenced postzygotic differentiation and development of proembryos.
Berlyn, 1962	pines	Knop and White medium. Mature and immature embryos showed growth and development, and needle elongation was enhanced by autoclaved coconut milk.
Gjonnes, 1962	spruce	White medium. Embryos had excellent growth when cultured on the medium with coconut milk, sucrose and corn starch.
Howe, 1962	pine	White medium. IPA at 0.1 and 0.01 mg/l and GA at 0.1 mg/l did not enhance embryo growth.
Lee and Chang, 1963	pine	Nitsch medium. Best growth was obtained by embedding cotyledons in the medium with 5% sucrose and radicles in the medium free of sucrose.
Brown, 1964	pines	Gautheret macronutrients and Nitsch micronutrients. Best growth of embryo was obtained on the media without any growth regulators.
Engvild, 1964	pine	With complete nutrients the chlorophyll obtained was 2/3 that of a normally germinating seedling.
Berlyn and Miksche, 1965	pine	Knop macronutrients and Nitsch micronutrients. The embryos did not require a haustorial function of cotyledons for growth.
Konar and Oberoi, 1965	arbor-vitae	White medium. Embryoids were regenerated from the cotyledons.
Berlyn, 1967	pine	Nitsch solution. Described the morphology of a mature embryo and a developing seedling after germination.
Bulard, 1967a	ginkgo	Heller medium. Description was given on the effect of different growth regulators on growth and development of embryos.
Bulard, 1967b	ginkgo	Knop solution and Heller micronutrients. GA_3, GA_4, and GA_7 at 1 µg/l or higher inhibited epicotyl growth.
Banerjee, 1968	pine	White medium. Irradiated zygotes had less abnormal differentiation but high rate in abortion than those in more advanced stages in culture.

Name	Species	Summary
LePage-Degivry, 1968	yew	Heller medium. Break of dormancy was achieved by culturing the embryos in the liquid medium.
Banerjee and Radforth, 1969	pine	Crone solution. The older the proembryos the greater the potential for normal growth and development. Suspensor cells were induced to form secondary embryos.
Thomas, 1970a	pine	Heller, Halperin, and Murashige and Skoog medium. Precocious germination did not occur when immature embryos were cultured.
Thomas, 1970b	pine	Halperin medium. Immature or nearly mature embryos growing in the dark did not show any development comparable to those germinated naturally.
Kriebel and Shafer, 1971	pines	Nitsch medium. Proembryos and megagametophytes developed within the cultured cones.
Thomas, 1972	pines	Elongation of suspensor cells and division of embryonic cells were promoted by nursing culture with callus.
LePage-Degivry, 1973a	yews	Heller medium. Immature embryos or embryos from dormant seeds required either leaching and GA or leaching and chilling for germination.
LePage-Degivry, 1973b	yew	Heller medium. Embryos germinated in the liquid medium because inhibitors were leached out or diluted by the medium.
LePage-Degivry, 1973c	yew	Heller medium. Embryos contained ABA which was leached into the medium before germination.
LePage-Degivry and Garello, 1973	yew	Heller macronutrients and Gautheret micronutrients. Dormancy was removed by culturing embryos in liquid medium or by transferring to agar medium after 8 days in liquid medium.
Thomas, 1973	pines	Proembryos proliferated in embryonic cells and elongated in suspensor cells when callus was used in nursing culture.
Kriebel, 1975	pine	Limited success was obtained in culturing megastrobili.
Moore, 1976	pine	Embryo culture was used for morphological and histological study.

Name	Species	Summary
Berlyn and Beck, 1980	pines	Campbell and Durzan medium. In high mineral nutrient medium (1.5x), iron is inhibitory to embryo growth.
Haddock, 1981	pine	Embryo culture produced seedlings which grew to maturity.
Mapes and Zaerr, 1981	Douglas-fir	Murashige and Skoog medium. Best embryo growth was when 1 or 2% extract from megagametophyte was added to the medium.

8. *IN VITRO* CONTROL OF MORPHOGENESIS IN CONIFERS

TRAN THANH VAN K, D YILMAZ-LENTZ, and TH TRINH

1. INTRODUCTION

One of the important problems in forestry is the establishment of propagules of high quality for planting. Traditionally this implies the problem of selection through generations. However, due to their slow growth, long vegetative phase before the flowering phase, and their large size, breeding and selection of trees by conventional genetic methods are difficult. In breeding orchards, only female partners can be selected because open pollination relies on airborne pollen. Furthermore, delayed periods of flowering of the two parental partners strongly limit the possibilities for genetic selection. Finally, *in situ* controlled pollination is expensive.

Another way of obtaining high quality propagules is by cloning. This can be done by grafting mature scions onto young rootstock. However, grafts constitute a weak part of the tree and incompatibility problems often emerge later. Another method of asexual propagation is rooted cuttings. This method has been used successfully for propagation of a few conifer species. However, this method is limited because of the slow growth rate, the difficulties of rooting and the need to use relatively young plants.

Tissue culture techniques have shown to be effective in propagating plants from adventitious vegetative buds from organs of juvenile plants of a number of species, e.g., *Sequoia sempervirens* (10, 13, 33), *Pinus radiata* (3), *Pinus palustris* (22), *Pinus pinaster* (14), *Picea abies* (6, 10). However, difficulties are still encountered in propagation from tissues of selected mature trees.

No method has been found yet for the *in vitro* production of male and female cones to open new possibilities in breeding. It has been shown with Nicotiana that the degree of ovule receptivity and the duration of this receptivity are different if one compares in situ ovules with those cultured in vitro (31). The production of male and female cones of conifers in vitro would allow one to use a wider range of parental partners and to conduct successive generations of breeding in a shorter period of time. In vitro pollination would also open new prospects for the creation of new genotypes. Furthermore, it has been shown that, at least in Nicotiana, in vitro produced male and female gametophytes are more embryogenetic than the ones in vivo (30). For example, this method has been used to obtain an increased percentage of androgenic haploid plants of Nicotiana plumbaginifolia (20% of the in vitro produced anthers formed plantlets as compared to 3% of the in vivo anthers). Production of haploid conifers or other woody species would be an important tool for selection.

In this chapter we describe methods of regeneration of Pseudotsuga menziesii from tissues of mature trees and the production of viable female and male gametophytes of Sequoia sempervirens in vitro.

2. REGENERATION OF PLANTLETS FROM TISSUES OF MATURE DOUGLAS FIR (PSEUDOTSUGA MENZIESII (MIRB.) FRANCO)

In this species, the ability to form adventitious vegetative buds was until recently only expressed by tissues excised from embryos and from plants less than 2 years old (11, 12, 13, 32, 33). However, recently a technique for multiple shoot formation in cultures of explants taken from mature Douglas fir has been reported (16). There is a need to develop methods of propagation using tissues from adult trees because proper selection of trees for cloning can only be made after their specific wood properties and their growth characteristics are fully expressed, i.e., at the adult stage (5).

Because of the seasonal rhythm of growth of woody species one has to consider such experimental factors as temporal and spatial sequences when choosing explants, the exogenous supply of growth regulators or their trophic removal, and the environmental conditions to be applied to the donor tree and to the explants.

2.1. Material and methods

2.1.1. Nature of the explants. To compare the morphogenetic responses of organs of young and mature specimens, different types of explants were used: 1) embryo, cotyledon, and hypocotyl fragments from plants; 2) leaves of different ages, resting buds, fragments (1 cm long) of branches without axillary buds, and a thin layer of cortical and parenchymatous cells excised from these branches according to a method developed previously (24), all obtained from branches of an 80-year-old tree grafted onto 13-15-year-old trees.

All material (especially outdoor material) was washed thoroughly in water and detergent, and sterilized by a sodium hypochlorite solution (7%) for 20 or 30 minutes. Seeds were washed with water and sterilized for 15 minutes in sodium hypochlorite, then rinsed thoroughly in distilled water. Embryos were excised aseptically after 30 hours of imbibition in distilled water.

2.2. Culture medium

2.2.1. Juvenile material. Embryo, cotyledon, and hypocotyl fragments were cultured on a medium containing half strength of Murashige and Skoog (19) minerals, 5.10^{-6} M of indoleacetic acid (IAA), indolebutyric acid (IBA), N^6-benzyladenine (BA) and naphthaleneacetic acid (NAA), according to Cheng (11, 12, 13), 20 g/l of sucrose, and 8 g/l of agar.

2.2.2. Mature material. Two basal media containing either Murashige and Skoog (BM) (19) or Gamborg (G) (17) minerals were used with the addition of thiamin (0.4 mg/l), inositol (100 mg/l), sucrose (30 g/l) and agar (10 g/l). Phenylalanine and the growth substances NAA, IBA, Kinetin (Kin), BA and 6 $\gamma\gamma$ dimethylallylaminopurine (2-iP) were added to the basal media at the concentrations shown in Table 1. All media were adjusted to pH 5.6.

2.3. Environmental conditions

2.3.1. Donor plants. Plantlets were grown under the following conditions: 24-27°C, a photoperiod of 16 h of natural light complemented with fluorescent and incandescent lights to a minimal level of 80 $\mu Em^{-2}s^{-1}$, and a relative humidity of 65%. The potted mature donor grafts were grown either outdoors, or in growth chambers under the same environmental conditions mentioned above

Table 1. Callus and bud formation on explants from 80-year-old *Pseudotsuga menziesii*.

Culture medium			Explants		
				Stem	
Minerals	Growth substances (Mol)	Phenylalanine (Mol)	Leaves	Segment	Thin layer
BM or G	NAA 10^{-6}, Kin 10^{-6}		GCC	GCC	GCC
BM or G	NAA 10^{-6}, BA 10^{-6}		GCC	GCC	GCC
BM or G	2-iP10^{-6}, Kin 10^{-6}		GFC	GFC	GFC
G	NAA 10^{-6}, IBA 10^{-6}	10^{-6}	GCC	Vegetative buds	GCC

BM = Basal medium, G = Gamborg medium, GCC = Green compact callus, GFC = Green friable callus.

except that constant temperatures of 17°C, 22°C, 24°C and 27°C were used.

For outdoor trees, experiments were carried out with collections made in November and March.

2.3.2. Culture. Explants from juvenile material were cultured in a growth chamber at 19°C, 70% relative humidity, and a 16-h photoperiod. Mixtures of fluorescent and incandescent lights (80 $\mu Em^{-2}s^{-1}$) or mixtures of isoquantal monochromatic lights (450 nm, 550 nm, 620 nm and 680 nm) and white lights (80 $\mu Em^{-2}s^{-1}$) were used.

Explants from mature trees were cultured under 170 $\mu Em^{-2}s^{-1}$ of continuous cool white fluorescent and incandescent lights. Two temperatures were tested: 23° ± 1°C and 26° ± 1°C. All explants were cultured in 25 x 2.5 cm test tubes.

2.4. Results

2.4.1. Juvenile stage explants. The most dramatic developments were the following: monochromatic light of 550 nm, in comparison with other monochromatic lights and white light, increased the percentage of hypocotyl fragments that formed buds from 36 to 75% (Fig. 1). Cotyledon explants kept in the dark formed roots instead of shoots without first forming callus (Fig. 2).

FIGURE 1. Vegetative bud formation on a Pseudotsuga menziesii hypocotyl under 550 nm monochromatic light.

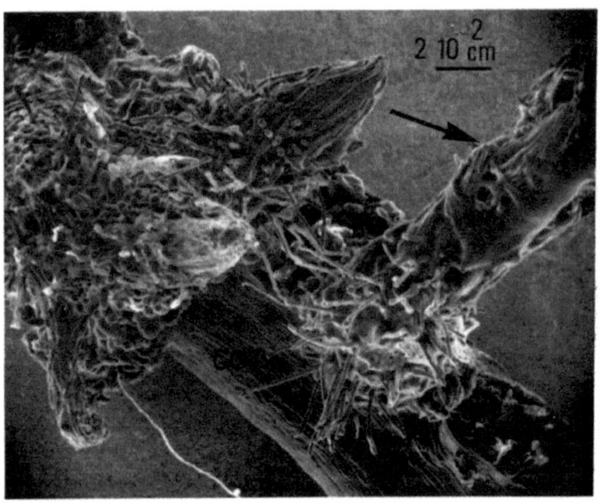

FIGURE 2. Direct root (arrow) formation on a cotyledon (Cot) of Pseudotsuga menziesii in the dark.

2.4.2. Mature tissues. For mature donors (grafts) grown in growth chambers, the results were as follows: different types of explants from mature material (Table 1) formed compact (if cultured with Kin) or friable green (if cultured with 2iP) callus. Callus also developed on thin cell layer explants (Fig. 3), but no organogenesis was observed in this callus. Fragments of branches cultured on Gamborg medium containing NAA 10^{-6} M, IBA 10^{-5} M and phenylalanine 10^{-6} M formed on average two or three axillary buds with a maximum of five buds after 4 weeks of culture at 23° 2°C (Fig. 4). Although these buds had a juvenile type of phyllotaxy, they stopped growing when the experiments were conducted with material collected in winter, spring, or autumn. However, continuous growth was obtained when the explants were excised in the summer and exposed to a constant temperature of 27°C instead of 23°C. Subsequent transfer onto basal medium (BM) of various strength (½, ¼), with or without activated charcoal did not result in the formation of a significant percentage of roots.

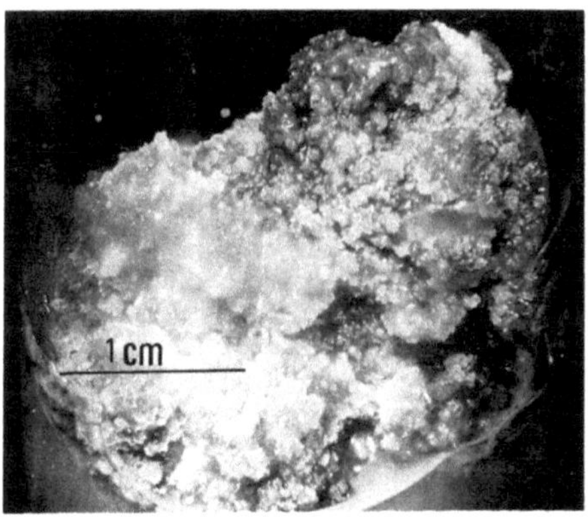

FIGURE 3. Callus from a thin layer of cells stripped from a stem of *Pseudotsuga menziessii*.
FIGURE 4. Vegetative bud development (arrow) from a stem section of an 80-year-old *Pseudotsuga menziessii*.

Fragments of branches excised from outdoor 80-year-old donor plants formed two or three buds which became macroscopically visible after 1 week of culture at 25°C on Murashige and Skoog basal medium deprived of growth substances (Fig. 5). In this medium, the base of 6% of the explants formed roots.

A histological study suggested that the buds obtained on medium supplied with growth substances, were formed de novo, while the buds obtained on basal medium originated from preexisting resting axillary cell territories (Fig. 6).

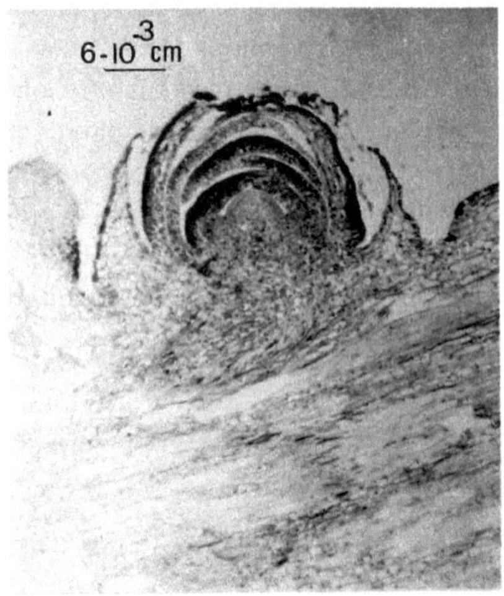

FIGURE 5. Formation of two vegetative buds (arrows) on a basal medium.
FIGURE 6. A longitudinal section of a preexisting axillary bud of a *Pseudotsuga menziesii* stem section.

2.5. Discussion

In vitro regeneration of vegetative buds from a mature tree was feasible. However, attempts to root these buds *in vitro* were successful only in a few cases. On the other hand, roots can be induced directly on cotyledons in the dark. With juvenile explants, monochromatic light of 550 nm had a positive effect. Future experiments, for example with sequential monochromatic radiation and dark treatments, could conceivably further improve budding and rooting in both juvenile and mature material. Pre-treatment of the donor tree by repeated spraying with growth substances or manipulation of the light/temperature regime could be useful (see Franclet, volume 1). Futhermore, preculture of mature stem fragments to initiate actively developing shoots in these fragments or the use of new culture media (16) could improve the reactivity of explants. In addition, successive grafts could be used to improve regeneration of plantlets from explants from mature trees (15).

3. MALE AND FEMALE CONE PRODUCTION IN *SEQUOIA SEMPERVIRENS* EXPLANTS

The production of haploid plants from male and female gametophytes is, potentially, a powerful tool in plant breeding programs. It has been shown that in some cases, the embryogenic capacity of male and female gametophytes produced *in vitro* is higher than those of gametophytes produced *in vivo* (30). The aim of our investigation is first to establish a tissue culture system to produce male and female gametophytes of gymnosperms *in vitro* and subsequently to study the characteristics and the embryogenetic capacity of these gametophytes. Although this is an important aspect of research, there are no reports on *in vitro* production of gametophytes of forest tree species.

Most of the literature deals with *in vitro* formation of vegetative buds from fragments of cotyledons or hypocotyls (9, 12, 13), or from fragments of young needles (1, 7, 21) or embryos (2, 4, 18, 20). In some cases, subsequent *in vitro* rooting of these vegetative shoots has not been well controlled. In this paper, we describe the successful *in vitro* production of viable male and female gametophytes of *Sequoia sempervirens*.

3.1. Material and methods

3.1.1. *Nature of explants*. Fragments, 1 to 2 cm long, with or without a terminal meristem, were excised either from lower branches of an approximately 70-year-old *Sequoia sempervirens* tree or from young sprouts developed from the base of the trunk of this tree. Experiments conducted in October-November, February, and April were repeated during three successive years. After excision, the explants were rinsed with detergent and sterilized with calcium hypochlorite (7%) for 30 min.

3.1.2. *Culture medium*. The basal medium (BM) contained half strength mineral nutrients of Murashige and Skoog (19), thiamin (0,4 mg/l), inositol (100 mg/l), sucrose (30 g/l) and agar (8 or

10 g/l). Basal media supplemented with 10^{-6} M of the following growth substances: gibberellic acid (GA_3 and $GA_{4/7}$), NAA, ethrel, abscisic acid, triiodobenzoic acid (TIBA) were tested. In addition, equimolar (10^{-6} M) combinations of GA_3 and NAA, of ethrel and NAA, of ethrel and IBA, and of ethrel and ABA were also tested.

As soon as male and female cones developed, they were transferred to BM containing Kin (10^{-7} M) and IBA (10^{-5} M) and sometimes Zeatin (2.10^{-6} M). After six weeks of growth, the explants were subcultured on fresh medium.

3.1.3. Environmental conditions. The explants were cultured in 25 x 2.5 cm test tubes at 25° ± 1°C under 170 $\mu Em^{-2}s^{-1}$ of fluorescent and incandescent light.

3.1.4. Germination test. Pollen grains from male cones collected in a breeding orchard and from male cones produced *in vitro* were submerged in an aqeuous solution containing 30 g/l sucrose and 1 g/l boric acid. Pollen germination percentages were determined after 24 h.

3.2. Results

3.2.1. Primary in vitro development of reproductive cones.

1) Influence of the position of the explant on the donor plant and of the time of excision of the explant. Four months after the beginning of the culture, the stem explants had increased in length from 1-2 cm to 8-12 cm. This increase was due to the elongation of the terminal bud or of the subterminal axillary bud. In explants excised in October from young stump sprouts these elongating buds developed into male and female cones (Figs. 7 and 8). In three replications of the experiment an average of 30% of the explants formed at least one male cone and 1% a female cone. In explants excised in February and April from the stump sprouts, or in explants excised from the lower branches of the crown at any time, reproductive organs did not develop.

2) Influence of the culture media. Reproductive cones were obtained only on basal medium. The growth substances tested, including TIBA and ABA, inhibited development of reproductive organs.

3) Morphology of cones formed in vitro. *In vitro* cones (Figs. 8, 11, and 12) were identical in shape and size to *in vivo* cones. At the axils of the bracts of the *in vitro* female cones ovules developed (Fig. 9) and in the male cones, yellow sacs containing pollen grains were formed (Fig. 10). The *in vitro* pollen grains, about half the size of *in vivo* pollen grains, were viable in the germination test. Cones formed *in vitro* were different from cones formed *in vivo* in some respects. Instead of senescing, male and female cones *in vitro* remained turgescent after maturation and could be maintained in a turgescent state for at least 10 months by subculture on fresh medium.

FIGURE 7. Male cones (arrows) from a *Sequoia sempervirens* stump sprout explant.
FIGURE 8. Male cone produced *in vitro*.

These long living functional gametophytes *in vitro* could possibly be used for; 1) pollination between relatively distant genotypes that are sexually poorly synchronized in time, and 2) improved production of male and/or female haploid plants by pollen or ovule cultures in a manner similar to that observed in tobacco (34).

3.2.2. Secondary shoot development from sexual cones in vitro. Some female and male cones *in vitro* developed secondary shoots

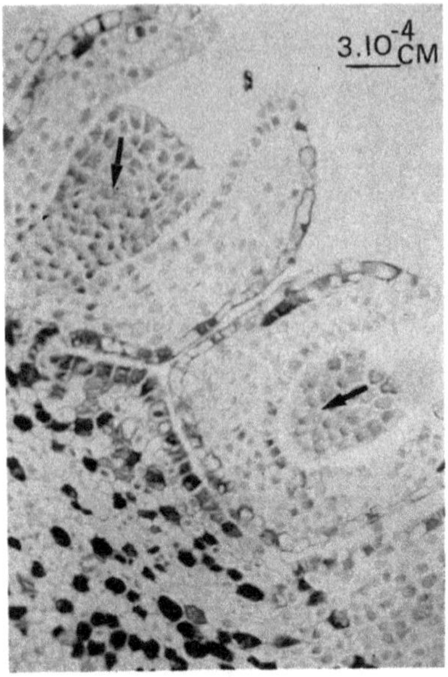

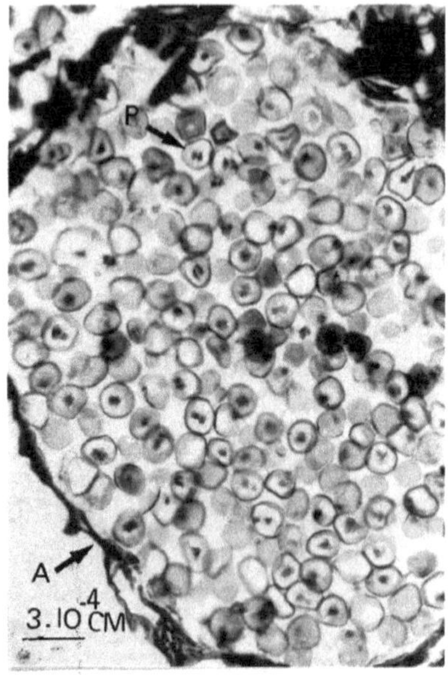

FIGURE 9. Histological longitudinal section of a female cone produced *in vitro* showing two ovules (arrows).
FIGURE 10. Pollen grains (P) in a pollen sac of a male cone produced *in vitro*.

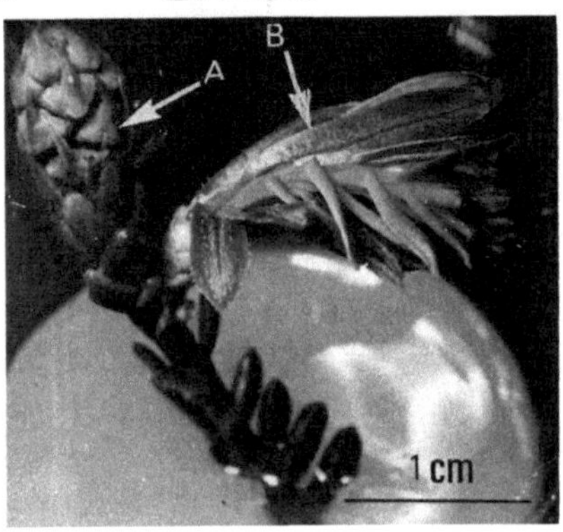

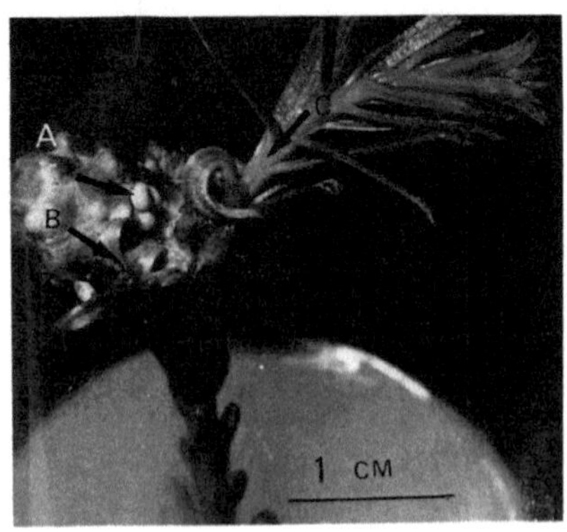

FIGURE 11. A shoot (B) developing from a bract axil at the base of a male cone (A).
FIGURE 12. A shoot developing from the base of a female cone.

(Figs. 11 and 12) about 2 weeks after their subculture on basal medium with zeatin (2.10^{-6} M) or zeatin (2.10^{-6} M) and 2-iP (10^{-6} M). After excision of these secondary shoots and their transfer to fresh basal medium the shoots derived from male cones formed male, and those derived from female cones formed female reproductive cones. These cone forming secondary shoots did not always arise from the base of the cones, sometimes they arose in a median position on the cone.

3.3. Discussion

Axillary buds of orthotropic sprouts at the base of the main stem of *Sequoia sempervirens* form vegetative shoots with a juvenile type phyllotaxy when cultured *in vitro* (8). Presumably, because normal apical dominance patterns do not prevail in these shoots, they can evolve reproductive cones, although only if the initial explant was excised from stump sprouts in the autumn. The inhibition of flowering by apical dominance, and the flowering of a certain defined type of juvenile meristem after its removal from apical dominance, were first demonstrated in *Geum urbanum* (23, 27, 29) and later in several orchids (25).

Due to separation from the correlative physiological controls within the donor plant, and due to the exogenous supply of growth substances and of nutrients, *in vitro* methods can induce a rejuvenation in the explant. This results in a reorganization of the network of internal correlations (i.e., of signals and messages) that can lead to the expression of different genetic programs. This is illustrated by the development of vegetative shoots from reproductive cones produced *in vitro* and the precocious formation of sexual cones on these. Besides being used in tree breeding, this *in vitro* production of gymnosperm gametophytes could be exploited to initiate new developmental patterns such as embryo formation. It has been shown that pollen grains resulting from *in vitro* meiosis in *Nicotiana tabacum* or in *Nicotiana plumbaginifolia* had a greater embryogenic ability than *in vivo* produced pollen grains (27, 30).

4. CONCLUSIONS

Our results have shown that in some woody species morphogenesis can be controlled by environmental factors (light, seasonal rhythm, culture medium) and by internal factors (alteration of correlations, positionnal effects). It has been shown recently that biologically active oligosaccharides can dramatically influence the morphogenetic pathways in tobacco thin cell layers, suggesting that cell wall components might release some specific signals required for cell division and/or organogenesis. This suggests that, in extension of the type of work described here, different oligosaccharides should be isolated and tested in organogenetically recalcitrant cultures of woody species.

Systematic studies at the biochemical and molecular levels are needed to characterize the different phases of the passage of the juvenile phase to the adult phase and of rejuvenation. These phases are critical in perennial woody plants. To advance knowledge in this field the metabolism of phenolic compounds and its effects on lignification and structure of the cell wall in woody species should also be studied. This could lead to new methods to activate cell division, cell reorganization, and cell elongation in arrested cells of resting or even of growing organs of these species.

5. REFERENCES

1. ABO EL-NIL M, Z WOCHOK 1977 *In vitro* developmental responses of wild full sib families of Douglas fir. Plant Physiol 59 (Suppl): 2
2. ARNOLD S, T ERIKSSON 1978 Induction of adventitious buds on embryos of Norway spruce grown *in vitro*. Physiol Plant 44: 283-287
3. BIONDI S, TA THORPE 1982 Growth regulator effects, metabolite changes, and respiration during shoot initiation in cultured cotyledon explants of *Pinus radiata*. Bot Gaz 143: 20-25
4. BONGA JM 1977 Application of tissue culture in forestry. *In* J Reinert, YPS Bajaj, eds, Plant Cell Tissue and Organ Culture. Springer Verlag, Berlin, pp 93-108
5. BONGA JM 1982 Vegetative propagation of mature trees by tissue culture. *In* AM Rao, ed, Tissue Culture of Economically Important Plants. COSTED and ANBS, Singapore, pp 191-196
6. BORNMAN CH, E JANSSON 1981 Regeneration of plants from the conifer leaf: Limitations imposed by needle morphology. *In* Proc IUFRO Sect S2 01 5. Int Workshop "In Vitro" Cultivation For Tree Species. Fontainebleau, France, pp 41-53

7. BORNMAN CH, E JANSSON 1980 Organogenesis in cultured Pinus sylvestris tissue. Z Pflanzenphysiol 96: 1-6
8. BOULAY M 1977 Multiplication rapide du Sequoia sempervirens en culture in vitro. Ann AFOCEL, 37-67
9. CAMPBELL RA, DJ DURZAN 1975 Induction of multiple buds and needles in tissue cultures of Picea glauca. Can J Bot 53: 1652-1657
10. CHAPULA V 1977 Development of isolated Norway spruce and Douglas fir buds in vitro. Commun Inst For Cech 10: 71-78
11. CHENG TY 1975 Adventitious bud formation in cultures of Douglas fir (Pseudotsuga menziesii (Mirb.) Franco). Plant Sci Lett 5: 97-102
12. CHENG TY 1976 Vegetative propagation of western hemlock (Tsuga heterophylla) through tissue culture. Plant Cell Physiol 17: 1347-1350
13. CHENG TY 1977 Factors affecting adventitious bud formation of cotyledon culture of Douglas fir. Plant Sci Lett 9: 179-187
14. DAVID H, K ISEMUKALI, A DAVID 1978 Obtention de plants de Pin maritime (Pinus pinaster Sol.) à partir de brachyblastes ou d'apex caulinaires de très jeunes sujets cultivés in vitro. CR Acad Sci 287: 245-248
15. DE LA GOUBLAYE DE NANTOIS T 1980 Rajeunissement chez le Douglas en vue de la multiplication végétative. Etudes sur la plagiotropie des parties aériennes et racinaires. DEA Physiol Veget, PARIS VI
16. GUPTA PK, DJ DURZAN 1985 Shoot multiplication from mature trees of Douglas fir (Pseudotsuga menziessii) and sugar pine (Pinus lambertiana). Plant Cell Rep 4: 177-179
17. GAMBORG OL, RA MILLER, K OJIMA 1968 Nutrient requirements of suspension cultures of soybean root cells. Exp Cell Res 50: 148-151
18. KADKADE PG, HJ O'CONNOR 1977 Influence of light quality on organogenesis in Douglas fir tissue cultures. In Proc For Biol Wood Chem Conf. TAPPI, Atlanta, pp 71-75
19. MURASHIGE T, F SKOOG 1962 Revised medium for rapid growth and bioassays with tobacco tissue cultures. Physiol Plant 15: 493-497
20. PATEL KR, TA THORPE 1984 In vitro differentiation of plantlets from embryogenic explants of lodgepole pine (Pinus contorta Dougl. ex Loud.). Plant Cell Tissue Organ Cult 3: 131-142
21. REILLY K, CL BROWN 1976 In vitro studies of bud and shoot formation in Pinus radiata and Pseudotsuga menziesii. Ga For Res Pap 86: 1-9
22. SOMMER HE, CL BROWN, PP KORMANIK 1975 Differentiation of plantlets in long leaf pine (Pinus palustris Mill.) tissue cultured in vitro. Bot Gaz 136: 196-200
23. TRAN THANH VAN K 1965 La vernalisation du Geum urbanum L. Etude expérimentale de la mise à fleur chez une plante vivace en rosette exigeant le froid vernalisant pour fleurir. Ann Sci Nat Bot 12, ser VI: 373-594
24. TRAN THANH VAN K 1973 Direct flower neoformation from superficial tissues of small explant of Nicotiana tabacum L. Planta 115: 87-92
25. TRAN THANH VAN K 1974 Growth and flowering of Cymbidium buds normally inhibited by apical dominance. J Am Soc Hortic Sci 99: 450-453

26. TRAN THANH VAN K 1977 Regulation of morphogenesis. In W Barz, E Reinhard, MH Zenk, eds, Plant Tissue Culture and its Biotechnical Application. Springer Verlag, Berlin, pp 367-385
27. TRAN THANH VAN K, H TRINH 1978 Plant propagation: Non-identical and identical copies. In KW Hughes, R Henke, M Constatin, eds, Propagation of Higher Plants through Tissue Culture. Univ Tennessee Symp Proc, pp 134-158
28. TRAN THANH VAN K 1981 Control of morphogenesis in *in vitro* cultures. Annu Rev Plant Physiol 32: 291-311
29. TRAN THANH VAN K 1985 *Geum urbanum*. In H Halevy, ed, Handbook of Flowering, Vol 3. CRC Press, pp 53-62
30. TRAN THANH VAN K, H TRINH 1980 Capacité embryogénétique des anthères des fleurs néoformees à partir de couches cellulaires minces et celle des anthères des fleurs prélevées sur la plante mère chez le *Nicotiana tabacum* L. et le *Nicotiana plumbaginifolia* Viv. Z Pflanzenphysiol 100: 379-388
31. VERONE P, TH TRINH, K TRAN THANH VAN, C DUMAS 1985 The control of ovule receptivity in the process of *in vitro* pollination. In DL Mulcahy and E Ottaviano, eds, Biotechnology and Ecology of Pollen. Springer Verlag, New York, in press
32. WINTON LL 1972 Callus and cell cultures of Douglas fir. For Sci 18: 151-154
33. WINTON LL, SA VERHAGEN 1977 Shoots from Douglas fir culture. Can J Bot 55: 1246-1250

9. MORPHOGENESIS (CYTOCHEMISTRY)

KAMLESH R. PATEL AND TREVOR A. THORPE

1. INTRODUCTION

Histochemical staining of cells and tissues is one of the major techniques currently applied to a range of problems in basic and applied biological research. A few decades ago application of staining techniques was limited to morphological studies, but in recent years the histochemical approach has become a common component of investigations in the fields of physiology and biochemistry. The principles and procedures of histo- or cytochemistry are essentially similar to those used in biochemistry, except that, in cytochemistry the colored end-product is precipitated within the cells. Thus, with tissues containing a variety of cell types, biochemical functions of a single cell type can be studied with the cells in a normal structural and presumably functional relationship with the other cell populations (39).

Plants and their organs are very complex with respect to their morphology and physiology. Cytochemical procedures help to unravel the biochemical complexity of such a system and provide data which can be interpreted in terms of cells and tissues. The in vitro organization and morphogenesis of multicellular plants depends upon the integration and mutual interaction of the various organs, tissues and cells. There is a need for much more intensive anatomical examination by both light and electron microscopy to trace the emergence of organization within the initially relatively uniform cell mass of callus or suspension culture aggregates (49, 52, see Simola and Pescrata first volume). Cytochemistry permits the determination of differences between cells which might otherwise appear to be a population of a single cell type.

In a chapter such as this where the topic is very broad, it is not feasible to cover all the available techniques. Our aim is to introduce tissue culturists to a few quantitative and qualitative cytochemical methods. For additional information the reader is referred to several excellent books on histochemical methods, development of techniques, principles, theoretical considerations and applications (6, 16, 24, 27, 28).

2. GENERAL METHODS FOR TISSUE PREPARATION FOR CYTOCHEMICAL STAINING

There are five essential steps in preparing cells or tissues before carrying out any histochemical staining, namely: 1. Preparation of tissue, 2. Fixation followed by washing, 3. Dehydration and clearing, 4. Embedding, and 5. Sectioning and mounting the sections. There are only a few set rules and, therefore, modification of the existing procedures is often required, depending on the experimental material and localization procedure.

2.1. Preparation of tissue

Callus tissue or a large explant should be cut into pieces no larger than several mm in thickness immediately before being placed in the fixative. For cells in liquid culture as well as protoplasts it is necessary to remove the culture medium by low force centrifugation before transferring them to the fixative.

2.2. Fixation

The fixation procedure must be adapted to the tissue to be examined and should be specifically designed to preserve the structure(s) or substance(s) which are intended to be localized. For example, the fixation of protoplasts requires artificial buffering of the fixative with an osmoticum to avoid their swelling and bursting. However, once fixation is achieved, the semipermiability of the protoplast membranes is abolished and further presence of osmotic stabilizers is not required (19). In some instances cells are examined without fixation or after only a brief fixation. This is done especially for enzyme localization, where tissue is often frozen at low temperature instead of being fixed with chemicals. Because of the profound effect of the fixation procedures on the stablization and retention of various substances, each staining method given in the following pages specifies the type of fixation necessary or recommended.

2.3. Dehydration, clearing and embedding procedures for paraffin and plastic sections

For a complete discussion of dehydration, clearing, embedding of tissue with paraffin, readers are referred to other sources (6, 27). Preparation of tissue for plastic embedding requires special procedures and solutions, and these are covered elsewhere (7, 13, 34). A preparative technique for histochemical localization of some enzymes in plant tissues in JB-4 plastic sections has also been described (56).

3. CYTOCHEMICAL STAINING PROCEDURES

As a general warning it is important to be aware that a certain amount of non-specific staining occurs with most histochemical reactions. Therefore great care must be exercised in carrying out any staining procedure, paying particular attention to the controls, so as to eliminate the possibility of spurious interpretations. In addition color stability is often limited in time, and specimens must frequently be photographed for analysis.

3.1. Localization of DNA by Feulgen reaction (6, 18)

3.1.1. Principle. Hydrolysis of tissue with HCl splits the purine-glycoside linkage in DNA and forms aldehyde groups on the deoxypentose sugar. When the tissue is stained with Schiff's reagent, each pair of aldehyde groups reacts with one molecule of the reagent and develops a magenta color.

3.1.2. Fixation. Fix the tissue blocks in cold (4°C) Carnoy's fluid (3:1, alcohol: acetic acid) for 30 min to 3 hrs depending upon the type of tissue.

3.1.3. Solutions. 1. Schiff's reagent. Dissolve 2g of stain (basic fuchsin) and 3.8 g of potassium metabisulfite in 200 ml of 0.15 N HCl and stir for 2 hrs. Decolorize by adding 1 g of activated charcoal, stir for 5 min, filter and store in the dark, in tightly stoppered bottles. 2. Bleach solution. Add 5 ml of 1 N HCl and 5 ml of 10% potassium metabisulfite to 90 ml of distilled water.

3.1.4. Procedure. 1. Deparaffinize and hydrate the sections. 2. Hydrolyze them in 5 N HCl at 25°C, generally for 30 min (Optimum time should be determined for each species). 3. Stop further hydrolysis by washing the sections in cold distilled water. 4. Stain with Schiff's reagent for 2 hrs in the dark at 25°C. 5. Wash three times in the bleach solution for 10 min each. 6. Rinse thoroughly in distilled water, dehydrate through an alcohol series, clear in xylene and mount.

3.1.5. Results. DNA appears reddish purple.

3.1.6. Controls. Extract DNA from the sections with 5% trichloroacetic acid at 95°C for 15 min or with DNase (0.5 mg/ml) in 0.1M Tris buffer (pH 5.7) containing 0.2M $MgSO_4$ at 37°C for 4 hrs. Stain as mentioned above, a negative reaction should be obtained.

3.1.7. Quantification. Under standard conditions the color produced is proportional to the amount of DNA. The absolute amount of DNA in plant cells can be measured microspectrophotometrically using chicken erythrocytes as an internal standard (10).

3.2. Localization of RNA with Azure B (14, 25, 46)

3.2.1. Principle. Azure B is a metachromatic stain used for localization of nucleic acids. The specificity of the stain has been demonstrated (14), and it was hypothesized that the difference in stainability of the nucleic acids was due to the number of phosphate binding sites on the respective acids, with RNA having more binding sites than DNA. However, in the absence of DNA and under the standard conditions mentioned below azure B stains RNA metachromatically.

3.2.2. Fixation. Use fresh frozen sections, or tissue fixed in Carnoy's fluid. Post-fixation of tissue in formalin helps retain most of the RNA which otherwise may be lost during the rehydration of the tissue (24).

3.2.3. Stain. Azure B. 0.4 mg/ml of dye in McIlvaine buffer pH 4.0.

3.2.4. Procedure. 1. Incubate the hydrated sections in DNase solution to extract DNA (see 3.1.6). 2. Rinse several times in distilled water. 3. Dehydrate and treat for 2 hrs in acetic anhydride at 27°C. This acetylation step eliminates possible blockage of dye-binding sites by proteins (46). 4. Hydrate the sections again and stain with azure B for 2 hrs at 40°C. 5. After staining rinse the slides in three 10 min changes of tertiary butyl alcohol (TBA). If necessary differentiate in TBA overnight, clear in xylene and mount.

3.2.5. Results. RNA is stained blue.

3.2.6. Control. Incubate previously treated DNase sections with 1 mg/ml RNase in veronal acetate buffer (pH 7.7) for 2 hrs at 37°C prior to staining.

3.2.7. Quantification. In stained tissue sections, azure B follows the Beer-Lambert's law. A method for *in situ* cytophotometric estimation of the absolute amount of RNA using azure B is available (46).

3.3. Localization of total proteins with Naphthol Yellow S (11, 25)

3.3.1. Principle. Naphthol Yellow S (NYS) is an acidic dye which binds electrostatically to the basic groups of the proteins and forms a colored NYS-protein complex at an acidic pH.

3.3.2. Fixation. Fix the tissue blocks or cells in absolute ethanol: acetic acid (5:1) or in absolute methanol: formaldehyde: acetic acid (17:2:1) for up to 2 hrs.

3.3.3. Solutions. 1% acetic acid, 0.1% Naphthol Yellow S in 1% acetic acid (pH 2.8).

3.3.4. Procedure. 1. Place the slides in NYS solution for 30 min at room temperature. 2. Rinse in 1% acetic acid (under continuous agitation) for 30 min renewing the bath at 30 sec, 5 and 15 min respectively. 3. Pass slides through three changes of TBA and two changes of xylene and mount.

3.3.5. Results. Sites of proteins appear yellow.

3.3.6. Control. Prior to staining, place the tissue in a mixture made up of 20 ml of $NaNO_2$ and 60 ml of 1% acetic acid at room temperature for 1-24 hrs. This deamination procedure effectively blocks at the $-NH_2$ moiety and tissue cations such as $-NH_3^+$ against undesired staining.

3.3.7. Quantification. The NYS staining for proteins is stoichiometric, follows Beer-Lambert's law and can be measured quantitatively (25, 50).

3.4. Localization of nuclear basic proteins (histones) with alkaline Fast Green (2,6)

3.4.1. Principle. Histones have a high content of basic amino acids. At pH 8.0, generally most proteins are above or near their isoelectric point, while histones are still below theirs. Thus, when the tissue is kept at pH 8.0 in an acid dye like Fast Green FCF, only histones possess groups capable of dye binding.

3.4.2. Fixation. Fix the tissue in 10% neutral buffered formalin [Na_2HPO_4 (anhyd) 6.5 g plus $NaH_2PO_4 \cdot H_2O$, 4.0 g in 10% formalin] for 24 hrs at room temperature.

3.4.3. Stain. 0.1% aqueous Fast Green FCF (pH 8.0-8.1). Adjust the pH with 1N NaOH.

3.4.4. Procedure. 1. Extract the nucleic acids from the sections with 1N trichloroacetic acid (TCA) for 3 hrs at 60°C. 2. Rinse thoroughly in three 10-min changes of 70% ethanol. 3. Stain for 30 min in alkaline Fast Green at room temperature. 4. Wash the slides for 5 min in distilled water followed by 95% and absolute ethanol. 5. Clear in xylene and mount.

3.4.5. Results. The nuclear basic proteins (histones) will appear green.

3.4.6. Control. After hydrolysis of sections in TCA, extract the histones with 0.25 N HCl for 1H at 4°C or use the deamination procedure (see the method for total proteins, 3.3.6.) prior to staining.

3.4.7. Quantification. The relative amount of histones in the nuclei can be determined using a scanning microdensitometer at a wavelength of 635 nm (12).

3.5. Localization of total carbohydrates by the periodic acid - Schiff's (PAS) reaction (27,30)

3.5.1. Principle. In this reaction, the periodic acid brings about the oxidation of the 1,2-glycol linkage within the sugar molecule, resulting in the formation of two aldehyde groups which develop a magenta colored complex when reacted with Schiff's reagent.

3.5.2. Fixation. Use unfixed cryostat sections or tissue fixed in 4% formaldehyde. In a fixed tissue the technique localizes only insoluble carbohydrates.

3.5.3. Solutions. 0.5% aqueous periodic acid, Schiff's reagent, 2.0% aqueous sodium bisulfite.

3.5.4. Procedure. 1. Place the hydrated sections in periodic acid solution at 25°C for 30 min. 2. Rinse the slides thoroughly in distilled water. 3. Stain the sections in Schiff's reagent for 30 min at 25°C. 4. Rinse the slides in distilled water and place them in sodium bisulfite solution for 2 min. 5. Wash the slides, dehydrate, clear in xylene and mount.

3.5.5. Results. The polysaccharides will stain an intense purple.

3.5.6. Control. Omit the oxidation step with periodic acid and directly stain with the Schiff's reagent. Several other control preparations confirm the specificity of the reaction (16).

3.5.7. Quantification. This method, although quite specific has not been frequently used for quantification of carbohydrates. However, it has been used for microspectrophotometric determination of cell wall free space in woody tissue (5).

3.6. Histochemical localization of lipids by Sudan Black B (4)

3.6.1. Principle. The basis of the action of Sudan dyes is that they are more soluble in lipids than the solvent in which they are initially dissolved (27). The interaction between the dye and the phospholipids may be due to a salt linkage (51).

3.6.2. Fixation. The most widely used fixatives for the preservation of lipids are formal-calcium (1% $CaCl_2$ in 4% neutral formalin) and chromium - formal (1% chromium tioxide and 4% neutral formalin, 1:1).

3.6.3. Solutions. 0.1% bromine water; filtered, saturated solution of Sudan Black B in 70% ethanol.

3.6.4. Procedure. 1. Place sections in freshly prepared bromine water overnight and wash thoroughly with distilled water. 2. Stain with Sudan Black B for 30 min at room temperature. 3. Wash thoroughly in distilled water and mount in glycerin jelly.

3.6.5. Results. Lipids stain black or dark blue. Some phospholipids may appear brownish black.

3.6.6. Control. Extract the sections with a lipid solvent (e.g. ethanol, methanol, chloroform) prior to staining.

4. ENZYME HISTOCHEMISTRY

Several techniques for enzyme localization have been developed, and these are widely applied to many diverse problems in the plant sciences. Localization procedures for some enzymes, which are frequently related to growth and differentiation in plants cells will be presented here. For additional methods and localization procedures for other enzymes, readers should refer to specialized texts (16, 28, 44).

4.1. Histochemical localization of succinate dehydrogenase (6)

4.1.1. Principle. In this reaction electrons are released as a result of oxidation of succinate by succinate dehydrogenase and the electrons reduce the colorless tetrazolium to form a colored tetrazolium product.

4.1.2. Fixation. All chemical fixation greatly reduces enzyme activity, but a brief fixation of 5-15 min in formaldehyde at low temperature (4°C) can be used. However, fresh free-hand or unfixed cryostat sections are preferable.

4.1.3. Solutions. 0.1 M buffered sodium succinate (phosphate buffer pH 7.8); 1.0 mg/ml nitro blue tetrazolium (NBT).

4.1.4. Procedure. 1. Incubate the sections in equal volumes of buffered sodium succinate and NBT for 30-60 min at 37°C. 2. Rinse in distilled water and post fix in 10% neutral formalin. 3. Dehydrate the sections rapidly, clear in xylene and mount.

4.1.5. Results. Deposition of blue formazan depicts the sites of succinate dehydrogenese activity.

4.1.6. Controls. Incubate the sections in medium without sodium succinate or containing sodium malonate to inhibit the enzyme activity.

4.1.7. Quantification. The relative amount of the colored end product of this enzyme (9) and other enzymes mentioned here can be measured at the cell level using integrating microdensitometry image analysis, plug microspectrophotometry or color film photometry (6,16).

4.2. Localization of peroxidase activity (22)

4.2.1. Principle. Peroxidase catalyses the reduction of hydrogen peroxide to water using electron donors other than hydrogen peroxide. Suitable donors include phenols, amines and alcohols. Some of these donors produce partially insoluble color products when oxidized.

4.2.2. Fixation. Peroxidase activity is relatively unaffected by a short period (about 2 hrs) of fixation in 1% formaldehyde or glutaraldehyde. Use fresh, cryostat or specially processed paraffin embedded tissue sections (15).

4.2.3. Solutions. 0.1 M Tris-HCl buffer (pH 7.6) containing 0.5 mg/ml p-phenylenediamine, 1 mg/ml pyrocatechol and 0.01 ml/ml of freshly prepared 1.0% H_2O_2 (from 30% stock).

4.2.4. Procedure. 1. Incubate the sections in above mentioned incubation medium for 2-20 min at room temperature. 2. Decant the incubation medium and rinse with distilled water. 3. Mount in glycerin jelly.

4.2.5. Results. The sites of peroxidase activity appear black.

4.2.6. Controls. Incubate the sections in incubation medium without H_2O_2 but containing 0.01% (ca. 50 units) catalase. Addition of catalase eliminates the endogenous H_2O_2 and the test yields a negative reaction for peroxidase.

4.3. Localization of acid phosphatase activity (18,45)

4.3.1. Principle. Acid phosphatase hydrolyzes sodium β-glycerophosphate and releases the phosphate group which reacts with lead ions in the incubation medium. The lead phosphate thus formed is converted to insoluble granules of lead sulfide when treated with ammonium sulfide.

4.3.2. Fixation. Unfixed, frozen sections are preferred, however, tissue fixed for 2 hrs in 3% glutaraldehyde can be used after a post-fixation wash (43). A processing technique for paraffin embedding has been described which permits demonstration of strong enzyme reaction in well maintained tissue architecture (15).

4.3.3. Solutions. 0.25 M Tris-maleate buffer (pH 5.0); 1.25% sodium β-glycerophosphate (pH 5.0), 0.2% $Pb(NO_3)_2$; 1% $(NH_4)_2S$.

4.3.4. Procedure. 1. Place the sections in a substrate made by combining 10 ml each of Tris-maleate buffer, sodium β-glycerophosphate and lead nitrate solutions. 2. Incubate the sections for 2-4 hrs at 37°C. 3. Stop the reaction by removing the sections from the incubation medium and washing them with several changes of tris-maleate buffer. 4. Treat the material in the ammonium sulfide solution for 1-5 min. 5. Rinse thoroughly in distilled water and mount in glycerin jelly.

4.3.5. Results. Black precipitates of lead sulfide indicate the sites of acid phosphatase activity.

4.3.6. Controls. Several factors can produce artifacts in this lead trapping procedure, hence two different controls are necessary (44). 1. Place the tissue sections in the incubation medium without the substrate, i.e., sodium β-glycerophosphate and 2. Add 10 mM sodium fluoride, an inhibitor of acid phosphatase activity to the medium and process as for normal incubation.

4.4. Localization of adenosine triphosphatase activity (21)

4.4.1. Principle. ATPase is a general term to describe a range of enzymes catalysing the hydrolysis of ATP and yielding inorganic phosphate. The mechanism of staining of these phosphate groups is the same as the lead method used for acid phosphatase activity.

4.4.2. Fixation. Unfixed thin free-hand sections or cryostat sections should be used. Post fixation of cryostat sections in 4% formaldehyde (pH 7.0) at low temperature improves the localization. Fixation in 1% glutaraldehyde for 2 hrs had little effect on activity (1).

4.4.3. Solutions. 0.2 M Tris-maleate buffer (pH 7.2), 1.25 mg/ml ATP, 0.1 M $MgSO_4$, 2.0% $Pb(NO_3)_2$, 1% $(NH_4)_2S$.

4.4.4. Procedure. 1. Place the sections up to 1 h at 37°C in the incubation mixture consisting of 10 ml ATP solution, 10 ml Tris-maleate buffer, 2.5 ml magnesium sulfate and 1.5 ml of lead nitrate (adjust to pH 7.2). 2. Incubate the sections for 1-3 min in ammonium sulfide solution. 3. Rinse the sections thoroughly in distilled water and mount in glycerin jelly.

4.4.5. Results. The brownish-black precipitates depict the sites of ATPase activity.

4.4.6. Control. Since ATP tends to be hydrolysed non-enzymatically during the prolonged incubation period at 37°C, it is necessary to incubate the heat-inactivated sections as one of the controls. Another control should be sections incubated in the test solution lacking ATP.

4.5. Starch substrate film method for the localization of amylase activity (32,48)

4.5.1. Principle. When a section is placed on a glass slide coated with a film of starch, the enzyme splits starch into soluble products, which are washed out. After the reaction when the starch film is stained with iodine or periodic acid- Schiff's reagent, the areas of starch digested by amylase remain clear, indicating the sites of the enzyme in the tissue. Four major starch hydrolyzing enzymes, α-amylase, β-amylase, debranching enzymes and phosphorylase, can be assayed with a recent technique using starch-agar gels containing different substrates and buffers (32).

4.5.2. Fixation. Unfixed cryostat sections or thin hand-cut sections should be used.

4.5.3. Preparation of substrate film. Prepare 1-5% solution of hydrolyzed starch in borate buffer (0.02 M boric acid and 0.01 M NaOH, pH 5.6) by heating the suspension. Dip the clean glass slides in a hot (70°C) starch solution for 1-2 min and allow to dry vertically at 25°C. Remove the starch film from one side of the slide.

4.5.4. Procedure. 1. Place the sections on the starch film slide and incubate in a moist atmosphere at 37°C for 3-15 min. 2. Remove the sections carefully after the incubation and place in the same relative position on another slide for morphological comparison. 3. Fix starch film and tissue for 1 hr in methanol: acetic acid: distilled water (50:10:50). 4. Rinse in running water for 1 min. 5. Stain with IKI (0.5% iodine in 5% aqueous KI) for 1 min or with periodic acid-Schiff's reagent.

4.5.5. Results. Sites of starch degradation by amylase remain unstained.

4.5.6. Control. Use heat killed sections as controls. Addition of 1 mM EDTA to the starch solution has been shown to inhibit α and β-amylase activity (32).

5. APPLICATIONS OF HISTO- AND CYTOCHEMICAL TECHNIQUES IN TISSUE CULTURE

There is a widespread misconception that plant cell suspension or callus cultures are composed of uniform and undifferentiated cells. In fact, variation occurs in the size, degree of vacuolation and cytoplasmic content, cell wall characteristics and shape of cells. The problem becomes even more important when it is realized that cells which are cytologically similar differ greatly in their biochemical competency (53). Since it is usually desirable to correlate any physiological and biochemical changes occurring in a tissue with morphological changes, a histochemical approach is often used. However, very few studies using histo- and cytochemical techniques have been undertaken to answer some basic questions on _in vitro_ differentiation. Some of these reports are outlined below:

Ball (3) was one of the first to correlate a cytochemical feature with organogenesis. He showed that in callus cultures of _Sequoia sempervirens_, the shoots were regenerated mainly by the division of a small group of cells which were essentially devoid of any tannin. He concluded that the tannin-containing cells probably had less regeneration potential than the clear cells. Root callus of sugar maple (_Acer saccharum_) exhibited differential metabolic activities and cellular morphologies at different depths (57). A thin layer of acid phosphatase-active cells enclosed a second layer of cells that reduced nitro-blue tetrazolium (NBT), i.e., they were metabolically very active. Mitosis occurred within this latter cell layer, and no acid phosphatase activity was detectable. The cells beneath this layer also showed strong acid phosphatase activity. The morphology of cells at the outer surface appeared tubular and loosely attached, while those in the NBT region were isodiametric. The cells on the outer surface were loosely attached to each other, while those below the NBT region showed a progressive degeneration with depth through loss of nuclei, cytoplasm and cell wall integrity.

In a comparative histochemical study of shoot-forming and non-shoot-forming tobacco callus increased levels of ribonucleic acid and proteins was observed prior to and during shoot initiation (54). Heavy accumulation of starch was also found in the shoot-forming tissue. The accumulation preceded shoot initiation and was mainly localized in the regions which ultimately gave rise to shoot primordia. In a later study nodules with lignified tracheary elements in the upper part of the callus were observed, while pronounced starch accumulation occurred in the lower part of callus in which protrusions of tissue into the medium occurred (29). In callus of *Nicotiana glauca* and *N. glauca* x *N. longsdorffii* hybrid, organized development began from a single parenchyma cell which divided and formed a growth center (33). The growth center was precociously separated from the surrounding callus by a layer of cells which were shown to be endodermis-like by position and histochemical reactions.

Cytochemical (38) and autoradiographic (55) studies in *Pinus radiata* showed a series of events leading to shoot initiation in the cotyledon explants cultured in the presence of a cytokinin. In shoot-forming cotyledons the incorporation of precursors of DNA, RNA and proteins preferentially occurred in the epidermal and subepidermal cell layers in contact with the medium. In contrast, in the cotyledons cultured in the absence of the cytokinin the incorporation was randomly distributed and the amount of incorporation declined with time. Cytochemical studies in cytokinin-treated embryonic explants of *Pinus coulteri* (36) and cotyledon explants of *P. radiata* (38) showed a decline in the reserves of proteins, lipids and starch with a concomitant increase in the staining intensities of nucleic acids and cytoplasmic and nuclear proteins in the areas of future shoot formation. The mobilization of reserves was characterized by high amylase and lipase activity in the explants, especially in the organogenic centers. At this point the enzymes involved in energy production, i.e., succinate dehydrogenase, and ATPase, became active in the shoot-forming regions. On the other hand, peroxidase activity was mainly localized in the cells forming the vascular tissue. It has been hypothesized that the initiation of organized development *in vitro* involves a shift in metabolism that precedes and is coincident with the process (52). In a similar study, increased staining intensities for nucleic acids and proteins was observed in the target cells and meristemoids during primary and secondary shoot formation in black and white spruce (42).

During *in vitro* differentiation of the crown in dormant embryonic shoots of *Picea abies* a band of sclerenchyma with heavy deposition of lignin, as indicated by phloroglucinol-HCl and other histochemical reagents, was found (26). However, when the same staining procedures were carried out on the material that had not been cultured in vitro, lignin could not be detected. It was concluded that retention of the crown in the explant had marked effects on the induction of adventitious buds from the needle primordia. It has been suggested that the crown acts as a translocation barrier to, or as a transducer or filter of nutritional and/or hormonal stimuli (8).

Studies on adventitious root formation in Pinus radiata hypocotyl explants revealed that it was possible to predict the site of root primordium formation by histochemically locating cells containing tyrosinase or polyphenol oxidase (47). A positive histochemical test for starch was also limited to the cells of the meristemoids. In tissue culture-derived plantlets of black and white spruce the continuity of the vascular systems of the shoot and root axis was demonstrated, after clearing in sodium hydroxide and staining histochemically (37). This observation is of direct importance to in vitro clonal propagation, as extensive callus formation at the base of the shoots and a lack of a proper vascular connection between the root and shoot could make field survival of the plantlets difficult.

Callus, cell suspensions and microspores cultured in vitro are the systems in which early events in embryogeny have been traced ultrastructurally and cytochemically and correlated with biochemical changes. In suspension cultures of carrot, the cells destined to become embryos were found to be highly cytoplasmic and to contain large diffusely staining nuclei each with a single intensely staining nucleolus (31). These cells also showed deep staining with the protein stains nigrosine and Coomassie blue. With the RNA stain gallocyanin, a high content of RNA was observed in the nucleoli and the cytoplasm. In addition, these embryogenic cells showed high dehydrogenase activity. On the other hand, the non-embryogenic cells of the clump gave weak or negative responses. The ultrastructural and stereological analysis of pollen grains of Hyoscymus niger showed that potentially embryogenic, uninucleate pollen could be identified within 6 hr of culture by an increased ratio of volume over density of the fibrillar zone and an increased ratio of dispersed to condensed chromatin in the nucleoplasm (41).

^3H-uridine administered as a one or two-hour pulse to these embryogenic pollen grains was autoradiographically localized in embryoids formed during a subsequent chase (40).

Quantitative cytochemical techniques have also recently been used to test the genetic fidelity of tissue culture-derived shoots and plantlets. Using Feulgen microspectrophotometry it was shown that genetic instability in multiple buds of Pinus coulteri regenerated from tissue culture occurred (35). On the other hand, cytofluorometric measurements of ethidium bromide stained protoplasts showed that the DNA profile for adventitious buds and shoots of Picea abies did not deviate from that of resting buds in situ (20).

The most recent application of cytochemistry is the use of fluorescent staining of cell parts, e.g., plant nuclei and chromosomes (27a), as well as fluorescence-activated cell sorting for rapid identification and selection of somatic hybrid cells produced by protoplast fusion (17,23). The latter technique involves chlorophyll auto-fluorescence or staining of two parental protoplast populations with different fluorescent dyes. The underlying principle is that the heterokaryons formed by protoplast fusion should contain both of the parental fluorescent markers which could be sorted using a flow cytometer-cell sorter. Further discussion of these approaches is beyond the scope of this chapter.

6. CONCLUSIONS

Histo- and cytochemical approaches in addition to regular biochemical techniques have been used to glean some information on metabolic and other aspects of in vitro differentiation. However, most of these studies are fragmentary and much more needs to be done utilizing these techniques for a better understanding of the emergence and development of organized structures in vitro in both woody and non-woody plant species.

7. REFERENCES

1. AL-AZZAWI MJ, JL HALL 1977 Effect of aldehyde fixation on adenosine triphosphatase and peroxidase activities in maize root tips. Ann Bot 41: 431-435

2. ALFERT M, II GESCHWIND 1953 A selective staining method for basic proteins of cell nuclei. Proc Natl Acad Sci US 39: 991-998
3. BALL E 1950 Differentiation in a callus culture of Sequoia sempervirens. Growth 14: 295-325
4. BAYLISS OB, CWM ADAMS 1976 Bromine-Sudan black (BSB). A general stain for tissue lipids including free cholesterol. Histochem J 4: 505-514
5. BERLYN GP 1969 Microspectrophotometric investigations of free space in plant cell walls. Am J Bot 56: 498-506
6. BERLYN GP, JP MIKSCHE 1976 Botanical Microtechnique and Cytochemistry. Iowa State University Press, Ames
7. BOTTI C, IK VASIL 1984 Plastic embedding for light microscopy. In IK Vasil, ed, Cell Culture and Somatic Cell Genetics of Plants. Vol. 1. Laboratory Procedures and Their Applications. Academic Press, New York, pp 684-688
8. CHALUPA V, DJ DURZAN 1973 Growth and development of resting buds of conifers in vitro. Can J For Res 3: 196-208
9. CHIECO P, P NORMANNI, PJ BOOR 1984 Improvement in soluble dehydrogenase histochemistry by nitroblue tetrazolium pre-uptake in sections: A qualitative and quantitative study. Stain Technol 59: 201-211
10. DHILLON SS, GP BERLYN, JP MIKSCHE 1977 Requirement of an internal standard for microspectrophotometric measurements of DNA. Am J Bot 64: 117-121
11. DIETCH AD 1955 Microspectrophotometric study of the anionic dye, Napthol Yellow S, by tissue sections and by purified proteins. Lab Invest 4: 324-357
12. DODDS JH, R PHILLIPS 1977 DNA and histone content of immature tracheary elements from cultured artichoke explants. Planta 135: 213-216
13. DOUGHERTY WJ 1981 Preparation of semi-thin sections of tissue embedded in water-soluble methacrylate for light microscopy. In G Clark, ed, Staining Procedures. Williams and Wilkins, Baltimore, London, pp 27-38
14. FLAX MH, MH HIMES 1952 Microspectrophotometric analysis of metachromatic staining of nucleic acids. Physiol Zool 25: 297-311
15. FUJIMORI T, T MOCHINO, M MIURA, I KATAYAMA 1981 Enzyme histochemistry on paraffin embedded tissue sections. Stain Technol 56: 355-359
16. GAHAN PB 1984 Plant Histochemistry and Cytochemistry - An Introduction. Academic Press, New York
17. GALBRAITH DW 1984 Selection of somatic hybrid cells by fluorescence-activated cell sorting. In IK Vasil, ed, Cell Culture and Somatic Cell Genetics of Plants. Vol 1. Laboratory Procedures and Their Applications. Academic Press, New York, pp 433-477
18. GOMORI G 1952 Microscopic Histochemistry, Principles and Practice. Univ Chicago Press, Chicago
19. GOULD AR 1984 Staining and nuclear cytology of cultured cells. In IK Vasil, ed, Cell Culture and Somatic Cell Genetics of Plants. Vol. 1. Laboratory Procedures and Their Applications. Academic Press, New York, pp 698-711

20. HAKMAN I, S VON ARNOLD, A BENGTSSON 1984 Cytofluorometric measurements of nuclear DNA in adventitious buds and shoots of Picea abies regenerated in vitro. Physiol Plant 60: 321-325
21. HALL JL 1971 Cytochemical localization of ATPase activity in plant root cells. J Microsc 93: 219-226
22. HANKER JS, PE YATES, CB METZ, A RUSTIONI 1977 A new specific, sensitive and non-carcinogenic reagent for the demonstration of horseradish peroxidase. Histochem J 9: 789-792
23. HARKINS KR, DW GALBRAITH 1984 Flow sorting and culture of plant protoplasts. Physiol Plant 60: 43-52
24. HOROBIN RW 1982 Fixation and other loss-limiting procedures. In RW Horobin, ed, Histochemistry. Gustav Fischer, New York; Butterworths, London, pp 19-55
25. JACQMARD A, JP MIKSCHE, G BERNIER 1972 Quantitative study of nucleic acids and proteins in the shoot apex of Sinapis alba during transition from vegetative to reproductive condition. Am J Bot 59: 714-721
26. JANSSON E, CH BORNMAN 1983 Morphogenesis in dormant embryonic shoots of picea abies: Influence of the crown and cold treatment. Physiol Plant 59: 1-8
27. JENSEN WA 1962 Botanical Histochemistry - Principles and Practice. WH Freeman, San Francisco
27a. LALOUE M, D COURTOIS, P MANIGAULT 1980 Convenient and rapid fluorescent staining of plant cell nuclei with '33258' Hoechst. Plant Sci Lett 17: 175-179.
28. LOJDA Z, R. GOSSRAU, TH SCHIEBLER 1979 Enzyme Histochemistry: Laboratory Manual. Springer-Verlag, Berlin
29. MAEDA E, TA THORPE 1979 Shoot histogenesis in tobacco callus cultures. In Vitro 15: 415-424
30. McMANUS JFA 1948 Histological and histochemical uses of periodic acid. Stain Technol 23: 99-108
31. McWILLIAM AA, SM SMITH, HE STREET 1974 The origin and development of embryoids in suspension cultures of carrot (Daucus carota). Ann Bot 38: 243-250
32. NOLAN MC, BD DAVIS 1984 A starch-agar gel method for the localization of starch hydrolyzing enzymes in the cotyledons and hypocotyls of beans. Am J Bot 71: 137-141
33. NUTIRONCHI V 1981 Histological study of organogenesis in vitro from callus cultures of two Nicotiana species. Can J Bot 59: 1969-1977
34. O'BRIEN TP, ME McCULLY 1981 The Study of Plant Structure: Principles and Selected Methods. Termarcarphi Pty Ltd, Melbourne, Australia
35. PATEL KR, GP BERLYN 1982 Genetic instability of multiple buds of Pinus coulteri regenerated from tissue culture. Can J For Res 12: 93-101
36. PATEL KR, GP BERLYN 1983 Cytochemical investigations on multiple bud formation in tissue cultures of Pinus coulteri. Can J Bot 61: 575-585
37. PATEL KR, C RUMARY, TA THORPE Plantlet formation in black and white spruce. III Histological analysis of in vitro root formation and the root-shoot union. N Z J For Sci (submitted)
38. PATEL KR, TA THORPE 1984 Histochemical examination of shoot initiation in cultured cotyledon explants of radiata pine. Bot Gaz 145: 312-322

39. PATTISON JR, L BITENSKY, J CHAYEN 1979 Quantitative Cytochemistry and its Applications. Academic Press, New York

40. RAGHVAN V 1979 Embryogenic determination and ribonucleic acid synthesis in pollen grains of Hyoscyamus niger (henbane). Am J Bot 66: 36-39

41. REYNOLDS TL 1984 An ultrastructural and steriological analysis of pollen grains of Hyoscyamus niger during normal ontogeny and induced embryogenic development. Am J Bot 71: 490-504

42. RUMARY C, KR PATEL, TA THORPE Plantlet formation in black and white spruce. II Histological analysis of adventitious shoot formation in vitro. Can J Bot (In Press)

43. SEXTON R, J CRONSHAW, JL HALL 1971 A study of the biochemistry and cytochemical localization of β-glycerophosphatase in root tips of maize and pea. Protoplasma 73: 417-442

44. SEXTON R, JL HALL 1978 Enzyme histochemistry. In JL Hall, ed, Electron Microscopy and Cytochemistry of Plant Cells. Elsevier/North-Holland, New York, pp 63-147

45. SHAYKH MM, LW ROBERTS 1974 A histochemical study of phosphatases in root apical meristems. Ann Bot 38: 165-174

46. SHEA JR 1970 A method for in situ cytophotometric estimation of absolute amount of ribonucleic acid using azure B. J Histochem Cytochem 18: 143-152

47. SMITH DR, TA THORPE 1975 Root initiation in cuttings of Pinus radiata seedlings. I Developmental sequence. J Exp Bot 26: 184-192

48. SMITH RJ, J FROMMER 1973 The starch substrate film method for the localization of amylase activity: Suggestions for a standardized procedure. J Histochem Cytochem 21: 189-190

49. STREET HE 1977 Old problems and new perspectives. In HE Street, ed, Plant Tissue and Cell Culture. Blackwell Sci Publ, Oxford, London, pp 501-511

50. TAS J, J JAMES 1981 Quantitative cytochemistry of nuclear and cytoplasmic proteins using the Napthol Yellow S and dinitrofluorobenzene staining methods. Histochem J 13: 711-716

51. TERNER YJ, J SCHNUR, J GURLAND 1963 Stable sudenophilia: Contributions to the histochemistry of Sudan dyes. Lab Inves 12: 405-411

52. THORPE TA 1980 Organogenesis in vitro: structural, physiological and biochemical aspects. In IK Vasil, ed, Perspectives in Plant Cell and Tissue Culture. Int Rev of Cytol Vol 11A. Academic Press, New York, pp 71-111

53. THORPE TA 1982 Callus organization and de novo formation of shoots, roots and embryos in vitro. In DT Tomes, BE Ellis, PM Harney, KJ Kasha, RL Peterson, eds, Application of Plant Cell and Tissue Culture in Agriculture and Industry. Univ Guelph, Guelph, Ontario, pp 115-138

54. THORPE TA, T MURASHIGE 1970 Some histochemical changes underlying shoot initiation in tobacco callus cultures. Can J Bot 48: 277-285

55. VILLALOBOS VM, MJ OLIVER, EC YEUNG, TA THORPE 1984 Cytokinin-induced switch in development in excised cotyledons of radiata pine cultured in vitro. Physiol Plant 61: 483-489

56. YEUNG EC 1984 Histological and histochemical staining procedures. *In* IK Vasil, ed, Cell Culture and Somatic Cell Genetics of Plants. Vol 1. Laboratory Procedures and their Applications. Academic Press, New York, pp 689-697

57. WEISMAN LF, MF MORSELLI 1981 Biochemical and morphological stratification in *Acer saccharum* root callus. In Vitro 17: 1111-1116

10. ROOT FORMATION

Th. GASPAR AND M. COUMANS

1. INTRODUCTION

The present review is exclusively limited to studies of adventitious rooting in sterile and non-sterile excised shoots, stems and leaves. A cutting is a plant suddenly lacking one of its physical and physiological support systems, the roots. Profound perturbation of normal metabolism is required in order to regenerate the root system, to restore the thermodynamics of the whole plant and to sustain life (41). Anatomical similarities in adventitious root development between species (24, 38) suggests that the supporting metabolism is similar in all of them. Complete understanding of this metabolism requires the description of biochemical differentiation in the entire cutting, in the rooting zone and in the initial cells of root primorida. Most of the literature describes the rooting zone but not the initiation of primorida. Consequently, there are only a few papers with histochemical and cytochemical data describing biochemical changes within the progenitor cells of root primordium initials (65).

Most studies primarily describe biochemical changes during the period from severance of the cutting to anatomical differentiation of the first root primordium, a period which can be very brief or may require weeks or months. Often it is not clear what developmental stage of rooting was sampled, nor was the physiological condition of the stock plant taken into consideration. Therefore, the literature of rooting is discrepant and has to be questioned. Most authors, up to recently, simply ignored the inductive or preparatory phase which precedes cell division, organization of the primordium, and growth and differentiation of the newly formed root. This inductive phase may be accomplished in the cutting already at excision (28). Distinction between the successive physiological phases is not always easy since specific biochemical markers are missing.

2. BIOLOGICAL, PHYSICAL AND CHEMICAL FACTORS AFFECTING RHIZOGENESIS. THEIR PUTATIVE ROLES

2.1. Biological factors

As stated by Haissig (41), there is plausible evidence for genetic control of adventitious rooting because of: 1) consistent differences between and within genera in adventitious rooting ability, 2) the occurrence of preformed adventitious root primordia in certain species, 3) the influence of genetic mutation on rooting ability, and 4) the dependence of plant regeneration *in vitro* on the presence of specific genes. However, it remains unknown how many genes or gene products are involved, and what is deficient in those instances where adventitious rooting does not occur.

Rooting is influenced by a variety of physiological factors including age of the parent plant, season when the cuttings are taken, type and size of cuttings, and presence of leaves and vegetative buds. Ease of adventitious root formation generally declines with increasing age of parent stock. Rooting of adult trees, however, may be facilitated by previous *in vitro* rejuvenation (26). A period of elongation following multiplication *in vitro* predisposes the shoots to root in response to auxin (70). Growth correlations within cuttings have often been noted as influential on rooting. The redirection of metabolic resources, particularly carbohydrates, may affect rooting, depending on the physiological status of the stock plant and its cuttings. Rooting may depend on maintenance of nitrogen metabolism (which itself relates to the metabolism of carbohydrates, nucleic acids, and proteins). Furthermore, breaking of bud dormancy and active (but not excessive) shoot growth have often been associated with a general physiological status that promotes rooting. The spurt in rooting generally coincides with the period of high cambial activity (15) during the period of vegetative growth in the spring. During this period a proper balance would exist between available auxin and nutrition (36). Cuttings taken from plants ready to flower root poorly. It has been assumed that rooting and flowering are two antagonistic phenomena from a hormonal point of view (28).

Sinnott's suggestion (77) that polarity is the first requirement for organized development in plants certainly holds true for rhizogenesis since roots always are formed at the physiological basal ends, irrespective of the orientation of cuttings. There is good evidence that polarized gradients of phytohormones are involved in these regeneration patterns (79, 80).

2.2. Physical factors

Adventitious rooting often occurs under conditions of water stress either at the base of cuttings (for instance, induced by exposure to polyethylene glycol solutions) or as water loss from leaves to the surrounding air (indirect effect on stomatal aperture by irradiance or CO_2 level). Available evidence indicates that water stress influences rooting partly through carbohydrate and hormone (abscisic acid, ethylene) metabolisms. The preparation of cutting itself causes water stress and probably reduces cellular ability to maintain optimum solute capacity. Furthermore, the rooting zone generally accumulates large concentrations of soluble sugars, nitrogenous compounds and phenolics, which leads to an abnormal cellular osmotic balance (41 + references therein). Additional investigation is needed to ascertain whether or not the control of rooting is caused by the resulting metabolic changes.

Higher temperatures (± 30°C) appear to favor primordium initiation. Lower temperatures (± 25°C) favor root elongation. The beneficial influence of higher temperature on initiation may be due to its effect on translocation of supportive (carbohydrates) factors and on the related increase in respiration (66) and in catabolism of simple sugars stored in starch at lower temperatures (83). Cheng and Voqui (13), however, found that rooting of excised Douglas-fir shoots rose from 3 to 80% when the temperature was lowered from 24°C to 19°C, and the sucrose and auxin content of the medium were reduced.

There is no consistent theory on the role of light in rooting, since the effect of light varies with the plant and with the method of propagation. The results, however, consistently show that etiolation or reduction of light intensity favors adventitious root formation (65 + references therein). A dark period is often required to induce root formation. Light-triggered control of organogenesis may be related, either directly or indirectly to the increased accumulation of starch in specific cells through the photosynthetic processes, but also presumably to the events mediated by phytochrome and the blue-light photoreceptor. The light effect could also be mediated by endogenous levels of free auxin (85).

2.3. Chemical factors

The literature does not suggest that root formation is markedly influenced by excesses or deficiencies of any particular mineral nutrient, with the possible exception of boron in some cases, and nitrogen in many species. Eliasson (20) suggested that applied nutrients may not generally be needed during rooting because endogenous nutrients are basipetally transported from the shoot. However, he suggested that applied calcium or boron may influence rooting because they are poorly transported in the phloem. Calcium could be needed for the increase in peroxidase activity before root initiation (31, see below). Similarly the relationship between manganese concentration in the leaves and rooting (73) can be associated with the activity of peroxidase binding to cellular structures (31) and activity in indoleacetic acid (IAA) and 1-aminocyclopropane-1-carboxylic acid (ACC) metabolism (31, 88).

In addition to its necessity for growth and differentiation at the whole plant level, the beneficial effect of boron on rooting (49, 59, 62) could be more specifically related to its control of membrane permeability and sugar transport, and of enzymes involved in the metabolism of carbohydrates, phenolics, lignin, auxins, and nucleic acids (58). This shows the difficulty in associating boron-related modification of rooting with any single, isolated metabolic process.

Application of carbohydrates to stock plants and cuttings can have beneficial effects on rooting but the studies often led to erroneous conclusions because they often invalidly assumed that the applied carbohydrates are taken up unchanged, i.e., they are not changed to metabolic inhibitors or physicochemically inactive products (41). Conversion of sucrose to reducing sugars, changes in osmotic pressures and in enzyme activities have been observed, which might indirectly modify rooting. Differential partitioning of carbohydrate into soluble, readily metabolized forms, the insoluble storage forms, may be another factor with a determining role in rooting (40).

Contrary to the generalization that high C/N ratios favor rooting, recent literature concludes that C/N ratio in stock plants is not related to subsequent rooting of cuttings (84). Haissig (41) details the reasons why C/N ratios in themselves may be unreliable predictors of rooting ability. The form of soluble nitrogen in both inorganic and organic compounds (purines and pyrimidines, amino acids and oligopeptides, polyamines, cytokinins and indole auxins) as well as total amounts may accumulate in the rooting zone of cuttings during propagation. However, like the redistribution and accumulation of carbohydrates, there does not appear to be any fixed pattern of spacial change of nitrogen within cuttings.

In addition to ammonium and nitrate nitrogen, endogenous and exogenous amino-acids, sometimes in complicated interactions with sugars, may influence rooting. However, no test has yielded general conclusions concerning the importance of such compounds, or their endogenous levels to the promotion or inhibition of rooting. Oligopeptides that have originated in the shoot have recently been proposed as rooting inducers (56). Such oligopeptides would contain tryptophan or phenylalanine to mimic the core structure of IAA or phenylacetic acids. Klambt (56) suggests that IAA mimics the function of the root-inducing oligopeptide, which recalls the root-inducing effects of auxin-amino acids conjugates. Since cytokinins appear to be inhibitors of rooting, an oligopeptide-cytokinin balance would determine whether or not rooting would occur.

Purine and pyridine bases have positive effects on rooting which indicates a need for nucleic acid synthesis and a dependence on the availability of specific active proteins (7, 19, 51). Purine and pyrimidine analogs suppress rooting (8).

Although applied polyamines do not increase rooting (3), there is evidence that polyamines and their metabolites are involved in rooting. Evidence is based on analogs of polyamine precursors (L-canavanine and L-canaline) and inhibitors of polyamine biosynthesis (methylglyoxal-bis-guanylhydrazone) (MGBG). These compounds lower the levels of spermine and spermidine, increase levels of putrescine, and block the auxin-induced increases in endogenous polyamines, and inhibit rooting (27, 50). With regards to rooting, polyamines may simply store and supply nitrogen (spermine and spermidine can be enzymatically oxidized to produce ammonium ions, 87), or possibly function in association with nucleic acids, or in stabilization of membranes, or in maintaining the structure and function of ribosomes.

There is substantial evidence that applied phenolics may favorably influence rooting. However, it has until recently proven difficult to study their effect (primarily related to auxin metabolism, see 41, 47 and references therein) since results could not be reproduced when cuttings of the same species were tested in different laboratories (compare 45, 46, with 89) and since rooting responses to a given phenolic may vary between species. Biochemical marking of the different rooting phases (28, 32) brought some light into these reported discrepancies in the literature (see section 3.). Leaf anthocyanin accumulation (2), conversion of aromatic amino acids into protein and of phenylpropanoids into lignin certainly indicate participation of endogenous phenolics in the rooting process. Qualitative and quantitative changes in enzyme activities during xylem differentiation in newly formed roots were observed (see section 3.). However, more research is needed to define the metabolism of simple phenolics that are substrates for lignification and anthocyanin biosynthesis, and of nonsubstrate phenolics (69).

In addition to polyphenolics, compounds as unrelated as indole, phenylacetic and phenylbutyric acids, α- and β-naphthol, dihydroasparagusic acid, and sodium metabisulphite have been classified as auxin-synergists (47). It is of course difficult to envisage that all synergists act in the same way. Several steroid hormones, cortisol in particular, stimulate adventitious root formation (34, 76). Vitamin D and its analogues also often stimulate rhizogenesis (10, 49, 62). In most cases examined, these sterol substances did not promote or only slightly promoted root formation by themselves but they acted synergistically with auxins. Their presumed interference with auxin metabolism (35) was confirmed by Moncousin and Gaspar (62), who showed that vitamin D2 favored root formation by modifying the pattern of peroxidase activity during the inductive phase and by Kevers et al. (55) who indicated that ergosterol, ergocalciferol and cholecalciferol affected calcium-mediated peroxidase secretion processes.

Mission et al. (61) have reviewed the beneficial effect of activated-charcoal on in vitro root formation. The stimulation is attributed to the removal of inhibitory phenolics as well as excess of the rooting auxin. Mineral analysis of the medium after treatment by activated charcoal indicated a specific absorption of Fe and Zn. Autoclaving slightly diminished this retention (61).

Promotive effects of oxygenation of the rooting medium on both primordium initiation and elongation were established long ago. All stages of mitosis require oxygen (1). Oxygen may serve chemical roles in rooting apart from respiration. Pingel (67) found a positive correlation among oxygen uptake by nodes, IAA-oxidase activity and rooting. Data were consistent with basic peroxidase and auxin level variations over the course of rooting.

2.4. Growth regulators

It is generally accepted that auxin has a central role in the initiation of adventitious roots. In so-called easy-to-root cuttings, auxins invariably induce the formation of a greater number of roots per cutting than other chemicals. Furthermore, in order to demonstrate the stimulatory influence of other chemicals it is often necessary to supply cuttings with auxin. Young leaves and active buds which are sources of auxin generally enhance rooting and often may be replaced by exogenous auxin. Inhibition of rooting by triidobenzoic acid (TIBA), which is known to interfere with auxin transport and action, confirms that auxin transport is basipetal and that root formation is associated with polarity. We already mentioned that phenolics, which influence auxin metabolism, may affect the rooting response of cuttings. Of the more commonly used auxins indolebutyric acid (IBA) and naphthaleneacetic acid (NAA) are usually more effective than IAA (33). They induce maximum rooting when high concentrations (just below those which induce symptoms of toxicity) are given immediately, or soon after, cuttings are made (48). Application of auxins, at concentrations which themselves do not influence root formation, partially offsets those processes which lead to diminishing root response with increasing delay of the root-inducing treatment (47). In both woody (37) and non-woody cuttings (22) there is a need for a constant supply of auxin throughout the initiation phase. High concentrations of auxin (54) are required to ensure initiation of cell division and organization of the primordium. However, application of a high auxin concentration for a long period inhibits growth of primordia or even initiation itself. These observations suggest that the second phase of auxin action as postulated by Went (86) may be associated with lower auxin-concentration than the first.

Adventitious root formation is generally inhibited by apical or basal application of cytokinin to cuttings. Some cytokinin, however, may be necessary for root formation since addition of small amounts of cytokinin is necessary in tissue culture and since root initiation in cuttings is inhibited by a cytokinin antagonist (78). The requirement for root initiation likely is a low cytokinin level and a favorable auxin/cytokinin ratio.

Gibberellins (GA) have been widely reported to inhibit adventitious root formation (47), particularly when supplied immediately, or soon after cuttings are made (32), while gibberellin antagonists or inhibitors of gibberellin synthesis stimulate root formation (23). Occasionally gibberellin will stimulate root formation. This stimulation is dependent upon irradiation conditions of the stock plants (42) and time of application (32). Both inhibitory and stimulatory effects of gibberellin may be indirect through enhancement of auxin supply either from the leaves to the site of regeneration, or through its well-known effects of carbohydrate metabolism, or these effects may be direct, i.e., gibberellin may directly affect the regeneration site (31, 81).

Leafy stem cuttings vary considerably in their response to basal application of abscisic acid (ABA). It is thought that ABA interacts with other growth regulators, or acts indirectly through its influence on lateral bud break during the rooting period (47).

The reports concerning the influence of ethylene on adventitious root formation and on ethylene production are contradictory (47). There appears to be no correlation between total ethylene production and the number of roots formed in response to auxins. Experiments using inhibitors of ethylene synthesis and action (23) and work on organogenesis in tissue cultures (44), suggest that endogenous ethylene is inhibitory to the early stages of regeneration but stimulatory or essential to a later stage of development.

As stated by Jarvis (47), given the central role of auxin and the more or less direct effect of the other regulators on its metabolism, it is not surprising that concentrations of GA_3, ABA and kinetin, which, in themselves are either inhibitory or without effect on root formation, actually enhance rooting when supplied with IAA.

3. METABOLIC CHANGES ASSOCIATED WITH ROOT FORMATION

3.1. Nucleic acids and proteins

Chloramphenicol, cycloheximide, and actinomycin-D, which inhibit protein synthesis either directly or indirectly, inhibit rooting (36). The recent research on nucleic acid and protein metabolism during rooting, as reviewed by Haissig (41) demonstrates: 1) that the promotive influence of auxin on primordium initiation and development is manifest through quantitative and qualitative changes in protein synthesis during an early part of rooting, after an initial lag; 2) that inhibitors of nucleic acid synthesis may lengthen the lag phase and thereby enhance root primordium initiation and development in some species; 3) that primordium initiation may require mRNA synthesis which may be enhanced by auxin treatments, through some action at the chromosomal level, possibly on transcription; and 4) that some factor in addition to auxin apparently triggers maximum RNA synthesis during the initiation of primordia.

3.2. Sugars, respiration, and photosynthesis

Adventitious rooting is endergonic, as indicated by increases in total cutting mass that may occur before root system development (40). Substantial evidence suggests that cuttings root best under conditions that yield optimum endogenous concentrations of non-structural carbohydrates before and during rooting (12, 57, 73). Auxin appears to mobilize carbohydrates in the rooting zone of cuttings (39, 59, 83). Controversies still exist about the cause-effect relationship between total soluble carbohydrate and starch levels in cuttings and rooting. Total carbohydrate levels may be related more to the formation of insoluble macromolecules such as cellulose, hemicellulose, lignin, proteins, and nucleic acids for root emergence than to root initiation (40, 41).

The respiration rate increases during rooting (66), especially in auxin-treated cuttings. Therefore, inhibitors or uncouplers of oxidative phosphorylation sometimes inhibit rooting (38). Stimulation of rooting by uncouplers of oxidative phosphorylation, which is observed occasionally, has been attributed to the resultant increase in respiration rate (64).

Photosynthesis is not always a requirement for rooting, as is demonstrated in chlorophyll-lacking tissue cultures and cuttings that produced roots while kept in continuous darkness. Furthermore, the net photosynthesis rate in leafy cuttings is much lower than in corresponding stock plants, and declines even during rooting (21). Nevertheless, there is ample direct and indirect evidence (16) that photosynthesis contributes substantially to the carbohydrate (sucrose, glucose, fructose, sugar alcohols, and starch) pools in the basal rooting zone.

Aerobic degradation of carbohydrates via the Embden-Meyerhof-Parnas (EMP) pathway of glycolysis and the pentosephosphate (PP) pathway is the likely source of energy in cuttings, i.e., glucose-6-phosphate, which is readily derived from simple sugars and starch, and is completely or partially catabolized (41). This would provide pentose, ATP and NADPH for nucleic acid and protein biosynthesis upon which rooting is strongly dependent.

3.3. Phytohormones

Initial pioneers' work suggested that the number of roots initiated per cutting may be a function of the amount of auxin-like substances in the zone of regeneration (5, 52). Despite the development of refined techniques for the identification and estimation of auxins and other phytohormones, relatively little progress has been made in recent years. By means of gas chromatography, Brunner (11) identified a high auxin concentration in the region of root regeneration during the phase of initiation, and found a declining auxin concentration prior to formation of the root primordium. These declining auxin concentrations coincide with increasing activity to peroxidase and IAA-oxidase frequently described prior to formation of root primordia (4, 6, 14, 18, 25, 28, 62). Gaspar (28) has suggested that root formation occurs only after total peroxidase has reached and passed a peak of maximum activity and when activity of basic isoperoxidases (of particular importance in IAA degradation) is declining. Experimental data (18, 62, 79) led Gaspar to propose that rooting is improved by conditions which first hasten and increase basic peroxidase activity and then induce a sharp decrease. This explains how substances such as GA_3 or polyphenols, which decrease activity of basic peroxidases (25, 75) or function as competitive inhibitors of their IAA-oxidase activity (68), may inhibit or enhance rooting (32) depending upon the time of application. That means that the auxin level would undergo an inverse variation during inductive and initiative phases (28, 31). In the absence of enough histological data related to auxin level, the question remains whether cell divisions occur before or after the passage through the minimum auxin level. Jarvis (47) is of the opinion that those high concentrations of auxins which initiate the rooting response would be inhibitory to subsequent growth, which demands a low auxin concentration (48). No visible primordia, however, were observed before the peaks of maximum peroxidase and IAA-oxidase activities, that is before the minimum auxin level was reached. Further increase in auxin concentration might stimulate primorida development and elongation. The question also can be posed in other terms: could the determining factor of tissue reentry in cell division and organization of the root primordium be a precisely determined, unstable transitory low auxin level? This unstable condition could be compared to the unstable transitory high auxin level in leaves which is required for flowering (30).

3.4. Phenolics and related enzymes

A correlation between the extent of rooting and the concentration of total phenol content, particularly diphenols, has been demonstrated in cuttings of several species (5,6). Increased rooting of cuttings of *Impatiens* seedlings, and an increase in auxin-like material, is associated with an increased content of anthocyanin-3-glucoside and 2-hydroxy-1,4-dihydroxynaphthalene-4-glucoside, the two major inhibitors of IAA-oxidase in this material. Overall data suggest a synthesis of phenolics throughout the cutting followed by basipetal transport to and greatest accumulation in the base of the stems. However, an increase in endogenous phenol level would only occur during the initiative phase since it varies inversely with the changes in peroxidase activity (18).

The basipetally transported phenolics may be involved in lignification of xylem cells for the future roots. The activity of acidic peroxidases specifically involved in lignification increases throughout the rooting process (28, 31). The activity of PAL which provides part of the lignin precursors (58) increases before xylem formation (41). PAL could also be responsible for the poorly understood anthocyanin biosynthesis that occurs during root formation. Both enzymes may be controlled by ethylene (29, 74). Polyphenol oxidase is another enzyme which increases before rooting (17). This increase could be a result of the preparation (wounding) of cuttings. Phenoloxidase catalysis of o-hydroxylation of monophenols and conversion of o-diphenols to quinones (82) may indirectly participate in an auxin-sparing mechanism and in lignification. Kaminski (53) suggested that phenoloxidase activity accompanies cellular proliferation and that peroxidase activity indirectly influences rooting through effects on cellular differentiation.

3.5. Other aspects

Nitrogen, phosphorous, and potassium are redistributed from mature to developing tissues, with the rooting zone primarily accumulating potassium. This redistribution, however, could occur only after primordium initiation and budbreak (41).

4. FACTORS INFLUENCING ROOT FORMATION *IN VITRO*

As already mentioned, root induction, root primordium initiation and root elongation each require different culture conditions. Rooting as a whole *in vitro* is enhanced by changing conditions after root induction (18, 32, 62). Transplantation to soil also normally produces the conditions which favor the last phase of root development.

In the absence of biochemical markers for the rooting phases, one can only suppose that the below mentioned cultural factors are beneficial for *in vitro* root formation. These factors may enhance the inductive and initiative phases successively, except that the generally recognized beneficial effect of previous shoot elongation (obtained with activated charcoal 5-10 g/l, with or without GA_3 1-10 mg/l) (60, 72) might be related to the inductive phase only.

The basal media of Murashige and Skoog (63) with half strength NO_3, diluted 2 to 5 times, and half strength Quoirin and Lepoivre medium (71) sometimes without any growth regulators, are the most commonly used for rooting. When auxins are needed, IBA or NAA (1 to 3 mg/l), or a mixture IBA (0,5 mg/l) and NAA (0,5 mg/l) are supplied for a period of ± 10 days before passing the rooting shoots to a medium deprived of auxins to obtain further root growth. In some cases a short soak of the cut basal ends of the shoots in a concentrated IBA (1 mg/ml) solution for ± 5 min is preferred as auxin treatment. Treatments and auxins that avoid a too rapid development of callus are generally preferred.

A low level of light, and a lower temperature than required for shoot proliferation are necessary for good rooting. Complete darkness (8 to 10 days) at the beginning of the rooting stage may be beneficial (18).

Tissue culture plantlets are equivalent to small rooting cuttings. Their transfer to conventional growing media involves adaptation of both the shoot and root systems to normal conditions. For good survival and growth of the plantlets it is necessary to transfer them to soil while they are actively growing and have roots which are not too much developed (3 to 5 mm long). An increase in the light intensity some days or weeks before transplanting increases the chances of survival; during this time provision should be made for adequate entry of CO_2 by slackening screw-top or other covers on the vessels that contain the plantlets. After planting in soil an increase in humidity, provided by mist or by a polyethylene cover is advisable until the roots have adapted (43).

5. CONCLUDING REMARKS

The initiation of root formation is a complex morphogenetic phenomenon, in which extrinsic and intrinsic factors play a role. The concept of a unique, specific rhizogenic substance or complex (9) has ceased to be valid, although its presumed components, auxin, polyphenols, oxygen and oxidases still are key elements. Instead we may conclude that quantitative interactions between diverse growth factors rather than specific morphogenetic substances provide a common mechanism for the regulation of the morphogenetic phenomenon. More research is needed to better define the successive physiological phases and their associated metabolisms and the organs or tissues involved in each of them. Therefore, as stated by Jarvis (47), it is of considerable importance that future studies should include the coordinated use of biochemical, histochemical and anatomical techniques. Micropropagation techniques need reliable criteria that can be used to determine the best explant origin, physiological state and size, and to compose the best culture media for the successive in vitro phases.

6. REFERENCES

1. AMOORE JE 1961 Dependence of mitosis and respiration in roots upon oxygen tension. Proc Roy Soc Brit 154: 95-108
2. BACHELARD EP, BB STOWE 1962 A possible link between root intiation and anthocyanin formation. Nature 194: 209-210
3. BAGNI N, SD FRACASSINI, P TORRIGIANI 1981 Polyamines and growth in higher plants. In CM Caldarera, ed, Advances in Polyamine Research, Raven Press, London, 3: 377-388
4. BALASIMHA D, N SUBRAMONIAN 1983 Role of phenolics in auxin induced rhizogenesis and isoperoxidases in cacao (Theobroma cacao L.) stem cuttings. Indian J Exp Biol 21: 65-68
5. BASTIN M 1966 Root initiation, auxin level and biosynthesis of phenolic compounds. Photochem Photobiol 5: 423-429
6. BHATTACHARYA NC, A KUMAR 1980 Physiological and biochemical studies associated with adventitious root formation in Phaseolus mungo L. in relation to auxin-phenol synergism. Biochem Physiol Pflanzen 175: 421-435
7. BHATTACHARYA S, KK NANDA 1978 Stimulatory effect of purine and pyrimidine bases and their role in the mediation of auxin action through the regulation of carbohydrate metabolism during adventitious root formation in hypocotyl cuttings of Phaseolus mungo. Z Pflanzenphysiol 88: 283-293
8. BLAZICH FA, CW HEUSER 1981 Effects of selected putative inhibitors of ribonucleic acid or protein synthesis on adventitious root formation in mung bean cuttings. J Am Soc Hortic Sci 106: 8-11
9. BOUILLENNE R, F WENT 1938 Recherches expérimentales sur la néoformation des racines dans les plantules et les boutures des plantes superieures. Ann Jardin Bot Buitenzorg 43: 25-202
10. BUCHALA AJ, A SCHMID 1979 Vitamin D and its analogues as a new class of plant growth substances affecting rhizogenesis. Nature 280: 230-231
11. BRUNNER H 1978 Einfluss verschiedener Wuchstoffe und Stoffwechselgifte auf wurzelregenerierendes Gewebe von Phaseolus vulgaris L. Veränderungen des Wuchstoffgehaltes sowie der Peroxydase - und der IES-Oxydase-Aktivatät. Z Pflanzenphysiol 88: 13-23
12. CHAMPAGNOL F 1981 Relation entre la formation de pousses et de racines par une bouture de vigne et la quantité d'amidon initialement présente. C R Acad Sci, Paris 67: 1398-1405

13. CHENG TY, TH VOQUI 1977 Regeneration of Douglas-fir plantlets through tissue culture. Science 198: 306-307
14. CHIBBAR RN, K GURUMURTI, KK NANDA 1979 Changes in IAA oxidase activity in rooting hypocotyl cuttings of Phaseolus mungo L. Experientia 35: 202-203
15. DAVIES FT 1984 Shoot RNA, cambial activity and indolebutyric acid effectivity in seasonal rooting of juvenile and mature Ficus pumila cuttings. Physiol Plant 62: 571-575
16. DAVIS TD, JR POTTER 1981 Current photosynthate as a limiting factor in adventitious root formation on leafy pea cuttings. J Am Soc Hortic Sci 106: 278-282
17. DHAWAN RS, KK NANDA 1982 Stimulation of root formation on Impatiens balsamina L. cuttings by coumarin and the associated biochemical changes. Biol Plant 24: 177-182
18. DRUART P, C KEVERS, P BOXUS, Th GASPAR 1983 In vitro promotion of root formation by apple shoots through darkness effect on endogenous phenols and peroxidases. Z Pflanzenphysiol 108: 429-436
19. EBRAHIMZADEH H, M AMIDE 1980 Evolution des protéines dans les fragments d'entre-noeuds de tige de Peperomia blanda cultivés in vitro au cours de la néoformation de racines et de bourgeons. Physiol Vég 18: 405-410
20. ELIASSON L 1978 Effects of nutrients and light on growth and root formation in Pisum sativum cuttings. Plant Physiol 43: 13-18
21. ELIASSON L, L BRUNES 1980 Light effects on root formation in aspen and willow cuttings. Physiol Plant 48: 261-265
22. ERIKSEN EN, S MOHAMMED 1974 Root formation in pea cuttings. II. The influence of indole-3-acetic acid at different developmental stages. Physiol Plant 30: 158-162
23. FABIJAN D, JS TAYLOR, DM REID 1981 Adventitious rooting in hypocotyls of sunflower (Helianthus annuus) seedlings. II. Action of gibberellins, cytokinins, auxins and ethylene. Physiol Plant 53: 589-597
24. FAVRE JM 1977 La rhizogenèse. Aspects divers d'un processus d'organogenèse végétale. Ann Univ Abidjan, Sér C Sci, ISBN No. 2-7166-0151-8
25. FOONG TW, MF BARNES 1981 Rooting cofactors in rhododendron: the fractionation and activity of components from an easy-to-root and a difficult-to-root variety. Biochem Physiol Pflanzen 176: 507-523
26. FRANCLET A 1983 Rajeunissement, culture in vitro et pratique sylvicole. Bull Soc Bot Fr, Actual Bot 130: 87-101
27. FRIEDMAN R, A ALTMAN, U BACHRACH 1982 Polyamines and root formation in mung bean hypocotyl cuttings. I. Effects of exogenous compounds on endogenous polyamine content. Plant Physiol 70: 844-848
28. GASPAR Th 1981 Rooting and flowering, two antagonistic phenomena from a hormonal point of view. In B Jeffcoat, ed, Aspects and Prospects of Plant Growth Regulators, Brit Plant Growth Regulat Group, Wantage, pp 39-49
29. GASPAR Th, C PENEL, FJ CASTILLO, H GREPPIN 1985 A two-step control of basic and acidic peroxidases and its significance for growth and development. Physiol Plant, In press
30. GASPAR Th, C PENEL, C RODUIT, C MONCOUSIN, H GREPPIN 1985 The role of auxin level and sensitivity in floral induction. Biol Plant, In press
31. GASPAR Th, C PENEL, T THORPE, H GREPPIN 1982 Peroxidases 1970-1980 A survey of their biochemical and physiological roles in higher plants. Univ Genève - Cent Bot
32. GASPAR Th, D SMITH, T THORPE 1977 Arguments supplémentaires en faveur d'une variation inverse du niveau auxinique endogène au cours des deux premières phases de la rhizogenèse. C R Acad Sci Paris 285: 327-330
33. GENEVE RL, CW HEUSER 1982 The effect of IAA, IBA, NAA and 2,4-D on root promotion and ethylene evolution in Vigna radiata cuttings. J Am Soc Hortic Sci 107: 202-205

34. GEUNS JMC 1974 Physiological activity of corticosteroids in etioled mung bean plants. Z Pflanzenphysiol 74: 42-51
35. GEUNS JMC 1978 Steroid hormones and plant growth and development. Phytochem 17: 1-14
36. GURUMURTI K, BB GUPTA, A KUMAR 1984 Hormonal regulation of root formation. In SS Purohit, ed, Hormonal Regulation of Plant Growth and Development, Agro Bot Publ, India, 1: 387-400
37. HAISSIG BE 1972 Meristematic activity during adventitious root primordium development. Plant Physiol 49: 886-892
38. HAISSIG BE 1974 Origins of adventitious roots. NZ J For Sci 4: 299-310
39. HAISSIG BE 1982 Carbohydrate and amino acid concentrations during adventitious root primordium development in Pinus banksiana Lamb cuttings. For Sci 28: 813-821
40. HAISSIG BE 1984 Carbohydrate accumulation and partitioning in Pinus banksiana seedling and seedling cuttings. Physiol Plant 61: 13-19
41. HAISSIG BE 1985 Metabolic processes in adventitious rooting. In MB Jackson, ed, New Root Formation in Plants and Cuttings. Martinus Nijhoff/ Dr W Junk Publ, The Hague, In press
42. HANSEN J 1976 Adventitious root formation induced by gibberellic acid and regulated by the irradiance of the stock plants. Physiol Plant 36: 77-81
43. HUSSEY G 1980 In vitro propagation. In DS Ingram, JP Helgeson, eds, Tissue Culture Methods for Plant Pathologists, Blackwell Sci Publ, Oxford, pp 51-61
44. HUXTER TJ, TA THORPE, DM REID 1981 Shoot initiation in light- and dark-grown tobacco callus: the role of ethylene. Physiol Plant 53: 319-326
45. JAMES DJ, IJ THURBON 1981 Shoot and root initation in vitro in the apple rootstock M 9 and the promotive effects of phloroglucinol. J Hortic Sci 56: 15-20
46. JAMES DJ, IJ THURBON 1981 Phenolic compounds and other factors controlling rhizogenesis in vitro in the apple rootstocks M 9 and M 26. Z Pflanzenphysiol 105: 11-20
47. JARVIS BC 1985 Endogenous control of adventitious rooting in non-woody cuttings. In MB Jackson, ed, New Root Formation in Plants and Cuttings. Martinus Nijhoff/Dr W. Junk Publ, The Hague, In press
48. JARVIS BC, AHN ALI, AI SHAHEED 1983 Auxin and boron in relation to the rooting response and ageing of mung bean cuttings. New Phytol 95: 509-518
49. JARVIS BC, A BOOTH 1981 Influence of indole-butyric acid, boron, myo-inositol, vitamin D2 and seedlings age on adventitious root development in cuttings of Phaseolus aureus. Physiol Plant 53: 213-218
50. JARVIS BC, PRM SHANNON, S YASMIN 1983 Involvement of polyamines with adventitious root development in stem cuttings of mung bean. Plant Cell Physiol 24: 677-683
51. JORDAN M, L ITURRIAGA, W FEUCHT 1982 Effects of nitrogenous bases on root formation of hypocotyls from Prunus avium L. "Mericier" and "Bing" grown in vitro. Gartenbauwiss 47: 46-48
52. HEMBERG T 1954 The regulation between the occurrence of auxin and the rooting of hypocotyls in Phaseolus vulgaris L. Physiol Plant 7: 323-331
53. KAMINSKI C 1971 Variations des activites peroxydasique et phenoloxydasique au cours de la croissance de Coleus blumei Benth var Automne. Planta 99: 63-72
54. KANTHARAY GR, S MAHADEVAN, G PADMANABAN 1979 Early biochemical events during adventitious root initiation in the hypocotyl of Phaseolus vulgaris. Phytochem 18: 383-387
55. KEVERS C, L STICHER, C PENEL, H GREPPIN, Th GASPAR 1983 The effect of ergosterol, ergocalciferol and cholecalciferon on calcium-controlled peroxidase secretion by sugarbeet cells. Physiol Plant 57: 17-20

56. KLAMBT D 1983 Oligopeptides and plant morphogenesis: a working hypothesis. J Theoret Biol 100: 411-435
57. LEAKEY RRB 1983 Stock plant factors affecting root intiation in cuttings of Triplochiton scleroxylon K. Schum., an indigenous hardwood of West Africa. J Hortic 58: 277-290
58. LEWIS DH 1980 Lignification and the origin of vascular plants - a unified hypothesis. New Phytol 84: 209-230
59. MIDDLETON W, BC JARVIS, A BOOTH 1980 The role of leaves in auxin and boron-dependent rooting of stem cuttings of Phaseolus aureus Roxb. New Phytol 84: 251-259
60. MILLER NF, LE HINESLY, FA BLAZICH 1982 Propagation of Fraser fir by stem cuttings: effects of type of cutting, length of cutting, and genotype. HortScience 17: 827-829
61. MISSION JP, P BOXUS, M COUMANS, P GIOT-WIRGOT, Th GASPAR 1983 Role du charbon de bois dans les milieux de culture de tissus végétaux. Med Fac Landbouww Rijksuniv Gent 48(4): 1151-1157
62. MONCOUSIN C, Th GASPAR 1983 Peroxidase as a marker for rooting improvement of Cynara scolymus L. cultured in vitro. Biochem Physiol Pflanzen 178: 263-271
63. MURASHIGE T, F SKOOG 1962 A revised medium for rapid growth and bioassays with tobacco tissue culture. Physiol Plant 15: 473-497
64. NANDA KK, R SETHI, S KUMAR 1982 Some paradoxical effects of metabolic inhibitors in root initiation and floral bud initiation. Ind J Plant Physiol 25: 1-16
65. NOUGAREDE A, P RONDET 1983 Bases cytophysiologiques de l'induction rhizogène en réponse à un traitement auxinique dans l'épicotyle du Pois nain. Ann Sci Nat Bot Paris, 13e ser 5: 121-149
66. OOISHI A, H MACHIDA, T HOSOI, H KOMATSU 1978 Root formation and respiration of the cuttings under different temperatures. J Jap Soc Hortic Sci 47: 243-247
67. PINGEL U 1976 Der Einfluss phenolischer Aktivatoren und Inhibitoren der IES-oxydase-Aktivität auf die Adventivwurzelung bei Tradescantia albiflora. Z Pflanzenphysiol 79: 109-120
68. PILET PE, Th GASPAR 1968 Le catabolisme auxinique. Masson et Cie, Paris
69. POESSEL JL, JJ MACHEIZ, L MOSELLA-CHANCEL 1982 Importance physiologique des composés phénoliques chez les arbres fruitiers. Applications au microbouturage in vitro. 2° Coll Rech Fruit-Bordeaux, pp 61-71
70. QUOIRIN M, P BOXUS, Th GASPAR 1974 Root intiation and isoperoxidases of stem tip cuttings from mature Prunus plants. Physiol Veg 12: 165-174
71. QUOIRIN M, P LEPOIVRE 1977 Etudes de milieux adaptés aux cultures in vitro de Prunus. Acta Hortic 78: 437-442
72. REILLY K, J WASHER 1977 Vegetative propagation of radiata pine by tissue culture: plantlet formation from embryonic tissue. NZ J For Sci 7: 199-206
73. REUVENI O, M RAVIV 1980 Importance of leaf retention to rooting of avocado cuttings. J Am Soc Hortic Sci 106: 127-130
74. ROBERTS LW, AR MILLER 1983 Is ethylene involved in xylem differentiation? Vistas Plant Sci 6: 1-24
75. RUNKOVA L, Th GASPAR 1976 Modification du spectre des isoperoxydases par l'acide gibbérellique et les réducteurs de croissance. C R Acad Sci Paris 282: 545-548
76. SCHMID A 1978 Influence de l'hydrocortisone (cortisol) sur la rhizogenèse de boutures chlorophylliennes de Populus tremula L., en association avec l'acide indolylbutyrique ou l'acide abscissique. C R Acad Sci Paris 186: 1499-1502
77. SINNOTT EW 1960 Plant Morphogenesis. McGraw Hill, New York

78. SKOOG F, RY SCHMITZ, RM BOCK, SM HECHT 1973 Cytokinin antagonists: synthesis and physiological effects of 7-substituted 3-methylpyrazolo (4,3-d) pyrimidines. Phytochem 12: 25-37
79. THORPE T, M TRAN THAN VAN, Th GASPAR 1978 Isoperoxidases in epidermal layers of tobacco and changes during organ formation in vitro. Physiol Plant 44: 388-394
80. VARDJAN M, JP NITSCH 1961 La régénération chez Cichorium endiva L.: étude des auxines et des kinines endogènes. Bull Soc Bot Fr 108: 363-374
81. VARGA M, EC HUMPHRIES 1974 Root formation on petioles of detached primary leaves of dwarf beans (Phaseolus vulgaris) pretreated with gibberellic acid, triiodobenzoic acid and cytokinins. Ann Bot 38: 803-807
82. VAUGHAN KC, SO DUKE 1984 Function of polyphenol oxidase in higher plants. Physiol Plant 60: 106-112
83. VEIERSKOV B, AS ANDERSEN 1982 Dynamics of extractable carbohydrates in Pisum sativum. III. The effect of IAA and temperature on content and translocation of carbohydrates in pea cuttings during rooting. Physiol Plant 55: 179-182
84. VEIERSKOV B, AS ANDERSEN, EN ERIKSEN 1982 Dynamics of extractable carbohydrates in Pisum sativum. I. Carbohydrate and nitrogen content of pea plants and cuttings grown at two different irradiances. Physiol Plant 55: 167-173
85. WEIGEL U, W HORN, B HOCK 1984 Endogenous auxin levels in terminal stem cuttings of Chrysanthemum moriforium during adventitious rooting. Physiol Plant 61: 422-428
86. WENT FW 1939 The dual effect of auxin on root formation. Am J Bot 26: 24-29
87. WHITE A, P HANDLER, EL SMITH 1968 Principles of Biochemistry. Blakiston Div, McGraw-Hill Book Co, New York
88. YANG SF, NE HOFFMAN 1984 Ethylene biosynthesis and its regulation in higher plants. Annu Rev Plant Physiol 35: 155-189
89. ZIMMERMAN RH, OC BROOME 1981 Phloroglucinol and in vitro rooting of apple cultivar cuttings. J Am Soc Hortic Sci 106: 648-652

11. CORRELATIONS WITHIN THE TREE

PW EVERS

1. INTRODUCTION

The growth and form of trees are largely the result of an innate pattern of development (26). Under field conditions the form is greatly modified by influences such as environment and growth correlations. The complexity of the form does not disappear when such influences are removed in experimental conditions (18). Tissue excised from a certain part of the tree retains its local plan for next year's development (14). Under special circumstances, such as in tissue culture or when taking cuttings (6, 14), part of this basic plan for development at the organism level, also called architecture (21), finds expression. The innate form of architecture is obscured by the physiological status of the tree as imposed by climate.

Recent reviews of tissue culture of trees (5) and horticultural crops (1) do not thoroughly cover the factors operating before the establishment of the cultures. The fundamental importance of these factors was emphasized during a conference on tissue culture of woody species in 1978 in Gembloux (4) and at a symposium of juvenility in woody perennials in 1975 (2). Recent studies on growth and development deal primarily with differentiation of cells rather than with the physiology of a differentiated part of a plant or with synchronized morphogenesis within the plant. This chapter discusses the theory underlying the influence that the physiological status of the original tissues, before isolation *in vitro*, has on morphogenesis outside the complex interactions in the tree. If this status is not studied in greater detail, progress in some sectors of tissue culture technology may stop, because standard techniques of tissue culture that have been developed for the most important tree species, are based only on incomplete premises.

2. CONSEQUENCES OF TREE PHYSIOLOGY

If the complex, gradually changing topophysical differentiation of buds, were better understood we could adapt media and physical factors in tissue culture to the specific needs of the various types of organs in the tree. Obviously, endogenous growth regulators whose ratios fluctuate over time will play an important role in the establishment of tree architecture. Elongation growth, diameter growth, and tree form are probably controlled mainly by gibberellic acids (GA) and auxins (22, 25). In Douglas fir, the optima of growth regulators in tissue culture media varied among buds taken from different topophysical positions (12). This was the consequence of the physiology at the site of excision. Optima were established for indole acetic acid (IAA) but not for GA_3. Because there are as yet no indicators of the gradually changing state of juvenility and/or physiology, it will remain difficult to define appropriate optima in tissue culture for all physiological ages, all periods of the year, and all sections of the architecture of the tree.

Carbon fixation is the most basic process for canopy growth if water is not a limiting factor. This process is very complex in trees, because feedback mechanisms regulate relative growth and because meteorological fluxes are distributed within the canopy (19). These fluxes further complicate the prediction of the physiological state of the initial tissue culture material. This shows how necessary it is to control the physical factors when preparing the trees for tissue culture. Control is needed to reduce the variation resulting from topophysical and physiological parameters.

The critical physiological threshold state cannot usually be expressed in the tissue culture because of the low light intensity (below 50 $\mu Em^{-2}s^{-1}$) in most culture rooms. Probably, most tree tissues are cultured below the CO_2 compensation point. This parameter of photosynthesis is another innate characteristic obscured by experimental design. The topophysical pattern of compensation points can be studied by determining this parameter in developing buds excised from various sites in the tree (10). A problem is that in the tree this parameter varies with the ontogeny (20).

In the field, the form of the tree is often the result of the predetermined growth pattern of the buds as influenced by water stress (18). In tissue culture, we usually deal with an abundance of water, which, at the organism level, would be called waterlogging. This introduces physiological aberrations (cf. Gaspar *et al*. volume 1).

3. JUVENILITY AND TOPOPHYSIS

The retention of leaves during winter in some broadleaved tree species is generally considered to be a juvenile characteristic (26). In older trees, this phenomenon occurs only in the lower branches. In conifers, when buds are isolated from the lower branches of young Douglas fir trees and grown *in vitro* into shoots, the rooting potential of this material is higher than that in cultures derived from buds in higher parts of the tree (14). Because rootability is associated with juvenility, there seems to be a relation between the physiological repercussions of topophysis and juvenility (3). It has often been observed that a tree that has been propagated as a cutting can retain some of the characteristics of the portion of the tree from which the cutting was taken (23, 24).

For rootability, differences in juvenility may occur in a group of cuttings coming from the same tree. From the above, there appears to be no physiological age of the organism as a whole except in an integrated sense. The age of the different parts of the tree varies. This explains why, the topophysical pattern (and, related to this, juvenility) changes in relation to the chronological age of the tree. It is difficult to ascertain the state of juvenility in organs excised for tissue culture. Thus a better understanding of the fluxes that occur during topophysical differentation should have a high priority and it would be helpful to have an indicator of the state of juvenility in a defined position in the canopy.

4. MOTHER TREE PRETREATMENTS

When trees are prepared for tissue culture, cold and forcing treatments should be applied to break dormancy. Severe treatments may lower the amplitude of topophysical variation in morphogenesis among the excised tissues (11). In adult trees, topophysis will normally be more differentiated and thus more variable than in juvenile trees. However, it is impossible to apply any of the above-mentioned treatments to large trees to reduce variability in subsequent tissue cultures.

Pretreatments such as hormonal sprays, pruning, and grafting, modify the topophysical predisposition of tissue in trees, with varying intensity. Recently, the effects of growth regulators, applied exogenously in various ways to change bud development and consequently the form of the tree, have been reviewed (25). The growth rate in general, the apical inhibition of lateral buds and shoots, the branch angle, and plagiotropic growth can all be altered by exogenously applied growth regulators. Several of these effects in trees have been described after application of TIBA, BA, or GA's (25). The consequence of pretreatments with growth regulators on the morphogenesis of excised buds *in vitro* has been described for Douglas fir (12). In these experiments, IAA in lanolin paste was applied to the wounds of decapitated trees and BA was applied in an aqueous solution around damaged roots. Morphogenesis of excised buds of these treated trees *in vitro* showed that, compared with the control trees, the growth pattern had been changed by the pretreatments. Comparable effects can be obtained by spraying mother trees with growth regulators during the first week of bud flushing, one week before excision (15). When *Quercus*, *Alnus*, *Platanus*, and *Salix* trees of various ages were sprayed in a greenhouse, GA_3 (1 mg/ml ethanol 50%, 20 ml/tree) stimulated elongation growth on the tree but not after the material has been isolated *in vitro*. BA (20-90 mg/tree in ethanol 50%) severely disturbed the growth pattern of the trees. Spraying with BA changed the branch angle and plagiotropic growth habit of *Alnus* and *Platanus* trees, because hyponasty occurred. In all genera, the elongation growth of buds was inhibited. Apparently, this was a consequence of the induction of axillary bud development, which consumed a major part of the available energy. These observations show that most growth regulator pretreatments force the tree to develop buds and, consequently, shoots for the next growing season, without a cold period. In all cases, the earliest flushing buds were influenced more by the pretreatments than shoots that develop later. This response changed the form of the tree. Mechanical distortions of the tree can also result in the tree revising its growth priorities (21). Sometimes the stimulation of growth of certain bud types as a result of pruning is associated with some degree of rejuvenation (16).

5. CASE STUDY: DOUGLAS FIR

In the research on the morphogenesis of buds on Douglas fir *in vitro*, the time of isolation, chronological age, and topophysis proved to be crucial for the development into microcuttings (8,9). On 2-year-old trees, 10 topophysical positions were defined (Fig. 1). The optimal light intensity at the explant level, and the concentrations of nitrate and sucrose in the medium for morphogenesis varied according to the topophysical origin of the buds on the tree. An example of topophysis-dependent elongation growth is shown in Figure 2. The graph indicates that apart from the difference in growth potential, the optimum concentration of the medium, in this case sucrose, is also influenced by topophysis. The optimum concentrations change during the year and, unfortunately, not always in the same manner in buds from different topophysical origins. Topophysis also influences the gas exchange of the cultures. Developing buds of Douglas fir *in vitro* demonstrate a pattern of CO_2 consumption that is comparable with photosynthesis in trees of the same age (10). A study of the influences of topophysis on photosynthesis of developing buds *in vitro* probably reflects the variability in the same process in the tree after flushing.

For growth, e.g., elongation (13), *in vitro* studies on topophysis are less revealing because the scale is not comparable with that *in situ*. Dark respiration, net photosynthetic maximum, light compensation point and photochemical efficiency were determined for developing buds from each topophysical position (Table 1). This indicated that the buds "remembered" the local light situation in the tree. In general, shoots from the least exposed group of positions (P4, P8, P9, P10) seemed to have a higher net maximum (P_{Nm}) and efficiency (α_n) and a lower compensation point (I_c) than those of the more exposed group. This coincides with high relative growth rates of these least exposed developing buds. The parameters of photosynthesis were influenced by the tissue culture conditions preceding the measurements. A high sucrose concentration in the medium generally inhibited the P_{Nm} and α_n; it also resulted in a higher I_c, except when the topophysical origin was P9 or P10. Shoots grown at a high light intensity (215 $\mu Em^{-2}s^{-1}$) demonstrated a lower dark respiration and P_{Nm} than shoots grown at a lower light intensity (110 $\mu Em^{-2}s^{-1}$). The hypothesis proposed to explain why photosynthesis was inhibited by a high sucrose level was that a high sucrose level in the medium might compensate for a low autotrophic C-fixation. This was not observed when morphogenesis of shoots on media with high or low concentrations sucrose (15 g/l or 45 g/l) at a high or a low light intensity (40 or 110 $\mu Em^{-2}s^{-1}$) was compared. The inhibition of elongation growth at a low light intensity could not be compensated for by a high sucrose concentration. However, P10 shoots grew faster on a medium containing 45 g/l sucrose than on a medium with 15 g/l sucrose at this low light intensity (Fig. 3). Apparently, photosynthesis cannot be translated directly into growth. At a low light intensity (40 $\mu Em^{-2}s^{-1}$), topping the trees stimulated the growth of the shoots compared with the growth of shoots from untopped trees; this was predominantly the case in buds from "shaded" positions (P4, P8, P10). This effect was not found at a high light intensity (22 Wm^{-2}).

The optimum sucrose concentration of the medium for elongation growth of shoots from the 10 topophysical positions was found to depend on the method used to break dormancy (Figs. 4, 11).

The top part of the shoot initials grew faster after the trees were forced than after natural breaking of dormancy; this was particularly noticeable in shoot initials isolated from the

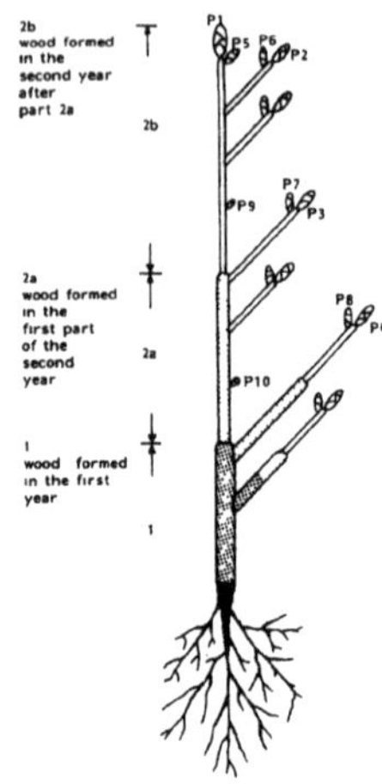

A. Terminal positions.

Position 1(P1) · Terminal buds of the tree
" 2(P2). " buds of the highest branch on stem part 2b
" 3(P3) " buds of the highest branch on stem part 2a
" 4(P4) · " buds of the highest branch on stem part 1

B. Axillary positions close to terminal buds:

Position 5(P5): Axillary buds close to the P1 buds
" 6(P6) " " " " " P2 buds
" 7(P7) · " " " " " P3 buds
" 8(P8). " " " " " P4 buds

C. Positions on the stem.

Positons 9(P9) Axillary buds on the stem part 2b, closest to part 2a (Figure 2)
" 10(P10) . Axillary buds on the stem part 2a, closest to part 1

FIGURE 1. Positions of buds in selected 2-year-old Douglas fir trees (1a1), displaying Massart's model of architecture (7).

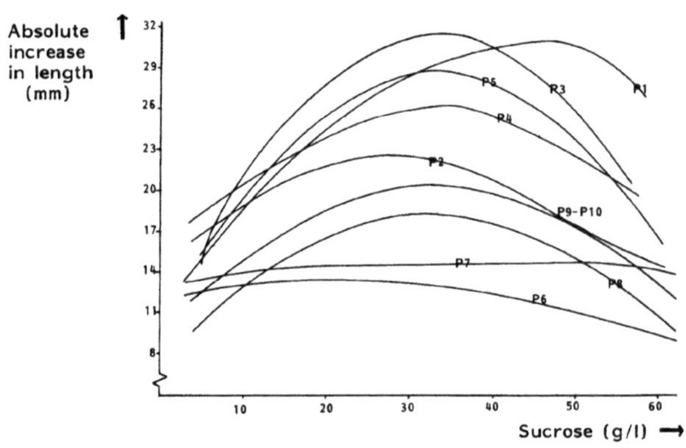

FIGURE 2. Influence of the sucrose concentration on extension growth of shoot initials isolated in April from buds in the positions P1-P10.

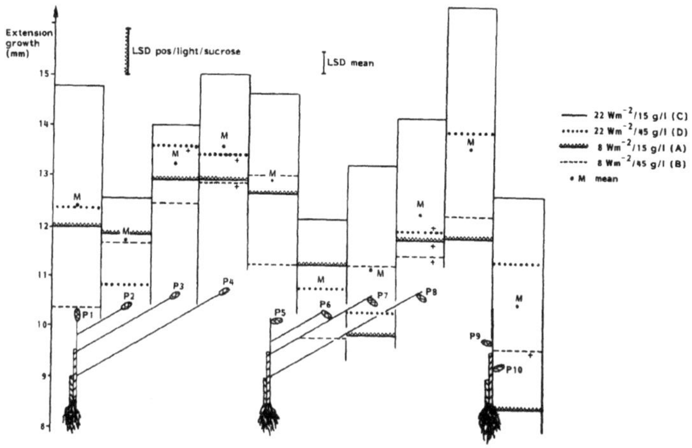

FIGURE 3. Extension growth in vitro of shoot initials of buds taken from intact trees after 42 days. Explants were taken from 10 positions (p 1 to 10) on intact trees as influenced by 4 light intensity/sucrose concentration treatments (A to D); LSD = Least significant difference.
+ = significantly higher extension growth was observed in shoots from the same position in decapitated trees ($p < 0.05$).

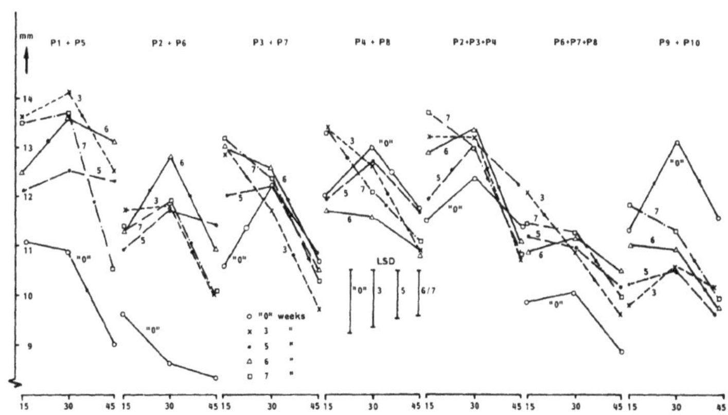

FIGURE 4. Influence of the duration of forcing (0 to 1 week) and the sucrose concentration on the mean extension growth of shoot initials isolated from buds in horizontally and vertically distributed groups of positions (p).
LSD = Least significant difference.
. = significant increase or decrease.
"O" = natural breaking of dormancy 15,30,45 = g/l sucrose.

highest buds in the tree (P1, P2, P5, P6). For buds on the stem (P9, P10), natural breaking of dormancy resulted in a higher rate of elongation growth than after forcing.

In conclusion, in Douglas fir, topophysical expression *in vitro* is a useful tool for studying tree architecture and the physiological status of the tree. Results are modified by hormonal pretreatments and the method used to break dormancy. Once in tissue culture, medium factors (predominantly sucrose and nitrate) and physical factors (light intensity) reveal the variations in the topophysically based morphogenetic strategy.

6. ACKNOWLEDGEMENTS

Part of this work was supported by the "Wood as a Renewable Raw Material" program of the EEC. The Douglas fir project was a joint program of the Dorschkamp Institute and the Departments of Horticulture and Silviculture of the Agricultural University of Wageningen. Thanks are also due to J Burrough for editing the English.

7. REFERENCES

1. BHOJWANI SS, RAZDAN MK 1983 Plant Tissue Culture: Theory and Practice. Elsevier, Amsterdam
2. BORCHERT R 1976a The concept of juvenility in woody plants. *In* RH Zimmerman, ed, *Symposium on Juvenility in Woody Perennials*, Acta Hortic 56: 21-36
3. BORCHERT R 1976b Differences in shoot growth patterns between juvenile and adult trees and their interpretation based on systems analysis of trees. *In* RH Zimmerman, ed, *Symposium on Juvenility in Woody Perennials*, Acta Hortic 56: 123-142
4. BOXUS P (ed) 1978 Proc Round Table Conference "*In Vitro*" Multiplication of Woody Species. CRA, Gembloux, *Belgium*
5. DODDS JH 1983 Tissue Culture of Trees. Croom Helm, London
6. DURAND-CRESSWELL R 1977 The vegetative propagation of Eucalyptus by organ culture. PhD Thesis Univ New Engl NSW, Australia
7. EDELIN D 1977 Images de l'architecture des conifères. Thèse Biol Veg Acad Montpellier
8. EVERS PW 1981a Growth and morphogenesis of shoot initials of Douglas fir, *Pseudotsuga menziesii* (Mirb) Franco, *in vitro*. I Plant, nutritional and physical factors. Uitvoerig Versl

Rijksinst Onderzoek Bos- Landschapsbouw "De Dorschkamp," Wageningen 16(1)

9. EVERS PW 1981b Growth and morphogenesis of shoot initials of Douglas fir, Pseudotsuga menziesii (Mirb) Franco, in vitro. II Growth factors, topophysis and seasonal changes. Uitvoerig Versl Rijksinst Onderzoek Bos- Landschapsbouw "De Dorschkamp," Wageningen 16(2)
10. EVERS PW 1982a Growth and morphogenesis of shoot initials of Douglas fir, Pseudotsuga menziesii (Mirb) Franco, in vitro. III Photosynthesis in vitro. Uitvoerig Versl Rijksinst Onderzoek Bos- Landschapsbouw "De Dorschkamp," Wageningen 17(1)
11. EVERS PW 1982b Growth and morphogenesis of shoot initials of Douglas fir, Pseudotsuga menziesii (Mirb) Franco, in vitro. IV The influence of topping, forcing, the sucrose concentration and the light intensity. Uitvoerig Versl Rijksinst Onderzoek Bos- Landschapsbouw "De Dorschkamp," Wageningen 17(2)
12. EVERS PW 1983a Growth and morphogenesis of shoot initials of Douglas fir, Pseudotsuga menziesii (Mirb) Franco, in vitro. V The influence of growth regulators applied in vivo and in vitro and their interaction with topophysis. Uitvoerig Versl Rijksinst Onderzoek Bos- Landschapsbouw "De Dorschkamp," Wageningen 18(1)
13. EVERS PW 1983b Growth and morphogenesis of shoot initials of Douglas fir, Pseudotsuga menziesii (Mirb) Franco, in vitro. VI Comparison of growth and development in vivo and vitro. Uitvoerig Versl Rijksinst Onderzoek Bos- Landschapsbouw "De Dorschkamp," Wageningen 18(3)
14. EVERS PW 1984 Growth and morphogenesis of shoot initials of Douglas fir, Pseudotsuga menziesii (Mirb) Franco, in vitro. PhD Thesis Agric Univ Wageningen, The Netherlands
15. EVERS PW, DONKERS J 1985 Growth regulator preconditioning of Platanus mother trees for micropropagation. Symp Gembloux (In Press)
16. FRANCLET A 1983 Rajeunissement, culture in vitro et pratique sylvicole. Bull Soc Bot Fr 130: 87-98
17. HOUTZAGERS G 1954 Houtteelt in de Gematigde Zone. Tjeenk Willink, Zwolle
18. KOZLOWSKI TT 1971 Growth and Development of Trees. Vol 1. Academic Press, New York
19. LEDIG FT 1976 Physiological genetics, photosynthesis and growth models. In MRG Cannell, FT Last, eds, Tree Physiology and Yield Improvement. Academic Press, London, pp 21-54
20. LUUKANEN O 1976 Relation between the CO_2 compensation point and carbon fixation efficiency in trees. In MRG Cannell, FT Last, eds, Tree Physiology and Yield Improvement. Academic Press, London, pp 119-138
21. OLDEMAN RAA, HALLE F, TOMLINSON PB 1978 Tropical Trees and Forests, an Architectural Analysis. Springer-Verlag, Berlin
22. PHARIS RP, KUO CG 1977 Physiology of gibberellins in conifers. Can J For Res 7: 229-325
23. ROBINSON LW, WAREING PF 1969 Experiments on the juvenile-adult phase change in some woody species. New Phytology 68: 67-78
24. ROMBERGER JA 1976 An appraisal of prospects for research on juvenility in woody perennials. Acta Hortic 56: 301-317

25. ROSS SD, PHARIS RP, BINDER WD 1983 Growth regulators and conifers their physiology and potential uses in forestry. In LG Nickell, ed, Plant Growth Chemicals. Vol 2, CRC Press, Boca Raton, pp 36-78
26. ZIMMERMAN MH, BROWN CL 1971 Trees Structure and Function. Springer Verlag, Berlin

Table 1. The influence of the sucrose concentration and the light intensity on the parameters of the photosynthesis of Douglas fir shoots during their culture: R_d (dark respiration, mg CO_2 g^{-1} dry weight h^{-1}), P_{Nm} (maximum net photosynthesis at saturating irradiance, mg CO_2 g^{-1} dry weight h^{-1}), l_c (compensation irradiance, Wm^{-2}) and α_n (photochemical efficiency, µg CO_2 J^{-1}). THSD = Turkey's honest significant difference ($p<0.05$), M1 = morning measurement, M2 = afternoon measurement

Light intensity during the culture of shoots	Low (22 Wm^{-2}) - Gro-lux light												High (43 Wm^{-2}) - Son T light							
Sucrose conc. (g/l)	15				30				45				30 (M1)				30 (M2)			
Position	R_d	P_{Nm}	l_c	α_n	R_d	P_{Nm}	l_c	α_n	R_d	P_{Nm}	l_c	α_n	R_d	P_{Nm}	l_c	α_n	R_d	P_{Nm}	l_c	α_n
1	4.82	8.32	7.67	0.50	6.88	6.49	7.61	0.61	6.57	2.61	13.01	0.25	3.00	3.88	10.87	0.20	2.78	4.00	8.69	0.24
2	4.43	9.40	5.90	0.61	5.94	5.84	9.41	0.44	7.14	3.38	15.79	0.24	3.69	1.49	23.71	0.08	4.17	1.51	19.74	0.10
3	6.52	5.84	8.31	0.53	5.92	5.55	10.75	0.38	7.29	3.50	16.14	0.24	3.03	2.54	12.47	0.16	2.96	2.83	11.17	0.18
4	5.23	7.02	5.82	0.67	5.90	6.98	6.68	0.63	5.52	2.19	14.52	0.19	2.87	3.62	8.10	0.26	2.62	4.02	7.90	0.26
5	5.20	10.67	4.75	0.89	5.86	3.90	9.88	0.36	5.72	1.98	21.92	0.12	2.95	2.24	11.70	0.16	2.33	2.32	11.43	0.14
6	9.51	2.94	17.15	0.25	7.80	1.59	23.19	0.12	9.92	3.09	17.69	0.25	2.45	0.26		0.02	2.14	0.28	83.22	0.001
7	8.00	5.98	10.87	0.47	9.84	4.04	8.19	0.61	10.35	3.00	13.93	0.32	2.78	1.72	10.77	0.15	2.51	2.40	13.28	0.13
8	5.34	12.25	4.52	0.98	10.63	5.19	10.84	0.53	8.18	6.24	10.87	0.48	2.77	1.42	21.87	0.07	2.30	1.44	20.97	0.06
9	7.00	7.54	5.35	0.93	7.25	8.25	6.44	0.81	8.58	5.82	8.82	0.64	3.00	1.15	19.11	0.08	2.62	1.75	15.59	0.10
10	5.30	10.71	3.98	1.11	6.34	7.19	6.81	0.67	9.22	6.17	9.98	0.57	2.90	0.99	21.61	0.06	2.96	1.30	14.76	0.10

THSD (22 Wm^{-2}) R_d: 3.42; P_{Nm}: 1.64; l_c: 5.01; α_n:0.310
(43 Wm^{-2}) R_d: 0.88; P_{Nm}: 1.01; l_c: 3.82; α_n: 0.049 (P6 shoots excluded)
(Total) R_d: 2.74; P_{Nm}: 1.53; l_c: 4.86; α_n: 0.250

12. HAPLOIDS (GYMNOSPERMS)

R. ROHR

1. INTRODUCTION

In vitro culture of male and female gametophytes of gymnosperms is generally carried out with two objectives in mind.

The first, the historic one, is to follow the development of a single microgametophyte *in vitro*. The slow development and inaccessibility of the microgametophyte make it difficult to study *in situ*.

The second objective is to initiate sporophyte development from haploid tissues. This would allow the mass production of haploid and homozygous diploid plants, which would be of interest for genetic studies that are facilitated by the haploid state (32, 40), and for the mass multiplication of selected genotypes.

The first efforts at *in vitro* culture were initiated towards the end of the last century. These efforts resulted in observation of the first pollen mitosis in pollen cultured on a medium containing simple sugars (2, 57). Subsequently, completion of microgametophyte development *in vitro* has been obtained with pollen of more than a dozen species (10, 20, 24, 27, 28, 29, 61, 64). These pollen were cultured on media supplemented with natural additives, such as ovule extracts or micropyle excretions (12, 38), or of more complex artificial media.

Experimental androgenesis has taken its greatest strides with angiosperms, primarily herbaceous ones, but good results have also been reported for a few tree species (reviews, 32, 53, 67). With gymnosperms morphogenesis in microsporophyll or pollen cultures has so far been limited (*Picea*, 54, 55). However, gymnosperms offer an alternative approach because they allow megagametophyte culture *in vitro*. Megagametophytes are easily excised and cultured, and regeneration of plantlets from these explants (gynogenesis) is envisaged. In Ginkgo and Cycadales the full development of the endosperm can occur in the absence of fertilization. Therefore, these gametophytes enable the study of the properties of the endosperm in vitro without interference by the embryo.

This chapter will emphasize the normal as well as abnormal ontogeny of micro- and megagametophytes in vitro, and discuss the regeneration of plantlets from these tissues.

2. SOURCES OF HAPLOID CELLS

2.1. Microgametophytes

The male cones of gymnosperms are homologous with the male flowers of angiosperms. Microsporophylls with pollen sacs are stationed around a short axis. At anthesis the pollen each contain from one to a dozen nuclei or cells. This variation in number is partly due to the existence or absence of prothallial cells, and depends on the level of maturation of the gametophyte at the time of anthesis. The prothallial cells are missing in some gymnosperms (Taxaceae, Cupressaceae). In others there is only one prothallial cell (Cycadaceae), in some there are two (Ginkgo, Pinaceae), and in a few there are many (Araucariaceae). The most simple pollen are those of the Taxaceae and Cupressaceae, which are shed at the unicellular stage, or at the bicellular stage with one generative and one pollen tube cell (Torreya). In most cases the pollen are tetracellular, i.e., they have two prothallial cells, one generative cell and one tube cell. Finally, the generative cell sometimes has already divided into the sperm mother cell and stalk cell, thus providing the pollen with five cells (Pseudotsuga menziesii, Ephedra foliata).

Generally, the entire male cone or microsporophylls are excised from the cone and cultured in vitro (Table 1). Occasionally pollen is cultured directly, either after removal from the pollen sacs by squashing (22) or by natural shedding under aseptic conditions (61, 64). Very young microspores of Pinus resinosa have developed callus in vitro (3). We have observed that young pollen of Taxus baccata can produce callus composed of a tangled mass of partially developed pollen tubes (45).

Callus development is common when mature pollen are cultured (Table 1). These calli all originated from several microsporophylls or pollen. Because of genetic differences in the pollen, due to genetic segregation in meiosis, these calli are non-isogenic (6). To avoid such genetic chimera one has to culture single isolated pollen, which is far more difficult with gymnosperms than with some of the angiosperms (review, 67).

2.2. Megagametophytes

The megagametophytes of gymnosperms are not as independent of the sporophyte as those of Pteridophytes, either *in vitro* or *in vivo* (68). They are large, and sometimes complex (*Ginkgo*) structures which feed the young embryo with reserves accumulated either before (Cycadales, *Ginkgo*) or after fertilization (conifers).

Because the megagametophyte was formed from a single cell, i.e., the megaspore, its tissues are isogenic. Therefore, these tissues are eminently suitable for the production of isogenic haploid or dihaploid plants. At present, megagametophyte tissues of only 7 species have shown an organogenetic capacity *in vitro* (Table 2). This organogenetic capacity was expressed in megagametophytes from developing ovules (15, 16) as well as from mature seeds (30, 34). Differentiation of organs was observed in gametophytes excised just before fertilization (*Zamia*) (35, 37, 42), as well as in gametophytes with archegonia (*Ephedra foliata*, 56). At present, gametophytes excised at a more immature stage, i.e., taken at a stage of development between the megaspore (Fig. 1) and the archegonial stage, have not been responsive in culture. Another problem with these early stage explants is our inability to remove all surrounding somatic, diploid cells from the haploid tissue. Megaspores and coenocytes (Fig. 2) can be obtained free of diploid cells when material killed in fixative is used (52). However, such a process is not yet available for living tissues, and the risk of introducing diploid somatic cells into culture is considerable.

The developmental stage that is most suitable for *in vitro* culture is different for each species. Some gametophytes are capable of rapid growth *in vitro* when excised before fertilization.

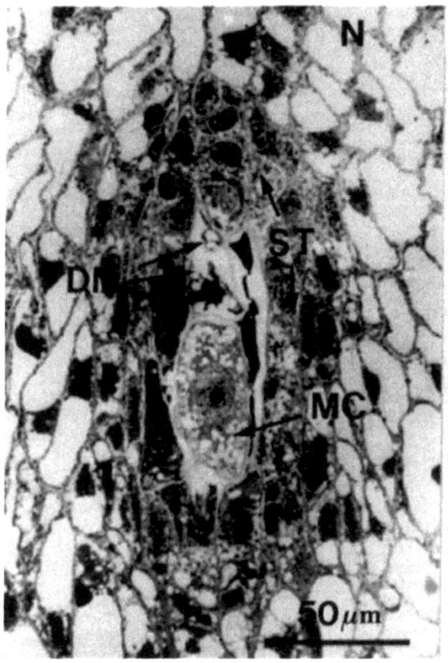

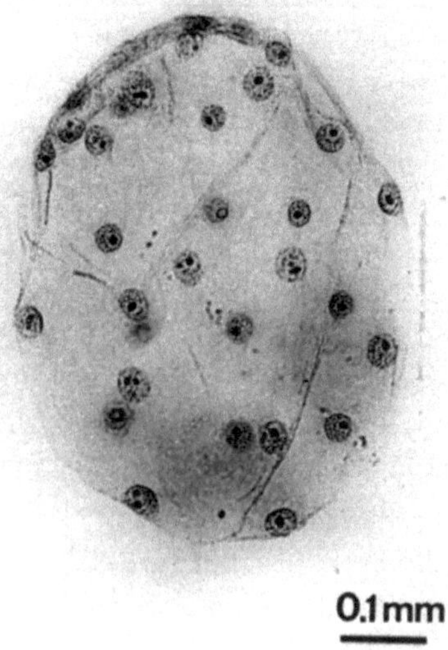

FIGURE 1. Megaspore cell (MC) in situ in the nucellus (N) of *Taxus baccata* (ST = sporogenous tissue, DM = degenerated megaspore).
FIGURE 2. A female gametophyte of *Larix decidua* at the free nuclear stage (32 nuclei) isolated from fixed nucellar tissue (52).

Others still have a remarkable growth potential when the endosperm has reached maturity.

2.3. Gametophyte formation in vitro

Ylmaz and Tran Than Van (69) (also this volume) have recently discovered a novel source of haploid tissue of *Sequoia sempervirens*. They cultured sections of stump sprouts containing either the shoot apex or axillary buds. After 4 months 30% of the cultures produced male cones and 2% female cones. This induction of sexual cones provides a new source of aseptic pollen and female gametophytes for *in vitro* culture. This would be of great benefit in the culture of pollen, which, in earlier procedures, have always been more exposed to surface sterilization agents than the better protected female gametophytes.

3. BEHAVIOR OF GAMETOPHYTES *IN VITRO*

3.1. Normal ontogeny of the microgametophyte

The first experiments in which normal spermatogenesis was obtained, were carried out with natural supplements, of mostly undefined chemical composition. These supplements included micropylar drops (38), ovule extract (12), tomato juice (37), yeast extract and/or coconut milk (see Table 1). Coconut milk has been used until recently for the production of microgametes of *Encephalartos* (14), *Ceratozamia* (17), *Juniperus* and *Sequoia* (24), *Taxus* (46), and *Ginkgo* (50). Synthetic media containing mineral salts, sucrose, vitamins and growth regulators have been used for *Torreya* (64), a simple mineral medium with sucrose for *Tsuga* (28) or without sucrose for *Larix* (27). These defined chemical media supported normal development in those cases where the pollen had few nutritional demands and required only a few days *in vitro* for maturation and sperm cell production (27).

Different media were required for spermatogenesis of pollen of different gymnosperm species. On the other hand, sometimes one particular medium would support different types of development of pollen in species that are taxonomically remote from each other, e.g. androgenesis in tobacco pollen (9) and spermatogenesis in *Taxus* (46) and *Ginkgo* (50) pollen.

The mass culture of pollen has permitted study of the relationship between the physiology and structure of spermatogenesis. Some phenomena, that have always been difficult to observe *in situ* have now been studied in detail with the aid of *in vitro* culture. For example, *in vitro* culture has made it possible to understand the process of release of the spermatogenous cell of gymnosperms from pollen tubes. This release has been followed by photomicroscopy (21) and was studied at the ultrastructural level (47). These studies have answered many old questions and have put an end to half a century of speculations (25), at least regarding spermatogenesis in *Taxus* (46).

However, one has to be careful in interpreting results obtained with *in vitro* experiments because of possible artifacts. For example, by changing the medium, morphogenetic patterns are changed as well (22, 60). Even when spermatogenesis is complete *in vitro*, the chronology of events is often irregular (10, 23, 28, 45, 61). This underlines the function of the ovule in regulating the development of the male gametes. Conversely, the pollen also influence the development of the ovules, in part by producing and excreting growth regulating substances (31, 59).

3.2. Induced abnormalities

It is essential to establish whether tissues formed *in vitro* are truly of gametophytic origin, or whether somatic cells of the explants have also participated in the formation of these tissues. The latter is possible when parts of microsporophylls (filament, connective tissue, and pollen sacs) are introduced in culture (4, 22, 63, 64). The female gametophyte, on the other hand, is large enough to be introduced in culture free of somatic ovular, embryo or suspensor tissue.

3.2.1. Microgametophytes. Study of the initial stages of pollen proliferation *in vitro* shows that this proliferation is initiated by either the vegetative cells of the gametophyte (*Ginkgo*, 60, 61) or the generative cell (*Cupressus*, 44), or by both (*Torreya*, 64). Sometimes it is not possible to establish the exact origin of this proliferation (*Ephedra*, 33). The cell colonies divide mitotically in *Ginkgo* and *Ephedra*. In *Taxus* the cells ressemble pollen tubes with the first mitosis being asymetric. In their further development the *Taxus* cell colonies show several abnormalities, e.g. branching tubes, and extrusion of the generative cell into the tube (Figs. 3, 4). Branching tubes were produced by pollen of *Pinus resinosa* (3), *Encephalartos* (14) and, *Zamia* (37). In *Torreya* cell multiplication occurred by budding, as in yeast (64).

The media that initiated these types of development included, with one exception (61), 2,4-dichlorophenoxyacetic acid (2,4-D) at concentrations between 0.05 mg/L (*Cupressus*, 44) and 5 mg/L (*Pinus*, 3) (Table 1).

Pollen calli were subcultured, but, with one exception (*Picea abies*, 54, 55), no differentiation beyond the parenchymatic or meristematic stage has been observed. In some cases tracheids were formed (44). The behavior of gymnosperm pollen is different in this respect from that of angiosperm pollen, in which androgenesis *in vitro* is a common phenomenon (review, 67). There are many

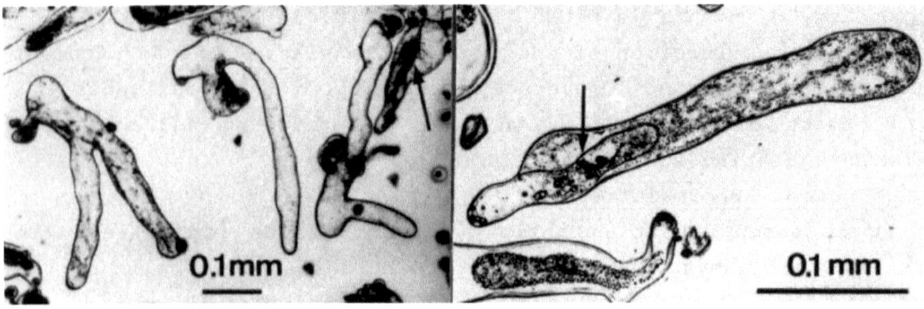

FIGURES 3 and 4. One and a half month old culture of Taxus baccata microspores. Some tubes branch (Fig. 3). Sometimes the generative cell extrudes into the pollen tube (Fig. 3 and 4, arrows).

possible reasons for the lack of androgenesis in gymnosperm cultures. One could be the improper choice of the stage of development of the microspores (5), another the lack of specific nutritional components. In some cases neighboring somatic tissues may be required to supply specific nutritional and other factors for normal androgenesis (22).

3.2.2. Megagametophytes. Mostly only tissue develops, although occasionally organogenesis has been observed, with the first cases of organogenesis having been reported many decades ago (13, 19). Tulecke obtained haploid tissue from juvenile megagametophytes of Ginkgo, excised between the time of pollination and fertilization (65). We have initiated proliferation of several highly differentiated cell types in Ginkgo megagametophytes that were excised 4-6 months after fertilization. In some cases the embryos were removed from the gametophytes before culture, in others they were left. Even gametophytic tissues with few cytoplasmic organelles and full of starch have grown by cell division (49) (Fig. 5). Proliferation of tissues from mature gametophytes was also obtained with Encephalartos umbeluziensis (15).

Gametophytes of several conifers, e.g. Pinus resinosa (3) and P. mugo (4), P. lambertiana (8), Araucaria araucana (11), Thuya occidentalis (37), Larix decidua (39), Taxus baccata in the presence (70) or absence (51) of an embryo, have produced calli or tissues. Generally these are parenchymatous. Occasionally tracheids develop in the calli or tissues, sometimes in association with organ development (1, 4, 65, 70). Xylogenesis is common in somatic tissues _in situ_ and _in vitro_. However, gametophytic tissues of gymnosperms never form vascular tissues _in situ_, and therefore, the capacity to do so could conceivably have been lost. However, the observations discussed above have shown that the genetic system responsible for xylogenesis is merely repressed in haploid cells _in situ_, and that exogenous stimuli supplied _in vitro_ can derepress this system. A similar situation appears to exist in ferns where callus from the prothallus has formed tracheids (18).

Many calli were capable of organogenesis, particularly when the initial culture medium, often rich in auxin and cytokinin, was replaced in subculture by media low in growth regulators (15, 34, 39, 42). The most simple media have often supported the most advanced apogamy (35, 36, 37) (Table 2). Norstog managed to obtain a high incidence (59%) of organogenesis in megagametophytes of _Zamia_ on a more complex medium, i.e., modified White's with additional amino acids (alanine, glutamine, asparagine), an auxin and cytokinin. Organogenesis was obtained with 2,4-D, but was more intense with indoleacetic acid (IAA). In megagametophyte cultures

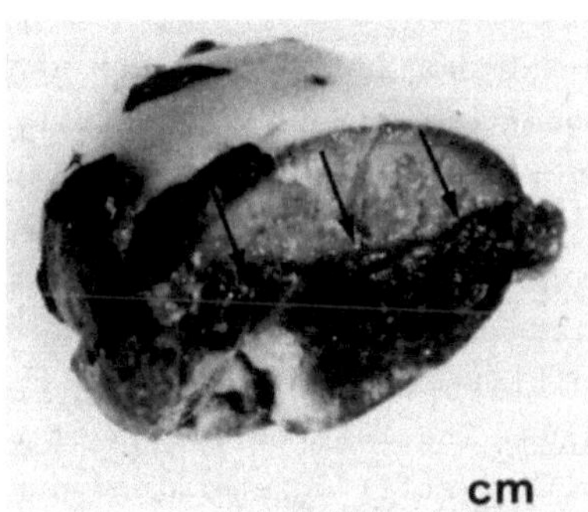

FIGURE 5. Basal end of a _Ginkgo_ megagametophyte removed from the medium after 45 days. A continuous row of outgrowths developed whereever the section was in contact both with agar and air (arrows) (49).

of *Picea abies* organogenesis was stimulated by a combination of the polyamines putrescine, spermidine, and spermine (30).

Organogenesis consisted of bud, root, leaf or stem formation, or embryogenesis (53). The rapidity of the response varied greatly between species. In *Ephedra foliata* roots developed in 2 weeks (56), while in *Cycas revoluta* roots did not appear until 2 years later (36).

These results show that in gymnosperms regeneration by gynogenesis is potentially possible. However, apogamous plantlets have remained small (41) and did not survive after transfer from the *in vitro* environment to soil (15, 43). One major problem is that proper vascular connections are often lacking between shoots and roots (30). Recently, Nagmani and Bonga (39) have obtained callus in cultures of immature megagametophytes of *Larix decidua* which produced a large number of embryos on hormone free medium. Small plantlets were obtained and transferred to soil. A few plantlets continued to grow and form a few new needles (Fig. 6), but all ceased growing after a few weeks on soil. At present we do not know if gynogenetically produced plantlets have the capacity of continued growth, or whether their growth will always be limited by deleterious recessive genes that are expressed in the haploid condition.

4. PLOIDY OF TISSUES AND ORGANS DERIVED FROM GAMETOPHYTES

Calli from microgametophytes were initially haploid (33, 61, 62, 64). Even after several years of subculture, haploid cells were still present in callus of *Taxus* (62), *Ginkgo* (61), and *Torreya* (64), although in each case diploid, tetraploid, and polyploid cells were also observed. Haploid and diploid cells were also found in the initial calli of *Juniperus communis* (22), *Pinus resinosa* (4), and *Taxus*. The initial callus of *Taxus* also contained polyploid cells with a still undetermined number of chromosomes (62).

According to Bonga (7) diploid cells, besides possibly being of somatic origin, may also have originated from multinucleate cells in which two of the haploid nuclei fused. The observation of bi- or multinucleate cells in cultures of microsporophylls of *Pinus*

FIGURE 6. A plantlet from subcultured callus from an immature megagametophyte of Larix decidua growing in soil (39).

resinosa (4) supports this view. In some cases diploidy may have resulted from a failure to halve the number of chromosomes in meiosis (7). It is further possible that diploidy and polyploidy have resulted from endomitosis, as is the case in cultures of anthers of some angiosperms (67).

In tissues of megagametophyte origin, the ploidy is often difficult to determine, in particular when the tissues only grow slowly and only a few metaphases are present for ploidy determination. The maturation level of the female prothallium may, to a degree, determine the ploidy of the callus derived from these. Callus of mature endosperm of Pinus lambertiana diploidizes spontaneously (8), that of Picea abies retained a mixture of haploid and diploid cells (30), and that of Taxus became polyploid (70). In contrast, Bonga (4) has noted that callus from juvenile prothalli of Pinus nigra and Pinus mugo, excised shortly after fertilization, remained haploid through several subcultures. Similarly, subcultured callus from juvenile megagametophytes of Ephedra foliata remained haploid (1). The polyploidy of Taxus could be due to the increasing complexity of the endosperm during its development. In thin microtome sections we have never found multinuclear cells in the young multicellular gametophyte, while in cells of the mature endosperm multinucleate cells are common. On the other hand, callus from mature endosperm only contains mononuclear cells, which suggests that nuclear fusion occurs in most multinucleate cells soon after transfer to in vitro culture.

One also has to be aware of the possibility of contamination of the cultures with cells of sporophytic (dipoid) origin (7). In our trials with Taxus this risk of contamination is avoided by using, in contrast to the method used by Zenkteler and Guzowska (70), only fragments of mature endosperm after removal of the embryo and suspensor (51).

Problems of interpretation did not arise when mature megagametophytes of Araucaria arucana were used, because the callus obtained from these gametophytes was haploid (11).

Callus of juvenile (66) and mature (48) endosperm of the Cycadales (42) and Ginkgo (66) remained haploid. The same was the case with organs regenerated from mature endosperm of Ceratozamia (15), or juvenile megagametophytes of Zamia (42). In the more evolved genera, organs derived from young gametophytes were haploid (Ephedra 1, 56), but the ploidy of organs of mature endosperm origin varied (Picea, 30).

In summary, it appears that the autonomy of the megagametophytes of Cycadales and Ginkgo is expressed in vitro by a maintenance of the original ploidy. This stability is also evident in the Ephedrales and conifers if the endosperm is excised prior to its total differentiation, but is lost if organized, mature endosperm is used.

5. CONCLUSION

Growth of gametophytes in vitro depends on many interacting genetic (species) and physiological (stage of development of the explant) factors. In addition, the nutrient and culture environment are important (58). Deviations in pollen development in vitro are probably caused more by cultural than by genetic factors. Pollen storage, and cold- (4, 61, 64) or centrifugation- (5) pretreatments may stimulate such development. However, these pretreatments are not essential since similar responses, though less frequent or intense, occur in cultures of fresh pollen.

After obtaining tissues from pollen of gymnosperms, Tulecke and Seghal wrote: "a problem that remains, however, is the induction of tissues from more highly evolved pollens of angiospems. Thus far, no tissue cultures are reported from flowering plants". In 1964, i.e., the year following this remark, Guha and Maheshwari discovered that anthers of Datura were embryogenic *in vitro* (26). Since then, the situation has reversed, and the number of Angiosperm species that turned out to be androgenic has increased steadily. For gymnosperms the culture of megagametophytes probably is a more promising approach than microsporophyll or pollen culture, since the capacity for organogenesis of megagametophytes has frequently been demonstrated.

To maximize the response of these cultures it is essential to determine more precisely the physiological state of the explants at the time of inoculation (c.f. chapter by Durzan, this volume). It appears that growth regulators are not always required for the initiation of andro- or gynogenesis. Presumably the endogenous supplies at the time of excision are generally sufficient. An analysis of these regulators *in situ*, and a determination of the physiological condition of the explants that is most favorable for *in vitro* growth, should be one of our research goals.

6. REFERENCES

1. BHATNAGAR SP, MN SINGH 1984 Organogenesis in the cultured female gametophyte of *Ephedra foliata*. J Exp Bot 35: 268-278
2. BELAJEFF WC 1891 Zur Lehre von dem Pollenschlauch der Gymnospermen. Ber Dtsch Bot Ges 9: 280-286
3. BONGA JM, DP FOWLER 1970 Growth and differentiation in gametophytes of *Pinus resinosa* cultured *in vitro*. Can J Bot 48: 2205-2207
4. BONGA JM 1974 *In vitro* culture of microsporophylls and megagametophyte tissue of *Pinus*. In Vitro 9: 270-277
5. BONGA JM, AH McINNIS 1975 Stimulation of callus development from immature pollen of *Pinus resinosa* by centrifugation. Plant Sci Lett 4: 199-203
6. BONGA JM 1977 Applications of tissue culture in forestry. *In* J Reinert, YPS Bajaj, eds, Applied and Fundamental Aspects of Plant Cell, Tissue and Organ Culture. Springer Verlag, Berlin Heidelberg, pp 93-108
7. BONGA JM 1981 Haploid tissue culture and cytology of conifers. *In* Proc IUFRO Sect S2 01 5. Int Workshop "In Vitro" Cultivation For Tree Species, Fontainebleau, France, pp 283-294

8. BORCHERT R 1968 Spontane Diploidisierung in Gewebekulturen des Megagametophyten von *Pinus lambertiana*. Z Pflanzenphysiol 59: 389-392
9. BOURGIN JP, JP NITSCH 1967 Obtention de *Nicotiana* haploides à partir d'étamines cultivées *in vitro*. Ann Physiol Vég 9: 377-382
10. BRANDSCHEIDT P 1939 Befruchtungsphysiologische Untersuchungen an *Taxus baccata* L. Ber Dtsch Bot Ges 57: 495-505
11. CARDEMIL L, M JORDAN 1982 Light and electron microscopic study of *in vitro* cultured female gametophyte of *Araucaria araucana* (Mol.) Koch. Z Pflanzenphysiol 107: 329-338
12. CHRISTIANSEN H 1969 On the germination of pollen of *Larix* and *Pseudotsuga* on artificial substrate, and on viability tests of pollen of coniferous forest trees. Silvae Genet 18: 104-107
13. COULTER JM, MA CHRYSLER 1904 Regeneration in *Zamia*. Bot Gaz 38: 130-139
14. DE LUCA P, S SABATO 1979 *In vitro* spermatogenesis of *Encephalartos* Lehm. Caryologia 32: 241-245
15. DE LUCA P, A MORETTI, S SABATO 1979 Regeneration in megagametophytes of Cycads. Giorn Bot Ital 113: 129-143
16. DE LUCA P, S SABATO 1980 Regeneration of coralloid roots on Cycad megagametophytes. Plant Sci Lett 18: 27-31
17. DE LUCA P, V LA VALVA, S SABATO 1980 Spermatogenesis and tissue formation in Cycad pollen grains. Caryologia 33: 261-265
18. DE MAGGIO A, R WETMORE, G MOREL 1963 Induction de tissu vasculaire dans le prothalle de fougère. CR Acad Sci (Paris) 256: 5196-5199
19. DUCHARTRE P 1888 Notes sur l'enracinement de l'albumen d'un *Cycas*. Bull Soc Bot Fr 35: 243-251
20. DUHOUX E 1973 Développement du gamétophyte mâle et gamétogenèse *in vitro* chez le *Juniperus communis* (Cupressacées). Premiers résultats. CR Acad Sci (Paris) 277: 2665-2668
21. DUHOUX E 1974 Sur le mode de division de la cellule reproductrice et la libération de ses produits dans les tubes polliniques chez le *Juniperus communis* L. et le *Cupressus arizonica* G. (Cupressacées). Rev Gén Bot 81: 193-204
22. DUHOUX E, B NORREEL 1974 Sur l'isolement de colonies tissulaires d'origine pollinique à partir de cônes mâles du *Juniperus chinensis* L., du *Juniperus communis* L., et du *Cupressus arizonica* G., cultivés *in vitro*. CR Acad Sci (Paris) 279: 651-654
23. DUHOUX E, L CHESNOY 1978 La gamétogenèse mâle chez le *Juniperus communis* L. (Cupressacées). Comparaison entre les cultures *in vitro* de grains de pollen et les observations dans l'ovule. Bull Soc Bot Fr, Actual Bot 1-2: 103-108
24. DUHOUX E 1979 Etude *in vitro* de la germination du pollen et de la gamétogenèse mâle chez quelques Gymnospermes. Thèse Doct Etat Univ Paris VII
25. FAVRE-DUCHARTRE M 1960 Contribution à l'étude des spermatozoïdes de *Taxus baccata*. Rev Cytol Biol Vég 21: 329-337
26. GUHA S, SC MAHESHWARI 1964 *In vitro* production of embryos from anthers of *Datura*. Nature 204: 497
27. HO RH, GE ROUSE 1970 Pollen germination of *Larix sibirica in vitro*. Can J Bot 48: 213-215
28. HO RH, O SZIKLAI 1971a Pollen germination of *Tsuga heterophylla in vitro*. Can J Bot 49: 117-119

29. HO RH, O SZIKLAI 1971b Germination and development of lodgepole pine pollen in vitro. Can J For Res 1: 12-19
30. HUHTINEN O, J HONKANEN, LK SIMOLA 1981 Effects of genotype and nutrient media on callus production and differentiation of Norway spruce endosperms cultured in vitro. In Proc IUFRO Sect S2 01 5. Int Workshop "In Vitro" Cultivation For Tree Species, Fontainebleau, France, pp 307-311
31. KAMIENSKA A, RP PHARIS 1975 Endogenous gibberellins of pine pollen. II Changes during germination of Pinus attenuata, P. coulteri and P. ponderosa pollen. Plant Physiol 56: 655-659
32. KARNOSKI DF 1981 Potential for forest tree improvement via tissue culture. Bioscience 31: 114-120
33. KONAR RN 1963 A haploid tissue from the pollen of Ephedra foliata Boiss. Phytomorphology 13: 170-174
34. KONAR RN, MN SINGH 1979 Production of plantlets from the female gametophytes of Ephedra foliata Boiss. Z Pflanzenphysiol 95: 87-90
35. LA RUE CD 1948 Regeneration in the gametophyte of Zamia floridana. Bull Torrey Bot Club 75: 597-603
36. LA RUE CD 1950 Regeneration in Cycas. Am J Bot 37: 664
37. LA RUE CD 1954 Studies on growth and regeneration in gametophytes and sporophytes of Gymnosperms. Brookhaven Symp Biol 6: 187-208
38. MEHRA PN 1938 The germination of pollen grain in artificial culture in Ephedra foliata Boiss. and Ephedra gerardiana Wall. Proc Indian Acad Sci 8: 218-230
39. NAGMANI R, JM BONGA 1985 Embryogenesis in subcultured callus of Larix decidua. Can J For Res 15: 1088-1091
40. NITZSCHE W, G WENZEL 1977 Haploids in plant breeding. In W Horn, G Röbbelen, eds, Advances in Plant Breeding. Suppl 8, J Plant Breeding. Verlag Paul Parey, Berlin, Hamburg, pp 1-101
41. NORSTOG K 1965 Induction of apogamy in the megagametophyte of Zamia integrifolia. Am J Bot 52: 993-999
42. NORSTOG K, E RHAMSTINE 1967 Isolation and culture of haploid and diploid Cycas tissues. Phytomorphology 17: 374-381
43. NORSTOG K 1982 Experimental embryology of Gymnosperms. In BM Johri, ed, Experimental Embryology of Vascular Plants. Springer Verlag, Berlin, Heidelberg, New York, pp 25-51
44. RAZMOLOGOV VP 1973 Tissue culture from the generative cell of the pollen grain of Cupressus ssp. Bull Torrey Bot Club 1: 18-22
45. ROHR R 1973a Production de cals par les gamétophytes mâles de Taxus baccata L. cultivés sur un milieu artificiel. Etude en microscopie photonique et électronique. Caryologia suppl 25: 177-189
46. ROHR R 1973b Ultrastructure des spermatozoides de Taxus baccata L. obtenus à partir de cultures aseptiques de microspores sur un milieu artificiel. CR Acad Sci (Paris) 277: 1869-1871
47. ROHR R 1974 Sur la rupture de la paroi de la cellule socle durant la gamétogenèse in vitro de Taxus baccata (Taxacées). CR Acad Sci (Paris) 278: 2433-2435
48. ROHR R 1978 Ultrastructure de tissu haploide de Ginkgo biloba L. Z Pflanzenphysiol 87: 365-378
49. ROHR R 1980a Dédifférenciation en culture in vitro des cellules à réserves dans l'endosperme de Ginkgo biloba. Z Pflanzenphysiol 96: 423-434

50. ROHR R 1980b Développement in vitro du pollen de Ginkgo biloba L. Cytologia 45: 481-495
51. ROHR R 1982 Activation et proliferation des celules du mégagamétophyte de Taxus cultuvé in vitro. Can J Bot 60: 1583-1589
52. ROHR R, JM BONGA 1984 A method to isolate the female sexual cells of Larix decidua for cytological studies. Can J Bot 62: 1651-1658
53. SANGWAN-NORREEL BS, E DUHOUX 1982 Les conditions de la gynogenèse et de l'androgenèse expérimentales in vitro chez les arbres. Rev Cytol Biol Bot 5: 171-187
54. SIMOLA LK 1981 Ultrastructure of callus cultures from trees. In Proc IUFRO Sect S2 01 5. Int Workshop "In Vitro" Cultivation For Tree Species, Fontainebleau, France, pp 201-209
55. SIMOLA LK 1982 Ultrastructure of callus cultures from Betula pendula and Picea abies. In A Fujiwara, ed, Plant Tissue Culture 1982. Proc 5th Int Cong Plant Tissue Cell Cult, Tokyo, pp 173-174
56. SINGH MN, RN KONAR, SP BHATNAGAR 1981 Haploid plantlet formation from female gametophytes of Ephedra foliata Boiss. in vitro. Ann Bot 48: 215-220
57. STRASBURGER E 1982 Ueber das Verhalten des Pollens und die Befruchtungsvorgänge bei der Gymnospermen. Histol Beitr 4: 1-158
58. SUNDERLAND N 1973 Pollen and anther culture. In HE Street, ed, Plant Tissue and Cell Culture, Botanical Monographs Vol 11. Blackwell Publ, Oxford, pp 205-239
59. SWEET GB, PN LEWIS 1969 A diffusable auxin from Pinus radiata pollen and its possible role in stimulating ovule development. Planta 89: 380-384
60. TULECKE WR 1957 A tissue derived fom the pollen of Ginkgo biloba. Science 117: 599-600
61. TULECKE WR 1957 The pollen of Ginkgo biloba: in vitro culture and tissue formation. Am J Bot 44: 602-608
62. TULECKE WR 1959 The pollen cultures of CD La Rue: a tissue from the pollen of Taxus. Bull Torrey Bot Club 86: 283-289
63. TULECKE WR, LG NICKELL 1960 Methods, problems and results of growing plant cells under submerged conditions. Trans NY Acad Sci 22: 196-206
64. TULECKE WR, N SEHGAL 1963 Cell proliferation from the pollen of Torreya nucifera. Contrib Boyce Thompson Inst 22: 153-163
65. TULECKE WR 1964 A haploid tissue culture from the female gametophyte of Ginkgo biloba L. Nature 203: 94-95
66. TULECKE WR 1967 Studies on tissue cultures derived from Gingko biloba L. Phytomorphology 17: 381-386
67. VASIL IK 1980 Androgenetic haploids. Int Rev Cytol, suppl 11A: 195-223
68. WALP RL 1951 Fern prothallia under cultivation for twelve years. Science 113: 128-129
69. YLMAZ D, K TRAN THAN VAN 1985 Production de gamétophytes mâles et femelles in vitro en vue de l'amélioration des aptitudes androgénétiques et gynogénétiques de Conifères. Coll Sect Fr IAPTC, Orsay, France, Bull Soc Bot Fr, Actual Bot (in press)
70. ZENKTELER MA, I GUZOWSKA 1970 Cytological studies on the regenerating mature female gametophyte of Taxus baccata L. and mature endosperm of Tilia platophyllos Scop. in in vitro culture. Acta Soc Bot Pol 39: 161-173

Table 1. Callus and tissue formation in *in vitro* culture of the male gametophyte of gymnosperms. YE: yeast extract, CM: coconut milk.

Species	Reference	Explant Source	Stage of Development	Medium	Results
Ginkgo biloba	60-61	male strobilus or pollen grains	4-celled gametophyte	White + YE and IAA	haploid tissue originally
Taxus brevifolia	62	microsporangia	microspores	White + CM and 2,4-D	tissue, evidence of polyploidy
Ephedra foliata	33	unopened microsporangia	mature pollen	Reinert + CM and 2,4-D	subculturable haploid tissue
Torreya nucifera	64	male cones, pollen	binucleate shedding stage	White + CM and 2,4-D	subculturable callus
Taxus baccata	45	male cones, microsporangia	young microspores	Bourgin & Nitsch 2,4-D, IAA, CM	callus made of pollen tubes
Cupressus benthamii	44	unopened microsporangia	mature pollen	modified White's, 2,4-D	callus with tracheids
Cupressus arizonica *Juniperus chinensis* *Juniperus communis*	22	undehiscent male cones, pollen	mature pollen	Brown & Lawrence White or Heller + 2,4-D, NAA, CM	haploid and diploid callus
Pinus resinosa	4	microsporophylls	5-12 d before anthesis	Brown & Lawrence + 2,4-D	callus (haploid + diploid cells)
Ceratozamia mexicana	17	unopened microsporangia	mature pollen, 3-celled stage	Tulecke 1957 + 2,4-D	tissue

Table 2. Organogenesis and plantlet regeneration in *in vitro* culture of the female gametophyte of gymnosperms. CM: coconut milk, AA: amino-acids, *k: kinetin*.

Species	Reference	Stage of Development	Medium	Results
Cycas revoluta	36-37	mature endosperm	wet sand	bud and root formation after 2 yr
Zamia floridana	35-37	pre-fertilization	various minerals	bud and root formation in 3 mo
Zamia integrifolia	42	pre-fertilization	White + 2,4-D k, various AA	haploid organs
Picea abies	6	mature seed	Brown & Lawrence + IAA	haploid plantlet, no further growth
Ceratozamia mexicana	15	mature endosperm	White	adventitious embryos
Ephedra foliata	34	mature seed ↓	Murashige & Skoog + CM, 2,4-D	callus ↓
		subcultures	Murashige & Skoog + CM, k	shoot and roots
Ephedra foliata	56	archegonial stage	Murashige & Skoog + CM, k	haploid shoots and roots in 13 d
Cycas revoluta	16	mature endosperm	White + 2,4-D, k	coralloid roots
Picea abies	30	mature seed, embryo excised	N6 (macro-) MS (micro- + vit) + 2,4-D, polyamines	shoots, roots and plantlets (n, 2n and mixaploid)
Sequoia	Ball (see Vol. 3)		unpublished	plantlets

13. INDUCTION OF ANDROGENESIS IN HARDWOOD TREES

CHEN ZHENGHUA

1. INTRODUCTION

Over the past decade much progress has been made in the induction of haploid plants in economically important forest and fruit trees in China. This has resulted in the production of "pollen plantlets," i.e., plantlets derived from pollen *in vitro*, of 9 families, 11 genera, and more than 20 species (Table 1).

Table 1. List of pollen plantlets in tree genera

Common name	Genera	Reference
Poplar	*Populus* (13 species or *interspecific hybrids*)	13, 16, 20, 21, 29, 31
Rubber tree	*Hevea*	5
Horse-chestnut	*Aesculus*	19
Orange	*Citrus*	3
Grape	*Vitis*	28
Chinese crab-apple	*Malus*	22, 23
Apple	*Malus*	11
Litchi	*Litchi*	12
Longan	*Euphorbia*	26
Trifoliate orange	*Poncirus*	15
Chinese wolfberry	*Lycium*	10, 14, 27
Tea	*Cemellia*	4

The first pollen plantlets of poplar (*Populus nigra* L.) were produced in 1975 (21). Since then pollen plantlets from more than 10 poplar species and interspecific hybrids have been induced with a frequency of pollen callus induction of about 20% (a maximum of more than 50%) and a frequency of plantlet differentiation from these pollen calli of about 60% (13, 16, 20, 29, 31). The transplanted pollen trees are now 7-8 years old and some of them have bloomed (Fig. 1). Several promising, disease resistant clones have been selected. Moreover, a systematic investigation of the chromosome numbers of these was carried out for 8 years.

Several hundred pollen plantlets have been obtained from nine lines of rubber tree (Hevea brasiliensis Muell.-Arg.) via anther culture since 1977 (5). Optimal conditions for the development of microspores into embryoids have been worked out. With anther culture techniques an average of 100 (with a maximum of 140) embryoids are formed per 100 inoculated anthers. The frequency of viable pollen plantlets has been enhanced from 3% to 28-30% in some clones. The survival frequency of transplanted pollen plants (Fig. 2) was as high as 30%. Last year a test tapping was made on a 6-year-old pollen tree (6, 7, 9). Compared with the donor clones, four clones with superior growth vigour were obtained (see

FIGURE 1. Pollen tree of poplar (from Wu Kexian, 1983).
FIGURE 2. Pollen tree of rubber (from Baoting Institute of Tropical Crops).

section 6.). Crosses were made with several blooming pollen trees of varying ploidy (see section 5.2.), and the hybrid seeds were collected and sown (Xu Xuen *et al*. personal communication).

Since 1978 pollen plantlets have been obtained from 6 species of fruit trees. Among them the pollen plants of grape (*Vitis vinifera* L.) (Fig. 3) and apple (*Malus pumilla* Mill) have been transplanted into the field and grown normally (11). The pollen plants of Chinese crabapple (*M. prunifolia* (Willd) Borkh.) (22, 23) were cloned by top-budding and bud-grafting (Wu Jiangyun *et al*., personal communication). The remaining three species are subtropical trees, i.e., orange, litchi (3, 12) and Longan (*Euphorbia longan*) (26).

As for medicinal woody plants, pollen plants were obtained from trifoliate orange (15) and from one species of Chinese wolfberries (*Lycium halimifolium* Mill.) (27). Pollen plants from two other species of Chinese wolfberries (*L. barbatum* L. and *L. chinense* Mill.) have been growing normally in the field since 1981 (10, 14).

As regard to beverage crops, haploid trees of tea (*Cemellia stenesis* (L) D. Kuntze) have been induced by anther culture and transplanted to the field (4).

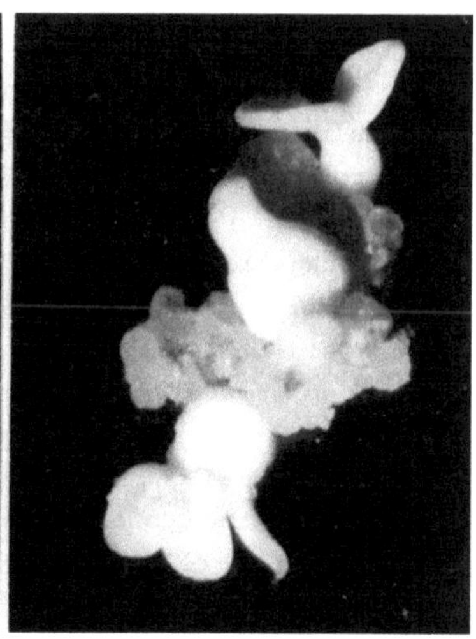

FIGURE 3. Pollen plantlet of grape (from Zou Changjie).
FIGURE 4. Pollen embryoids of rubber tree.

2. PROSPECTS OF IMPROVEMENT OF WOODY PLANTS BY USING HAPLOIDS

The haploids and homozygous diploids in woody plants are much more valuable than natural or artificial herbacious ones. Most trees do not have natural haploids and homozygous diploids. It is impractical to obtain pure homozygous diploid lines by successive generations of inbreeding, because a long period is required to obtain sexual progenies, and because in most woody plants seeds cannot be obtained by self-pollination.

2.1. Utilization of haploids in the study of genetics

So far, advances in genetics have been much slower in tree species than in annual herbaceous plants, mainly because of the lack of favourable material for study. The genetics of many economically important characters has not been clearly illustrated. Therefore, it is still unknown whether the desirable characters of the parents will or will not appear in their progenies. Only when crossing between different homozygous diploid plants, obtained from haploids, is carried out can we gain a clear idea of the dominance of genes controlling various characters and whether these characters are monogenic or polygenic, etc.

2.2. Utilization of haploids for woody plant improvement

2.2.1. Utilization in breeding new varieties. It is possible to obtain haploid and homozygous diploid plants with superior genotypes by anther culture and unpollinated ovule culture. Pollen trees with superior genotype and chromosomal variation, which could be retained by vegetative propagation, are extremely valuable for improving varieties, especially because valuable chromosomal variants in plants are often eliminated by meiosis and generative propagation.

Several rubber tree clones with good growth vigor and a clone with a high yield of latex have been produced. In addition, poplar clones with good growth vigor have been obtained. We used the best avaliable varieties as material for inoculation and selected subsequently for improved gametoclonal variation. Some clones significantly surpassed the donor clone in some economic characters, because of rigorous elimination of inferior genotypes at cell, embryoid or plantlet level during culture and after transplantation.

2.2.2. Utilization of pure lines for heterosis. Heterosis is much higher in good combinations of pure lines than in crosses between two different varieties. However, because of the long life cycle of trees, it is imposible to obtain pure lines by means of successive generations of inbreeding. The establishment of anther culture techniques of woody plants enables us to obtain pure lines of different genotypes in a short time and to pave the way to utilize heterosis of good combinations rapidly. We plan to produce self-rooting plants or artificial seeds from trees of good combinations by means of micropropagation. Production and planting would be much easier if such self-rooting clones could replace clones obtained by bud grafting.

2.2.3. Gene mutation at the haploid cell level. In general, the accumulation of variability is correlated with the duration of culture. In haploid plants the dominant and the recessive characters express themselves simultaneously, which will help to separate mutants. Moreover, the haploids are also very good material for studies of the mutation frequency for a spectrum of traits. Microbe and fungus breeding has been very effective because these organisms have only one set of chromosomes and because the induction and selection of mutants is carried out at the cell level. If the mutation and selection of crops could similarly be carried out at the haploid, single cell level in vitro, the efficiency of genetic improvement could be markedly enhanced.

Recent experiments have revealed that differences in resistance to mineral stresses and diseases are expressed in the same manner in cultured cells as in whole plants. It is reasonable to expect that genotypes differing in resistance to mineral stress and diseases can be identified and isolated in cultures of haploid or homozygous diploid cells and that this resistance will be expressed in haploid or homozygous diploid plants regenerated from these cells.

2.2.4. Genetic manipulation by haploid cell hybridization. Genetic manipulation by somatic hybridization of haploid protoplasts offers fascinating prospects for woody plants. For example, we can regenerate heterotic plants by hybridization between different pure lines of the same species and estimate their combining ability. We can select the best combiners, produce self-rooting trees by tissue culture, and plant the heterotic self-rooting trees on a large scale.

It is reasonable to assume that new, stable, amphidiploid plants can be obtained by means of haploid protoplast fusion. Thus new experimental systems that would speed up the procedure of breeding in woody plants, may be established. However, little success has been reported to date in plant regeneration from protoplasts in woody plants.

3. MORPHOGENESIS IN ANTHER CULTURE

3.1. Duration of culture

Anther cultures of woody plants are slower than those of herbaceous plants. For example, for the woody plants used it required 100-200 days from anther inoculation to plantlet formation while for herbacious plants it required only 36-88 days (Table 2). Furthermore, the induction of pollen plantlets of trees requires three or four steps. The embryogenesis of trees goes through two stages, i.e., the young and mature embryo-stage, before they can be transferred to the plantlet-forming medium. Embryoids of herbacious plants can be transferred to that medium when still young, thus saving one step. Special attention should be paid to the ontogenesis of embryoids and plantlets as well as to asynchronism in embryoid formation, to help avoid embryoid abnormalities.

3.2. Ontogenesis of embryoids and plantlets

There are four ways of producing embryoids and plantlets:

3.2.1. A pollen grain develops directly into an embryoid and then into a plantlet. In this case each embryoid or plantlet has its own genotype and the chromosome number in the embryoids is relatively stable (3).

3.2.2. A pollen grain develops into a callus which then differentiates into a cluster of embryoids. Almost all the embryoids (Fig. 4) originate from one callus from one pollen grain and have the same genotype. A few of the embryoids have a different chromosome number and in some the chromosome number has doubled (12, 28).

Table 2. Days of culture required for obtaining pollen plantlets of different species.

Plant	Formation of						
	Callus	Embryoids Young	Embryoids Mature	Shoots	Plantlets	Roots	Total
Rubber tree	50	20-30	40-60		30-40		140-180
Orange		70	20		20-30		110-120
Litchi	60	30	30-40		40		160-170
Grape	30		20-30		60		110-120
Chinese crabapple	30			90-120	30	20	180-200
Logan	30-40	30-80	60-130		20-30		120-270
Tobacco		21			15-20		36-41
Rice	21-28				45-60		66-88
Rape		20-30			20-30		40-60

3.2.3. A pollen grain develops into an embryoid, the hypocotyl and cotyledon primordia of which produce several secondary embryoids. In this case the secondary embryoids have the same haploid, or sometimes diploid genotype as the primary embryoid (Fig. 7) (10, 14).

3.2.4. A pollen grain develops into a callus and then the latter differentiates into a bud or a cluster of buds having the same genotype. The experiment showed that several pollen plants (H_1) of poplar, that had originated from one callus, had uniform characters. A few of these pollen plantlets showed variation in chromosome number (25).

To be able to predict the genotype and the extent of chromosome variation of the pollen plants it is important to identify the origin of the embryoids and plantlets. Transplanted pollen plants that have originated from different pollen grains can give many different genotypes for utilization in breeding. Since the formation of pollen embryoids and plantlets often is asynchronous, conditions more favorable for synchronous morphogenesis should be created.

Recent reports in the literature indicated that the pollen plantlets of poplar and Chinese crabapple were formed by pathway 3.2.1. and 3.2.2. and 3.2.3.; grape and litchi by 3.2.2. and

3.2.3.; Chinese wolfberry by 3.2.1. and 3.2.3. and 3.2.4.; trifoliate orange and horsechestnut by 3.2.1. only.

3.3. Asynchrony of organogensis in the embryoid

Organogenesis in most pollen embryoids is asynchronous. In embryoids obtained in anther cultures of rubber tree the main root and lateral root primordia, as well as the stem, cotyledon, and terminal bud primordia do not differentiate simultaneously. About 15 days after the transfer of the callus-forming anthers onto the differentiation medium, embryoids without obvious primordia became visible. After about 30 days the main root and cotyledon primordia formed. After 30-60 days the cotyledon primordia and the verticillate lateral root primordia differentiated. At this time the embryoids were transferred to the plant-forming medium, and formed a main root, lateral roots and green cotyledons, but failed to form a shoot. After 60-90 days the size of the embryoids markedly increased and the stem and terminal bud primordia formed. Therefore, it is recommended to transfer the embryoids after about 90 days from the differentiation to the plant forming medium to increase the induction frequency of intact plantlets. During the first 90 days of culture it was necessary to transfer the embryoids two to three times to fresh differentiation medium (8).

In anther cultures of woody plants the frequency of plantlet formation by pollen embryoids was usually low. Furthermore, most of the embryoids developed abnormally. The induction frequencies of plantlets in the first three species in Table 3 were low (3, 12, 26), presumably because of the asynchronism of the developing primary and secondary embryoids. This asynchrony may be caused by a rather large difference in the nutrition and hormone supply for each pollen grain and embryoid within each anther, i.e., only a small part of the pollen grains and embryoids were exposed to favourable conditions.

3.4. Causes of abnormal embryoid formation

High levels of hormones may have caused the abnormal embryoids. In anther cultures of the rubber tree the adverse effects of unfavourable levels of hormones were especially serious at high temperatures.

Table 3. The induction frequency of pollen callus, embryoids and plantlets in some tree species.

Species	Callus	Pollen embryoids	Plantlets %
Orange		2-3	<1
Litchi	20-30	about 1000	<1
Longan	14	7	<1
Rubber tree	40-60	100-140	10-30

At continuously high temperatures (about 30°C) most of the buds of poplar embryoids were deformed and many of the embryoids of rubber tree remained small in size or developed without cotyledons. Anatomical observations showed that the root pole of the embryoids of rubber tree developed quickly and vigorously but the shoot pole was retarded in development. The embryoids formed roots but not shoots after transfer to the plant-forming medium (8).

Because of these abnormalities the media for rubber tree culture were modified. Clusters of young embryoids were separated and transferred to the new media and better development was obtained. By these measures the induction frequency of intact plantlets of rubber tree was enhanced from 3 to 30 plantlets per 100 embryoids (see Table 3 and Fig. 5).

4. ANDROGENESIS AND CULTURE MEDIUM

The MS medium (18) supplemented with both cytokinin and auxin appears to be the most useful and widely used medium for anther culture of woody plants. The MS medium and modified MS medium are adequate for induction of pollen calli, small embryoids and plantlet differentiation of most of the genera mentioned in Table 1 (except grape and orange). MS medium contains a high level not only of total amount of nitrogen (840.7 mg/l) but also of NH_4NO_3 and mineral salts, which in some species, is necessary for the development of embryoids or calli from microspores. The addition of both cytokinin and auxin to the basic medium is necessary for this development of pollen embryoids and pollen calli, but the kind and level of hormones required varies with different species of trees.

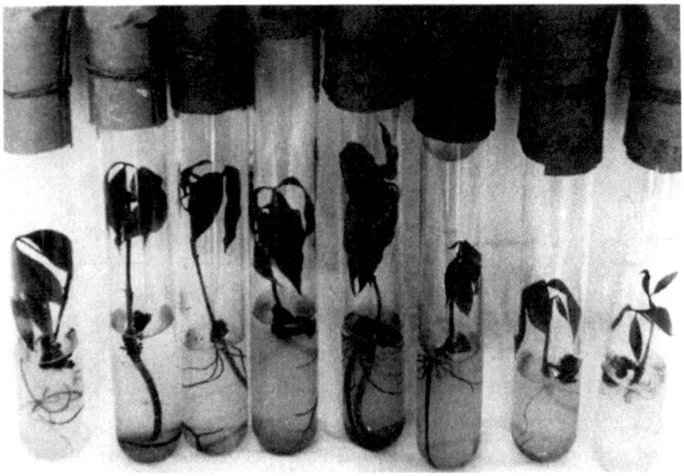

FIGURE 5. Numerous pollen plantlets of rubber tree.

4.1. Poplar

As reported by Zhang et al. (30), on the 2nd day after inoculation of poplar anthers on the dedifferentiation medium the microspore nuclei divided equally. On the 7th day of culture the number of equally divided microspore nuclei increased to peak value. After 20 days, a few multicellular masses ruptured the extine, while at the same time some calli emerged from the anther wall. These anther wall calli were eliminated. Many more pollen calli emerged from the anthers after 30-40 days of culture (30). These pollen calli were transferred to a differentiation medium supplemented with N^6-benzylaminopurine (BA) and α-naphthaleneactic acid (NAA) or indole-3-acetic acid (IAA). A large number of thin and delicate buds differentiated on the calli. Clusters of these buds were then transferred to a differentiation medium supplemented with kinetin (KT) and NAA (or IAA) for strong shoot formation. The survival frequency, after transfer to the field, of plantlets treated with KT was higher than that of plantlets not treated with KT. The culture medium that was used is listed in Table 4.

4.2. Chinese crabapple and apple

The MS medium was used as dedifferentiation medium for anther

Table 4. Culture medium for anther culture of poplar (13).

Step	Basic medium	Hormones(mg/l)	Sucrose (%)	Culture product
1	MS	KT 2, 2,4D 2	3	calli
2	1/2MS[a)]	BA 1, NAA or IAA 0.2-0.5	2	clusters of buds
	1/2MS	KT 1, NAA or IAA 0.2-0.5	2	strong shoots
3	1/2MS	NAA 0.8, IAA 0.2	2	roots

[a)] 1/2MS = amount of macroelements is half of that of MS medium

culture of the two species of Malus. On the 5th day after inoculation on this medium the microspores divided into binuclear microspores. On the 14th day multicellular masses formed in some of the swelling anthers. After the calli had grown to a few mm in diameter, they were transferred to the differentiation medium to promote plantlet formation. The culture media are listed in the Table 5. Biotin and insulin promoted formation of crabapple plantlets. For apple a low level of hormones was used in the dedifferentiation medium and the pollen embryoids formed directly in this medium (see Table 5). Plantlets formed after transfer to the MS medium supplemented with BA, indole-3-butyric acid (IBA) and gibberellic acid (GA_3).

4.3. Chinese wolfberry and trifoliate orange

In anther culture of these two species the MS medium was used as basic medium. A low concentration of hormones (Table 6) depressed the anther wall growth and thus promoted the development of pollen embryoids. Pollen embryoids developed directly from microspores.

4.4. Horse-chestnut and litchi

The modified MS medium, in which pollen embryoids of horse chestnut were induced, was supplemented with thiamine-HCl 2 mg/l, nicotinic acid 5 mg/l, Ca-pantothenate 10 mg/l, myo-inositol 100 mg/l and casein hydrolysates 200 mg/l.

Both the differentiation and plantlet-forming media for anther culture of litchi required L-glutamine 500-1600 mg/l and lactoprotein hydrolysates (LH) 200-500 mg/l. Without these the embryoids

Table 5. Media for anther culture of Malus (11, 23).

Species	Step	Basic medium	Hormones and organic substances (mg/l)	Sucrose (%)	Product
Chinese crabapple	1	MS	KT 2, 2,4D 2, LH 100	3	calli
	2	MS	BA 1, NAA 0.5, biotin 5, insulin 8	2	shoots
	3	1/2MS[a]	IAA 1, IBA 0.2	2	roots
Apple	1	MS	KT 0.2, 2,4D 0.4, IAA 4	3-8	embryoids
	2	MS	BA 1, IBA 0.5, GA_3 0.1	3-8	plantlets
	3	1/2MS	IAA 1.5	1.5-2	roots

[a] 1/2MS = amount of macroelements is half of that of MS medium.

Table 6. Culture media for Chinese wolfberry and trifoliae orange (10, 14, 15).

Species	Step	Basic medium	Hormones (mg/l)	Sucrose (%)	Product
Chinese wolfberry	1 or	MS	KT 0.5-1, 2.4D 0.2-2	3-5	calli
		MS	KT 0.2-1, NAA 0.1, IAA 1	3-5	embryoids
	2	MS	IAA 0.1, GA_3 0.1-0.5	3	plantlets
	3	MS	IAA 0.1	3	roots
Trifoliate orange	1-2	MS	KT 0.2-2, IAA 0.2	5	embryoids
	3	MS	hormone free	2	plantlets

and plantlets did not form. Addition of 400 mg/l royal jelly to the differentiation medium had a positive effect on the differentiaton of litchi pollen embryoids (12).

4.5. Rubber tree

The rubber tree medium (RT) (called modified MB medium before) was designed by us as follows (mg/l): $KNO_3$950, NH_4NO_3 1650, KH_2PO_4 510, $MgSO_4.7H_2O$ 370, $CaCl_2.2H_2O$ 440, Na_2EDTA 37.2, $FeSO_4.7H_2O$ 27.8, $MnSO_4.4H_2O$ 10, H_3BO_3 12, $ZnSO_4.7H_2O$ 8.6, $CuSO_4.5H_2O$ 0.025, $Na_2MoO_4.2H_2O$ 0.25, thiamine-HCl 0.5, pyridoxine-HCl 0.5, glycine 2, nicotinic acid 0.5, folic acid 1 (8) (Table 8).

KT and coconut water (CW) are added to the dedifferentiation medium. The simultaneous addition of NAA and KT significantly improved the development of small embryoids into large ones. Adding GA_3 stimulated further growth of embryoids and cotyledon formation (6).

Table 7. Culture medium for horse-chestnut and litchi (12, 19).

Species	Step	Basic medium	Hormones and organic substances (mg/l)	Sucrose (%)	Product
Horse-chestnut	1-2	Mod. MS	KT 1, 2,4D 1	2	embryoids
	3	Mod. MS	Hormone free	2	plantlets
Litchi	1	MS	KT 2, 2,4D 2, NAA 0.5	3	calli
	2	MS	KT 0.5,NAA 0.1,LH 500, royal jelly 400	3	small embryoids
		MS	LH 500, royal jelly 400, glutamine 500, LH 500	3	mature embryoids
	3	White (macro- x 2.5)	KT 0.1, IAA 0.01, LH 500, glutamine 1600	1	plantlets

Five days after inoculation of anthers on the dedifferentiation medium, only a few of the nuclei divided equally. On the 20th day after inoculation, about 12% of the pollen had developed into multicellular masses within the extine. After 25 days of culture the multicellular masses within the ruptured extine increased in size, and after 30 days some of them developed into small embryoids. Histological investigation of paraffin sections of anthers after 35-50 days of culture revealed that some large embryoids had ruptured the dehiscent layer of the anther. Some embryoids protruded through the rupture and some of the pollen calli proliferated quickly. At the same time embryoids of various size could also be seen. About 50 days after inoculation the callusing anthers were transferred to the differentiation medium and many more haploid embryoids were obtained.

4.6. Longan

There were three steps in the induction of haploid plants in anther cultures of longan. First step; the pollen calli and pollen embryoids were induced. The addition of 0.5% active carbon had a positive effect on the increase of induction frequency of these calli. Second step; the pollen calli differentiated into embryoids. The macroelements in the MS medium had to be reduced (Cuckoo medium, Table 9) for differentiation of pollen embryoids. Third step; the pollen embryoids were transferred to the plant-forming medium on which they developed into intact plantlets.

Table 8. Culture media for rubber tree.

Step	Basic medium	Hormones and organic substances (mg/l)	Sucrose (%)	Product
1	RT	KT 1, 2,4D 1, NAA 1, 5% (v/v) CW	7	callusing anthers
2	RT[a]	KT 0.5-1, NAA 0.2-0.3, GA_3 0.5	7-8	embryoids
3	MS[b]	GA_3 1-4, IAA 1-2, 5-bromo uracil 1-2[c]	4-6	plantlets

[a] = Amount of microelements is doubled.
[b] = Amount of macroelements is 4/5 of that of MS medium.
[c] = 5-bromouracil inhibits callus formation on embryoids.

Table 9. Culture media for anther culture of longan (26).

Step	Basic medium	Hormones and organic substances (mg/l)	Sucrose (%)	Product
1	MS	KT 1, 2,4D 2, C[a] 5000	5	Calli
2	Cuckoo[b]	BA 0.5, NAA 0.1	3	Pollen embryoids
3	Cuckoo	BA 0.3, IAA 0.1	3	Plantlets

[a] C = Active carbon.
[b] Cuckoo = KNO_3 480, NH_4NO_3 400, $NAH_2PO_4.H_2O$ 380, $FeSO_4.7H_2O$ 55.7, Na_2-EDTA 74.5, the other elements are the same as in MS medium.

4.7. Grape, orange and tea

The N6 or B5 medium were used as basic medium for dedifferentiation in anther cultures of grape, orange and as differentiation medium for tea. The differentiation and plantlet-forming media were supplemented with lactoprotein hydrolysates or L-glutamine (see Table 10).

5. VARIATION OF CHROMOSOME NUMBER

Most of the pollen embryoids and plants were mixoploid. The process of anther culture, as well as the vegetative period, is much longer for trees than for herbacious plants. Furthermore, the variation of chromosome number in pollen embryoids and plants is more complex in trees than in annual plants.

5.1. Chromosome number in pollen embryoids and plantlets

Cytological observations on embryoids and plantlets in various trees revealed that they originated from pollen grains (Fig. 6).

Table 10. Culture media for orange, grape and tea (3, 4, 28).

Species	Step	Basic medium	Hormones and organic substances (mg/l)	Sucrose (%)	Product
Orange	1	N6	KT or BA 2, 2,4D 0.5-2	8	small embryoids
	2	MS	BA 2, IAA 0.1, LH 500	2	mature embryoids
	3	1/2 MS[a)]	IAA 0.1, GA_3 1-4, LH 500	2	plantlets
Grape	1	B5[b)]	BA 1.5-2, 2,4D 0.5	3	calli
	2	MS	BA 0.5-4,NAA 0.2,LH 500	2	embryoids
	3	1/2 MS[a)]	BA 0.1, NAA 0.1, LH 500	2	plantlets
Tea	1	MS[b)]	KT 2, 2,4D 0.5	10	calli
	2	N6	KT 2, 2,4D 0.5 L-glutamine 800, L-serine 100	10	shoots
	3	N6	Zeatin 2, adenine 20, lactoprotein hydrolysate 10	5-10	plantlets

[a)]Amount of macroelements is half of that of MS medium.
[b)]Modified.

There were some aneuploid and polyploid cells besides the majority of cells which had the haploid chromosome number. For example, most root-tip cells of grape pollen plantlets had 19 chromosomes but there were also aneuploid, diploid and polyploid cells. The ratio of various ploidies was different in every plantlet. In trifoliate orange there were some cells with X+1, X+2 and 2X chromosomes in the root-tip cells of pollen plantlets, besides cells with 9 chromosomes (X = 9). Most pollen plantlets were mixoploid. Among 238 metaphases from 46 embryoids of rubber tree, 79.8% of the metaphases had 18 chromosomes, and a few cells had more than 27 chromosomes. In most cases the chromosome numbers were multiples of 9, i.e., X = 9, 2X = 18, 3X = 27, 4X = 36 and 5X = 45. Among 146 metaphases of pollen plantlets of rubber tree, 32.4% had 18 chromosomes, 58.6% contained 27 chromosomes, 4.0% had 9 chromosomes, and 5% had aneuploid chromosome numbers. These results showed that during the culture of embryoids to plantlets the number of the cells with 27 chromosomes increased, while that with 18 chromosomes decreased (1, 9).

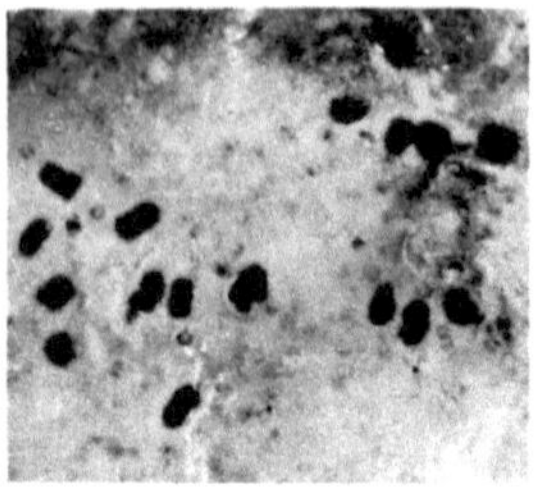

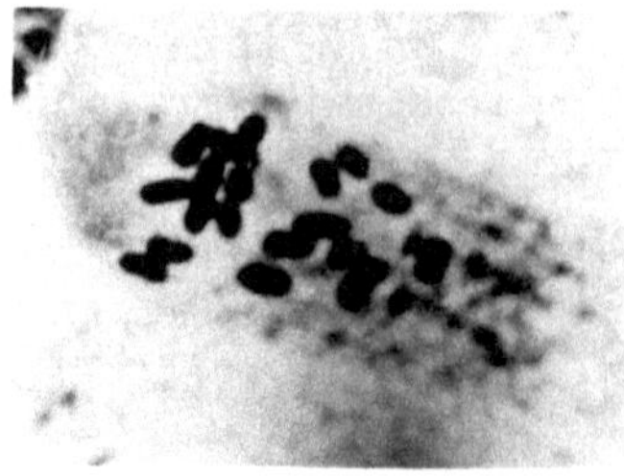

FIGURE 6. A root-tip cell of Chinese crabapple with 17 chromosomes.
FIGURE 7. A root-tip cell of rubber tree with 18 chromosomes.

5.2. Variation of chromosome number in transplanted trees

Systematical observations were carried out on transplanted pollen trees of poplar and rubber. Cytological studies of the apical cone of pollen trees of poplar showed that all of them were mixoploids. There were 5 types of somatic cells in each pollen tree: 1) diploid cells and aneuploid cells with a chromosome number close to 2n, 2) haploid cells and aneuploid cells with a chromosome number close to n, 3) aneuploid cells with chromosome numbers between diploid and haploid, 4) polyploid cells, and 5) cells with a chromosome number less than n. The ratio of these five types of cells varied in different pollen trees. Diploid cells prevailed in the first group of pollen trees, but haploid cells in the second and aneuploid cells in the third, while most of the cells in the control trees had a diploid chromosome number (11, 24). The ploidy of cells in the apical cone affected the vigor of the pollen trees. Some pollen trees grew vigorously because the majority of their cells had the diploid chromosome number, while others grew slowly because the majority of their cells were aneuploid.

For Populus berolinensis Dipp. 5 plants were observed in the 4th, 6th, and 8th year after transplantation. The results in Table 11 show that after the plantlets were transplanted into the nursery there was a tendency to increase the frequency of cells with the diploid chromosome number. The number of haploid cells gradually decreased and that of diploid cells gradually increased. The results also showed that spontaneous chromosome doubling was faster in the earlier stage than in the later stage of development of the tree. From 1975 to 1981 the frequency of diploid cells increased from 5.2% to 85.3%, with an average rate of 13.4% per year, while from 1981 to 1983 the average rate of increase was 2.8% per year. When the trees were 7-8 years old, the percentage of diploid cells was more than 90%, which is similar to that of the control (Lu Chihua et al. Personal communication).

Table 11. Chromosome count in metaphases of pollen plants of poplar[a] (Populus berolinensis Dipp).

Material	Year	No. of metaphases observed	Chromosome count (%) Haploid	Diploid	Aneuploid	Polyploid
Pollen plant	1979	181	23	50	26	1
	1981	581	13	75	12	0
	1983	516	2	92	3	3
Control	1979	30	0	100	0	0
	1981	109	0	100	0	0
	1983	95	0	100	0	0

[a] Transplanted in the field in 1975.

Propagation of pollen plants of poplar by rooting of cuttings increased the rate of chromosome doubling. In 1982, the number of diploid cells of the ramet averaged 80.5% (78.4-84.1%), while that of the ortet averaged only 72.8% (68.5-76.8%).

In pollen rubber trees the cells of young leaves of trees under 50 cm in height showed a considerable variation in chromosome number, most of the cells having 18 to 27 chromosomes. A larger tree of 160 cm in height had many leaf cells (60%) with chromosome numbers between 28-36 and relatively few cells (39%) with 18 to 27 chromosomes. There was also a gradual increase of cells with the diploid chromosome number during the growth of transplanted pollen plants. Buds of a pollen plant of 50 cm in height were grafted and grew into new plants. Chromosomes from cells of ten leaves from one bud-grafted plant were counted. Cells with 9 to 27 chromosomes were found. When this plant reached 170 cm in height, cells of young leaves near the top of the plant had chromosome numbers ranging from 18 to 36. This reveals a gradual increase in chromosome numbers during the growth period of the bud-graft of the pollen plant. This increase is similar to that in cells of the mother plant over that period of time (7).

Table 12. Chomosome counts in metaphases of young leaf cells from transplanted pollen plants and one grafted shoot of a pollen plant (7).

Material	Number of plants observed	Height of plant (cm)	Number of leaves observed	Number of metaphases observed	Chromosome counts 9-17	18-27	28-36	36
					number of cells			
Pollen plant	8	50	8	221	38	172	6	5
	1	160	10	69	1	26	41	1
Grafted shoot	1	50	2	11	3	8	0	0
	1	170	8	67	0	48	19	0

Therefore, the pollen trees showed a tendency to regain diploidy in cells of the apical cone. Consequently the pollen trees were mixoploids. The return of the chromosome number to the level of 2n may be advantageous to the woody pollen plant itself, i.e., the increased number of stable diploid cells suggests that these cells have a competitive advantage over those of other ploidy. It is conceivable that most of the pollen trees will finally become stable homozygous diploids. However, a few of them may produce stable variants with different chromosome numbers. Since the buds from such trees can be grafted, it will be possible to analyze the chromosomal variaton in shoots that originated from buds with various euploid and aneuploid constitutions. Both the homozygous diploids and chromosomal variants could be very useful in breeding work with woody plants.

6. METHODOLOGY FOR EVALUATION OF POLLEN TREES

Investigations of the morphology and vigor of pollen tree "H_1" of poplar, rubber tree and Chinese crabapple revealed that there was a great diversity in leaf morphology, growth vigor, circumference, etc. among pollen trees derived from the same donor tree. For example, the highest one-year old pollen tree of poplar was 80 cm in height, while the lowest was only 20 cm (25). The height of one-year-old pollen rubber trees from one donor ranged from 131 cm to 19.5 cm.

There may be some after-effects of hormones and other factors of culture on the pollen plants H_1. Wu et al. observed that among pollen plants of poplar, some types of morphological variation in leaf-shape disappeared gradually while other types of variation were retained in successive vegetative generations (25). A similar phenomenon was observed among pollen plants of rubber tree.

Differences in ploidy among pollen trees could lead to differences in morphology and growth rates of the pollen trees. On the other hand some morphological differences may be due to the techniques of in vitro manipulation. Important factors could be the stage of the pollen plantlet at transplantation, soil conditions, microenvironment, the length of time the plantlet grew in a pot (or basket) before transplantation to soil, etc.

In comparison with seedlings, the pollen plantlets usually lacked strong roots. Therefore, recovery time of transplanted pollen plants is much longer and their growth rate is slower (10-12 months for rubber tree) than that of seedlings derived from the same inflorescence. Therefore, it is difficult to choose a suitable control for comparison with the pollen trees. Consequently, there is an urgent need to design an accurate methodology, and a suitable control, for the study on transplanted pollen trees.

The comparison of clones of pollen tree H_1 with those of the donor tree were an ideal method for evaluation of pollen trees. This method was used in investigations of pollen trees of poplar and rubber tree and in the selection of elite clones from those pollen trees. For example, four vigorously growing pollen trees H_1 were selected on the basis of periodical measurement of their heights. Then they and their donor trees were simultaneously cloned by bud-grafting. The mean girth of clones of pollen trees and the donor tree (control) listed in Table 13 showed that the growth vigor of the clones of four pollen trees significantly surpassed that of the donor clone ($p<0.01$). Especially the mean girth of clone No. 3 was larger than the others ($p<0.05$). These results revealed that it was possible to select excellent, promising clones from pollen trees H_1. Moreover, pollen tree No. 9 was significantly taller than the others, but the mean girth of its clone was not the largest one. It is concluded that though the evaluation of pollen trees H_1 was important, this evaluation did not show all the genetic differences between pollen trees. The comparison of clones of pollen trees H_1 with that of their donor tree appeared to be a better method for evaluation of pollen trees. This method can minimize the variation due to cultivation techniques. It eliminates the aftereffect of culture conditions on pollen plants, as well as the difference in recovery time between pollen plantlets and seedlings used as control.

Table 13. Mean girth of two-years old clones of pollen rubber trees.

H_1 clone	Height (cm)	No. of cloned plants	Girth (mm) $\bar{X}$	S.E.	p value
No. 3	98	24	95.38	1.76	<0.001
No. 6	90	30	88.97	1.57	<0.001
No. 9	146	36	88.92	1.57	<0.001
No. 15	80	18	87.50	2.33	<0.01
Donor	---	24	76.46	2.26	---

7. REFERENCES

1. CEN MING, CHEN ZHENGHUA, QUIAN CHANGFA, WANG CHUANHUA, HE YOUNGTAO, XIAO YULING 1981 Investigation of ploidy in the process of anther culture of Hevea brasiliensis. Acta Genet Sin 8: 169-174. In Chinese, English abstract
2. CHEN HONGDIAO, DONG YAN, GUO YUFONG, DONG HONGHUI 1981 A study on the inheritance, selection and utilization of poplar pollen tree H_1. "Yangshu Keji Tongxun" 5: 7-15. In Chinese
3. CHEN ZHENGUANG, WANG MIAOQUING, LIAO HUIHUA 1980 The induction of Citrus pollen plants in artificial media. Acta Genet Sin 7: 189-191. In Chinese, English abstract
4. CHEN ZHENGUANG 1982 Obtaining plantlets by anther culture of tea. J Chinese tea 4: 6-7
5. CHEN ZHENGHUA, CHEN FAZU, CHIEN CHANGFA, WANG CHUANHUA, ZHANG SHIJIE, XU XUEN, OU XIAOHUI, HE YUNGTAO, LU ZUNMIN 1978 Induction of pollen plants of Hevea brasiliensis Muell.-Arg. Acta Genet Sin 5: 99-107. In Chinese, English abstract
6. CHEN ZHENGHUA, CHEN FAZU, QIAN CHANGFU, WANG CHUANHUA, ZHANG SHIJIE, XU XUEN, OU XIAOHUI, HE YUNGTAO, LU ZUNMIN 1979 A process of obtaining pollen plants of Hevea brasiliensis Muell.-Arg. Sci Sin 22: 81-90

7. CHEN ZHENGHUA, QIAN CHANGFA, CEN MING, XU XUEN, XIAO YULING 1982 Recent advances in anther culture of Hevea brasiliensis (Muell.-Arg.) Theor Appl Genet 62: 103-108
8. CHEN ZHENGHUA, QIAN CHANGFA, XU XUEN, DENG ZHONGTAO 1982 Anther culture of rubber tree and sugarcane. In A Fujiwara, ed, Plant Tissue Culture 1982. Proc 5th Int Cong Plant Tissue Cell Cult, Tokyo, pp 533-534
9. CHEN ZHENGHUA 1983 Microscopic observation of Hevea brasiliensis cultures. In Cell and Tissue Culture Techniques for Cereal Crop Improvement. Science Press Beijing China, pp 47-54
10. FAN YINGHAN, ZANG SHUYING, ZFHAO JINGFANG 1982 Induction of haploid plant in Lycium chinense Mill. and barbarum from anther culture. Hereditas (Beijing) 4: 25-26. In Chinese
11. FEI KAIWEI, XUE GUANGRONG 1981 The induction of Malus pumila Mill "Yuanshuai". Acta Hortic Sin 4: 41. In Chinese, English abstract
12. FU LIANFANG, TANG DAOYI 1983 Induction of pollen plants of litchi tree (Litchi chinensis Sonn). Acta Genet Sin 10: 369-374. In Chinese, English abstract
13. GROUP OF TREE BREEDING, NORTHE-EASTERN FORESTRY INSTITUTE 1977 Induction of haploid poplar plantlets from pollen. Acta Genet Sin 4: 49-54. In Chinese, English abstract
14. GU SHURONG 1981 Pollen plants of Lycium obtained from anther culture. Acta Bot Sin 23: 246-248. In Chinese, English abstract
15. HIDAKA T, YAMADA Y, SHICHIJO T 1979 In vitro differentiation of haploid plants by anther culture in Poncirus trifoliata (L.) Raf. Jpn J Breed 29: 248-254
16. LU CHIHUA, ZHANG XIAOFANG, LIU YUXI 1978 Preliminary observation on induction of poplar pollen plants and their genetic performance. In Proceedings of Symposium on Anther Culture (1977). Science Press Beijing, pp 195-197
17. LU CHIHUA, ZHANG XIAOFANG, LIU YUXI, ZHANG PEIGUO 1977 Observation of the chromosome of somatic cells of poplar pollen plants. J North-eastern For Inst 2: 14-22. In Chinese, English abstract
18. MURASHIGE, T.; SKOOG, F. 1962 A revised medium for rapid growth and bioassays with tobacco tissue culture. Physiol Plant 15: 473-497
19. RADOJEVIC, L. 1978 In vitro induction of androgenic plantlets in Aesculus hippocastanum. Protoplasma 96: 369-374
20. Tree Improvement Laboratory, Heilungkiang Institute of Forestry 1975 Induction of haploid poplar plants from anther culture in vitro. Sci Sin 18: 769-777
21. WANG CHINGCHU, CHU ZHINCHING, SUN CHINGSAN 1975 Induction of pollen plants of Populus. Acta Bot Sin 17: 56-62. In Chinese, English abstract
22. WU JIANGYUN 1980 Successful induction of plantlets from anther of crabapple. J North-eastern Agric Coll 2: 69. In Chinese
23. WU JIANGYUN 1981 Induction of haploid plants from anther culture of crabapple. J North-eastern Agric Coll 2: 105-108. In Chinese, English abstract

24. WU KEXIAN, XU MIAOZHEN, XIE QIMING 1980 A study on natural chromosome doubling of poplar pollen plants and technique of artificial doubling. "Linye Keji" 3: 6-10. In Chinese
25. WU KEXIAN, XU MIAOZHEN, MA HEPING, LI WEI, ZHU WENHUA 1981 Preliminary analysis on characteristic of pollen plants of *Poplus simonii* Carr. "Linye Keji" 2: 5-6 In Chinese
26. YANG YONGQING, WEI WENXIONG 1984 Induction of Longan haploid plantlets from pollens cultured in certain proper media. Acta Genet Sin 11: 288-293 In Chinese, English abstract
27. ZENKTELER, M. 1972 Development of embryos and seedlings from pollen grains in *Lycium halimifolium* Mill. in the *in vitro* culture. Biol Plant 14: 420-422
28. ZOU CHANGJIE, LI PERFEN 1981 Induction of pollen plants of *Vitis vinifera* L. Acta Bot Sin 23: 79-81 In Chinese, English abstract
29. ZHU XINGYU, WANG RUILING, LIANG YAN 1980 Induction of poplar pollen plantlets. Sci For 4: 49-54. In Chinese, English abstract
30. ZHANG XIAOFANG, LIU YUXI, LU CHIHUA 1979 Cytological observation of the development of the microspore in poplar anther culture. North-eastern For Inst 2: 5-13. In Chinese, English abstract
31. ZHANG LI-KUNG 1978 Induction and culture of regenerated plantlets from anther of Genus *Populus*. *In* Proceedings of Symposium on Plant Tissue Culture. Science Press, Beijing, pp 241

14. TRIPLOIDS

G. LAKSHMI SITA

1. INTRODUCTION

Many important agricultural species are polyploids. Polyploidy is less common in forest trees and rare in gymnosperms (12, 52). Among the polyploids, triploids occupy an important position in cultivated plants. However, due to sterility barriers they can only be propagated by vegetative means. Triploids are generally vigorous and the banana is perhaps an outstanding example of a triploid crop. In this species maximum vigor is associated with the triploid state. Moreover, since diploid bananas have hard seeds that make their fruit commercially unacceptable, the sterility that accompanies triploidy has the important commercial advantage of genetically deseeding the banana in an efficient and dependable way. Some poplars are also vigorous in the triploid state. The discovery by Muntzing (34) that certain naturally occurring giant types of European aspen are triploids, prompted others to try artificial production of triploid aspen. The normal practice to produce triploids is to cross tetraploids with diploids. Sometimes the crosses may not be successful and therefore the repeated production of triploid seeds would be difficult. Production of triploid plants via endosperm culture *in vitro* could be useful to overcome this problem.

2. ENDOSPERM

The angiosperm endosperm is a source of nutrition for the developing embryo. In contrast to the diversity in the patterns of embryo formation, a similarity exists in the mode of their nutrition. Endosperm is the product of double fertilization or triple fusion resulting in a triploid nucleus.

Nawaschin's (40) discovery of double fertilization in angiosperms posed the question – why do the zygote and primary endosperm nucleus behave so differently even when they are genetically identical. In nature, embryos are known to develop from integuments, nucellus, synergids and antipodal cells; but there is no confirmed report of embryos originating from the endosperm. Until recently it was assumed (15, 17) that the endosperm lacks the capacity of embryogenesis. However it is now known that the excised endosperm sometimes has the capacity of somatic embryogenesis *in vitro*. With the advent of tissue culture techniques successful attempts have been made to culture the endosperm of both angiosperms and gymnosperms. Although these endosperm cultures may have the same developmental potential as the sporophytic tissues within which they develop, they are unique in their nuclear contribution. In addition to explaining basic questions about its ability for embryo formation, endosperm culture can also be used to explain the complex and diversified metabolic interaction between the sporophyte and its nutritional milieu under controlled conditions. Endosperm culture can also play an important role in breeding. In the following pages I will discuss the early work in *in vitro* culture of endosperm and examine the recently reported (23, 29, 58) phenomenon of embryogenesis in detail. Advantages of somatic embryogenesis over organogenesis will also be discussed.

2.1. Ontogeny

In exalbuminous seeds, the endosperm is short lived and is used up during growth of the embryo. In albuminous seeds, it remains as a permanent storage organ of the mature seed and serves as a nurse tissue for the embryo during germination. During fertilization of the egg, one of the male gametes fuses wth the diploid polar fusion nucleus in the embryosac to form a triploid fusion nucleus (endosperm nucleus) which almost from its inception follows a developmental pattern sharply divergent from that of the zygote. In most angiosperms this triploid fusion nucleus divides by repeated mitotic divisions to form a free nuclear, cellular of helobial type of endosperm. In at least 80% of the angiosperm families the endosperm is triploid. There are however cases where diploid, tetraploid, pentaploid and higher ploidy endosperms are found.

There are also families where the endosperm does not develop, e.g., Trapaceae, Podostemonaceae and Orchidaceae.

3. *IN VITRO* CULTURE

3.1. Early work

The earliest report of endosperm culture *in vitro* was that of Lampe and Mills (26). Immature maize endosperm was grown on a medium containing potato extract or corn extract. Later LaRue (28) established maize endosperm callus in continuous culture. This work was further extended by Strauss and LaRue (55) and nutritional requirements, growth and morphology of the maize endosperm callus were established (53). This was followed by the work of Sternheimer (56) and Tamaoki and Ullstrop in 1958 (57). Although the use of yeast extract (YE), by itself or in combination with other substances, has served to bring cereal endosperm into culture (41, 56, 57), Strauss (54) was able to develop a completely synthetic medium containing mineral salts, vitamins and asparagine, which gave excellent growth of maize endosperm. Later Farrar and Ganugapati (11) reported that ammonium succinate could successfully replace asparagine to give excellent growth of endosperm. Successful culture of maize was followed by the endosperm culture of *Lolium* (41), *Cucumis* (38), *Asimina* (27), *Ricinus* (48) and *Santalum* (46). A detailed analysis of the above work was presented by Johri (15) and Raghavan (43).

3.2. Organogenesis in endosperm cultures

The early period was followed by the extensive work of the Delhi School and others with many herbaceous angiosperms and a few tree species (17, 51). Mature endosperm of several taxa yielded continuously growing tissue but only a few trees have given equally good results: *Jatropha* (49), *Putranjiva* (50), coffee (33), *Juglans* (7, 11), *Achras* (3), apple (36), coconut (10), and custard apple (35). There is no need at this time to review all previous research leading to plantlet formation *in vitro*. Several excellent reviews (16, 18, 51) are already available giving detailed accounts of these developments. While the tissue generally exhibited unlimited growth, organ formation was not achieved in all the above taxa. Among the trees, organogenesis was achieved with

Putranjiva, Jatropha and walnut (11, 50, 49). In Putranjiva, a woody euphorbiaceae, Srivastava found that casein hydrolysate (CH) was indispensible for the differentiation of shoot buds from the endosperm callus. The callus grew well with 2000 mg/l CH. With 3000 mg/l CH, the shoots did not grow beyond 0.8 cm, while 5000 mg/l CH caused profuse callussing. CH promoted shoot bud differentiation in 87% of the cultures. Individually valine could replace the effect of CH. In Jatropha, another woody species, friable callus was obtained after subculture on White's medium (WM) + Kinetin (Kin) + CH. Although shoot bud differentiation was observed in 22% of the cultures, these buds did not develop normal leaves. Addition of an auxin failed to influence bud differentiation.

The only reports of differentiation of roots and plantlets from the callus of the immature endosperm are those of Nakano et al. (37), in rice, and Mu et al., in apple (36). Immature endosperm of rice excised 4-7 days after pollination, proliferated on Linsmaier and Skoogs' medium (LS) supplemented with YE. Proliferation and callus formation were also observed on LS + 2,4-dichlorophenoxyacetic acid (2,4-D) + YE, LS + 2,4-D + coconut milk (CM), and modified WM + YE + 2,4-D. Differentiation of rootlets also occurred on WM + YE, with or without Kin or indole-3-acetic acid (IAA). Plantlets were formed on WM supplemented with YE or IAA, or YE + Kin. In apple, endosperm was isolated from young, developing seeds and cultured on MS medium containing Kin (1 mg/l) and 2,4-D (0.5-2.0 mg/l). When some of this callus was transferred to a similar medium containing BAP (0.1-1.0 mg/l) and naphtaleneacetic acid (NAA) (0.01-0.05 mg/l), one plantlet developed within 60 days. Roots and leaves also differentiated from the callus.

A literature survey indicates that organ formation was obtained mostly in callus from mature endosperm. In semi-parasitic angiosperms, differentiation of shoot buds occurred without callussing. In autotrophic species, the endosperm usually formed a callus mass before differentiation of shoot buds or roots occurred (51).

3.3. Organogenesis in naturally occurring triploids

Mathes (31) was the first to achieve growth of both shoots and roots in callus cultures of quaking aspen (Populus tremuloides).

Firm white callus was isolated from root sprouts. Shoots grew from the cut surface of that callus on medium with 0.05 mg/l N^6-benzylaminopurine (BAP). A few years later Wolter (63) showed that roots could be initiated on callus of P. tremuloides by adding BAP (0.14-0.3 mg/l) to basal medium without auxins. These shoots could be rooted in wolter's basal medium (64). Later, Winton (60) obtained shoots from two year old callus cultures of triploid quaking aspen. These shoots were produced on callus on Wolter's medium with 0.05-0.2 mg/l BAP. Roots were subsequently obtained with 0.04 mg/l 2,4-D and Kin 1 mg/l. These earlier reports (31, 60) left some doubt as to the viability and variability of the regenerated plants. Winton (61) demonstrated that normal tree seedlings could be obtained from callus cultures in vitro, although few in numbers. This was a significant observation, as this was the first time viable trees were produced.

Recently extensive work was done in another naturally occurring triploid fruit tree, i.e., banana (20). Berg and Bustamante (4) showed that cucumber mosaic virus-free plants of the economically superior Cavendish group AAA can be obtained from meristems and also from lateral buds of virally infected bananas by a combination of heat treatment and aseptic culture. Only one plant was obtained from each shoot apex that was cultured on Knudson's medium supplemented with Berthelot's (5) trace elements, thiamine HCl and CM 10% v/v and CH. Rooting of the excised shoot tips required the addition of 1 mg/l NAA to the medium.

This work led to the extensive commercial use of shoot tip culture techniques practiced now. Multiplication is induced by releasing dormant buds at the leaf bases and subculturing the proliferating mass of shoots. Protocorm-like bodies can also form at the shoot base. Multiple shoots that originate from these can, in turn, be separated by cutting, to produce individual plantlets. Functional roots can be initiated within a few days by using a combination of procedures (20).

The earliest efforts to establish callus and suspension cultures of the desert banana were made in 1950 and in subsequent years by Mohan Ram and Steward (32). Since then little work has been done with banana mainly due to the difficulty in obtaining subculturable callus. Mohan Ram and Steward used a variety of auxins to induce callus formation from mature and immature fruits. However, no differentiation was obtained.

3.4. Embryogenesis

Embryogenesis has not been reported until 1980 either from a naturally occurring triploid sporophyte or triploid endosperm. Lakshmi Sita *et al*. (23) have reported embryogenesis from endosperm cultures of *Santalum album*. Plantlets have developed from these embryoids and were established in soil. Cronauer and Krikorian (9) have reported somatic embryos from naturally occurring triploid plantain clones in liquid medium. Green somatic embryos formed only in the media containing 2,4,5-trichlorophenoxyacetic acid. These, after transfer to semisolid medium, produced roots, but no shoots. Recently somatic embryogenesis has been reported in walnut (unpublished; and Tulecke, this volume).

3.4.1. Embryogenesis in sandalwood. Rangaswamy and Rao (46) were the first to report callus initiation from endosperm tissue of sandalwood on modified white's medium (45) supplemented with 2 mg/l 2,4-D + 5 mg/l Kin and 0.25% YE. However, no differentiation was observed.

Callus induction, followed by differentiation, were reported later by Lakshmi Sita *et al*. in 1980 (23). Endosperm from fresh green fruits was cultured with or without embryo. Of the various media and growth regulators used, MS medium (35) supplemented with 2,4-D (1-2 mg/l), Kin (0.1-0.2 mg/l), BAP (0.5-2.0 mg/l) and NAA (1 mg/l) induced the most actively growing callus. Callus grew better on media supplemented with BAP and NAA than with 2,4-D. On 2,4-D media calus initiation was noticed within 8-10 days, but further growth was slow. Callus initiation on medium supplemented with BAP and NAA was slow initially but further growth was faster than with 2,4-D. Callus thus initiated was routinely maintained on MS medium supplemented with 1 mg/l 2,4-D. Experiments that were initially carried out without shoot callus and suspension tissues were repeated with endosperm tissue (21, 22). Numerous embryoids visible to the naked eye were observed from 3 weeks onwards on media with 1-2 mg/l gibberellic acid (GA). Differentiation of embryoids was also observed on MS medium supplemented with BAP (0.2-0.5 mg/l) and NAA or IAA (1 mg/l), GA (1 mg/l) and Kin (0.2-0.5 mg/l). Subculture of undifferentiated callus on MS supplemented with 0.2 mg/l Kin, 1 mg/l NAA and 10% CM showed intense embryogenesis (Fig. 1 and 2) Subculture on this medium for 2 weeks prior to subculture on MS + 1 mg/l GA gave excellent further growth. Mature embryoids developed into complete plantlets with a well developed shoot and root system after a final subculture on WM supplemented with 0.5 mg/l IAA (Fig. 3). These plantlets were established in soil as easily as diploid sandalwood plantlets. Triploid plantlets (Fig. 4) are similar to diploid plants except that their leaves are narrower and more linear in shape.

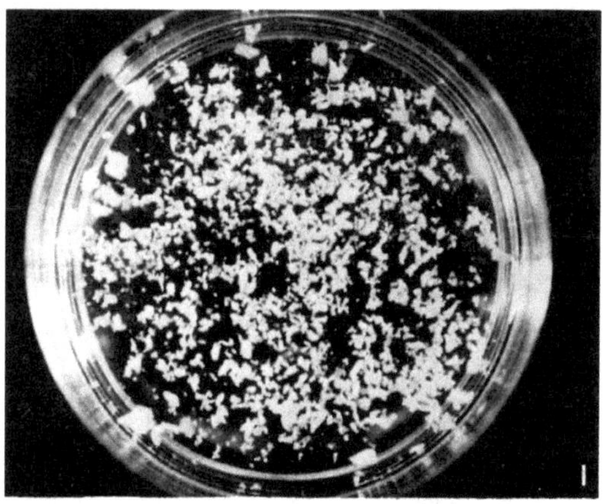

FIGURE 1. Differentiated embryoids from callus after 21 days (23) separated into a petri dish containing water to facilitate photography.

3.4.2. Nutritional requirements. In all the earlier reports of endosperm culture, successful proliferation of endosperm callus *in vitro* required an auxin, a cytokinin and a rich source of reduced organic nitrogen. For sandalwood it was possible to induce callus on media with auxin alone or with an auxin and a cytokinin. It was not necessary to use complex growth factors like CM, malt extract (ME), or YE for callus initiation or differentiation of embryoids.

FIGURE 2. Same as Fig. 1 (magnified x 16).
FIGURE 3. Plantlet developed from a triploid enbryoid on White's medium with 0.5 mg/l IAA. Note well developed shoot and root.
FIGURE 4. Four month old triploid plantlet in soil under natural conditions.

Contrary to earlier assumptions, contact with an embryo was not necessary for the initiation of endosperm callus. Low concentrations of NAA stimulate enlargement of endosperm. On a high concentration of NAA (1 mg/l) with low BAP (0.1-0.5 mg/l), the callus grew vigorously and appeared loose in texture and yellow in color. High concentrations of BAP (1-2 mg/l) induced a brownish, compact callus.

Initially organogenesis in endosperm callus was restricted to leaf bud initiation. Regeneration of roots and of viable plantlets was not possible until 1977 (16). Sandalwood is probably the only species where viable plantlets with a well-developed shoot and root system were formed from endosperm callus by embryogenesis. Recently, Wang and Chang (58) have reported embryogenesis but not viable plantlets from endosperm tissue of Citrus grandis. Their report confirms our finding that GA is required for the induction of embryogenesis. However, for sandalwood it was not necessary to use the high concentrations (15 mg/l) used by the Chinese group; 1 mg/l GA was suficient to bring about intense embryogenesis. At present, we have not yet defined the optimal conditions for embryogenesis.

A few generalizations can be made. Differentiation of sandalwood embryoids occurs on a medium supplemented with GA. Furthermore, GA stimulates elongation of embryos which have already differentiated. Failure of earlier workers to achieve embryogenesis can be attributed to one or two reasons. It is necessary for the cultures to be in an auxin medium in the presence of reduced nitrogen. WM which was used in most of the early work on endosperm cultures, had only a limited amount of reduced nitrogen in the form of glycine (3 mg/l) when it was not supplemented by additions such as CM, YE, CH or a defined mixture of amino acids. In the light of the need for reduced nitrogen for somatic embryogenesis (65), it is no wonder that only after the introduction of the MS basal medium, whisc is rich in reduced nitrogen (1650 mg/l), this phenomenon is observed more and more. The success of embryogenesis in sandalwood is due to the right combination of basal medium with appropriate growth regulators. Wang and Chang also reported that a combination of minerals, in high concentrations, and GA play an important role in inducing embryogenesis. Contrary to reports by others (13, 19, 39, 47), GA was not inhibitory. Johri and Srivastava (17) consider that the induction of endosperm callus and embryogenesis do not occur without the participation of the embryo, which they called the embryo factor. However recent results (23, 29) show that induction of endosperm callus and embryogenesis can be obtained with GA, without the embryo. In sandalwood, endosperm differentiation was also observed on other media containing auxin-cytokinin combinations. Recently, Lin et al. (29) have reported embryoid production and subsequent plantlet formation on MS medium supplemented with 1 mg/l zeatin.

As suggested by Raghavan (43) more information must be obtained before a comprehensive picture of the nutritional requirements for endosperm callus growth and subsequent embryogenesis can be defined. It would be of interest to know if there is a single hormone which regulates embryogenesis in the endosperm. Experiments designed to answer this question will be easier, once continuously growing cultures become available.

3.4.3. Advantages of embryogenesis over organogenesis. Embryogenesis has advantages over organogenesis since pantlets arising from embryoids will have a taproot which is superior to adventitious root systems. Even where traditional methods of vegetative propagation are available, as in rubber, the lack of a taproot often causes poor anchorage. Also, organogenesis often results in a limited number of shoots. On the other hand, induction of embryogenesis may result in numerous embryoids and a large number of plants.

Throughout the literature there are many claims of embryogenesis in cultures of diploid and haploid tissues. The term embryogenesis has been used for pro-embryo and embryo-like stuctures that did not necessarily develop into plantlets (62). Brown and Sommer (6) have discussed extensively the adventive embryogenesis in dicot trees. As with all species, the real breakthrough in mass propagation will come when embryogenesis in cell suspensions becomes a reality. Sandalwood is one of the few trees where embryogenesis can be routinely induced both in diploid and triploid callus and in suspension cultures. However, induction of embryogenesis was easier in callus cultures than in the latter. Plantlets measuring 5-8 mm are obtained in 5-6 weeks. These could be easily separated and transferred to soil after further growth in White's medium.

3.5. Protoplast culture

Protoplast from endosperm cultures of sandalwood can be isolated and cultured (25). Fusion of haploid, diploid and triploid protoplasts would enable production of various ploidies. Work is in progress towards this goal. Krikorian and Cronaur (unpublished) were able to prepare protoplasts reproducibly from various parts of desert banana and are working towards establishing cell suspensions from these preparations. No plants have yet been recovered from these protoplasts.

4. APPLICATIONS

4.1. Plant improvement

Polypoloidy breeding has been important in the development of new types of agricultural plants and promises to be of importance in forestry. The triploid condition, although sterile, stimulates growth in several members of Rosaceae (apples, pears, and *Prunus*). It is also vigorous in some poplars. Swedish and German experiments with two other genera, *Alnus* and *Ulmus* (14) corroborate the aspen work indicating promise for artificially produced triploids. Artificially produced tetraploids in these species were much slower growing than the diploids and triploids.

The culture of endosperm has opened up new avenues for plant breeders, who can now apply this technique to commercially important crop plants, especially where conventional breeding methods are ineffective. The most common method (*in vivo*) for raising triploids is crossing a tetraploid and a diploid. This cross, however, is not always successful due to the failure of the endosperm to develop and subsequent abortion of the embryo. Furthermore, production of triploids by crossing is a cumbersome process. Therefore the culture of endosperm followed by embryogenesis is a good means of obtaining large numbers of triploid plants.

Tissue culture should be exploited in the case of naturally occurring triploids like banana. Efforts to improve both bananas and plantains using conventional breeding techniques over the past 60 years have had limited results. Consequently, aseptic culture techniques should be used to induce variations and to recover novel plants (20). Production of callus and/or rapidly growing morphogenetically competent cell suspensions will be a major step towards the imporvement of banana as a crop. Fruit crops which form a major part of our edible foods can be exploited by endosperm culture especially to raise triploid seedless varieties. At a recent conference in China, there were preliminary reports of endosperm cultures of several fruit and forest trees (8, 29, 30, 59).

It has been suggested that freeze preservation of cultured tissues can be used for the prevention of ageing and long term conservation of germplasm (2). Bajaj (1) initiated rice endosperm callus from frozen seeds in 5-7 weeks. About 3% of the callus produced plantlets when subcultured on MS + Kin (2 mg/l) and IAA (1 mg/l). These plantlets differed little from normal plantlets.

4.2. Biochemical studies

Our understanding of the biochemical composition of endosperm has come from biological asays based on the ability of the endosperm to induce growth. Liquid coconut endosperm has been used extensively for this purpose. CM is a conglomerate of several inorganic ions, aminoacids, vitamins, growth hormones, sugars and sugar alcohols. Although subculturable endosperm of coconut was obtained by Eeuwens and Blake (14), its composition has not been established. As part of our investigations on sandalwood, both diploid (24) and triploid (unpublished) callus were analyzed for various aminoacids. Distribution of hydroxyproline containing protein is limited in plants, sandalwood being one of these. Crystalline hydroxyproline was isolated from callus cultures of both diploid and triploid tissues and identified by X-ray analysis as trans-4-hydroxyproline (28). Hydroxyproline isolated earlier (44) from leaves was found to be cis-hydroxy-L-proline, while the trans form occurred in the bound state. These differences may be due to cellular organization and secondary metabolism which is more complex in leaves than in callus. The absence of enzymes involved in isomerization pathways, could explain the variation.

Recently Pollard *et al.* (42), using suspension cultures of rye grass endosperm, found that hydroxyproline is rapidly converted into proline by a route which appears to be more direct than the pathways of hydroxyproline metabolism in mammalian and bacterial systems.

5. REFERENCES

1. BAJAJ YPS 1981 Growth and morphogenesis in frozen (-196°C) endosperm and embryo in rice. Curr Sci 50: 947-948
2. BAJAJ YPS 1983 Cryopreservation and international exchange of germplasm. *In* SK Sen, KL Giles, eds, Plant Cell Culture in Crop Improvement. Plenum Press, pp 19-41
3. BAPAT VA, S NARAYANASWAMY 1977 Mesocarp and endosperm culture of *Achras sapota* Linn. *in vitro*. Ind J Expt Bot 15: 294-296
4. BERG L, M BUSTAMANTE 1974 Heat treatment and meristem culture for the production of virus free bananas. Phytopathology 64: 320-322
5. BERTHELOT A 1934 Nouvelles remarques d'ordre chimique sur le choix des milieux de culture naturels et sur la manière de formuler les milieux synthétiques. Bull Soc Chim Biol 16: 1553-1557
6. BROWN CL, HE SOMMER 1984 Vegetative propagation of dicotyledonous trees. *In* JM Bonga, DJ Durzan, eds, Tissue Culture in Forestry. Martinus Nijhoff, The Hague, pp 109-148
7. CHEEMA GS, PN MEHRA 1982 Morphogenesis in endosperm cultures. *In* 5th Int Congr Plant Tissue and Cell Culture. Tokyo, pp 111-112
8. CHEN Z, S LIN, Q LIN 1984 A prelininary report on formation of plantlets from endosperm culture of loquat (*Tribatryyua japonica*). Abst Proc Conf Genetic Manipulations in Crop Plants, pp 75
9. CRONAUER S, AD KRIKORIAN 1983 Somatic embryos from cultured tissue of triploid plantain (*Musa ABB*). Plant Cell Rep 2: 289-291
10. EEWENS CJ, J BLAKE 1978 Culture of coconut and date palm tissue with view to vegetative propagation. Acta Hortic 78: 277-296
11. FARRAR KR, LS GANUGAPATI 1970 Growth of maize endosperm *in vitro*: 1. Nitrogen requirements of tissue grown on solid medium. J Sci Food Agric Sci (UK) 21: 329-332
12. GUSTAFFSON A 1960 Polyploidy and mutagenesis in forest tree breeding. *In* Proc 5th World For Cong. Seattle, US, pp 793-805
13. HALPERIN W 1970 Embryos from somatic plant cells. *In* HA Padykula, ed, Control Mechanisms in the Expression of Cellular Phenotypes, Symp Int Soc Cell Biol 9: 169-181
14. JOHNSSON H 1956 Auto and allotriploid *Betula* families, derived from colchicine treatment. Forst Genet Forstpflanzenzucht 5: 65-70
15. JOHRI BM 1975 Organogenesis in embryo and endosperm cultures of Angiosperms. 7th Philip R White Mem Lect Comm, Delhi, pp 1-24
16. JOHRI BM, SS BHOJWANI 1977 Triploid plants through endosperm culture. *In* J Reinert, YPS Bajaj, eds, Plant Cell Tissue and Organ Cultures. Springer/Verlag, Berlin, pp 398-411

17. JOHRI BM, PS SRIVASTAVA 1973 Morphogenesis in endosperm culture. Z Pflanzenphysiol 70: 285-304
18. JOHRI BM, PS SRIVASTAVA, AT RASTE 1980 Endosperm. In IK Vasil, ed, International Review of Cytology, Suppl XI. pp 157-182
19. KOCHBA L, P SPIEGELROY, N HANNA, S SAAD 1978 Stimulation of embryogenesis in Citrus ovular callus by ABA ethephon CCC and Alar and its suppression by GA_3. Z Pflanzenphysiol 89: 422-427
20. KRIKORIAN AD, SS CRONAUER 1984 Banana. In WR Sharp, DA Evans, PV Ammirato, Y Yamada, eds, Handbook of Plant Cell Culture, Vol 2. MacMillan & Co, pp 327-348
21. LAKSHMI SITA G, NV RAGHAVARAM, CS VAIDYANATHAN 1979 Differentiation of embryoids and plantlets from shoot callus of sandalwood. Plant Sci Lett 15: 265-270
22. LAKSHMI SITA G, J SHOBHA, CS VAIDYANATHAN 1980 Regeneration of whole plants by embryogenesis from cell suspension cultures of sandalwood. Curr Sci 49: 196-198
23. LAKSHMI SITA G, NV RAGHAVARAM, CS VAIDYANATHAN 1980 Triploid plants from endosperm cultures of sandalwood by experimental embryogenesis. Plant Sci Lett 20: 63-69
24. LAKSHMI SITA G, YV MAHALAKSHMI 1980 Isolation of hydroxyproline from tissue cultures of sandalwood. Ind J Biochem Biophys 18: 173
25. LAKSHMI SITA G, HS RANI 1983 Preliminary studies on isolation and culture of protoplasts from sandalwood. Experentia 45: 4-5
26. LAMPE L, CO MILLS 1933 Growth and development of the isolated endosperm and embryo of maize. Abstr Bot Soc Boston
27. LAMPTON RK 1952 Development and experimental morphology of the ovule and seed of Asimina triloba. Ph D Thesis, Univ Michigan, Ann Arbor
28. LA RUE CD 1949 Cultures of the endosperm of maize. Abstr Am J Bot 36: 798
29. LIN GY, GS RONG, TY XU 1984 Differentiation of organs in endosperm culture. Abstr Proc Conf Genetic Manipulation Crop Plants, pp 75
30. MASCARENHAS AF, S NAIR, RS IYER, PK GUPTA 1984 Genetic improvement of fruit crops through tissue culture. Abstr Proc Conf Genetic Manipulation Crop Plants, pp 13
31. MATHES M 1964 In vitro formation of plantlets from aspen tissue. Phyton 21: 137-141
32. MOHANRAM HY, FC STEWARD 1964 The induction of growth in explanted tissue of the banana fruit. Can J Bot 42: 15-59
33. MONACO K, Y KODA, Y OKAZAWA 1978 Application of tissue culture in the improvement of coffee. In J Reinert, YPS Bajaj, eds, Applied and Fundamental Aspects of Plant Cell Tissue and Organ Culture. Springer/Verlag, Berlin, pp 109-126
34. MUNTZING A 1936 The evolutionary significance of autopolyploidy. Hereditas 21: 263-378
35. MURASHIGE T, F SKOOG 1962 A revised medium for rapid growth and bioassays with tobacco tissue culture. Physiol Plant 15: 473-497
36. MU S, S LIU, Y QUIAN, P ZHANG, H XIE, F ZHANG, Z YAN 1977 Induction of callus from apple endosperm. Sci Sin 2: 370-377
37. NAKANO H, J TASHIRO, E MAEDA 1975 Plant differentiation in callus tissue induced from immature endosperm of Oryza sativa L. Z Pflanzenphysiol 76: 444-449

38. NAKAJIMA T 1962 Physiological studies of seed development especially embryonic growth and endosperm development. Univ Osaka Prefect Ser B 13: 13-48
39. NARAYANASWAMY S 1977 Regeneration of plants from tissue cultures. In J Reinert, YPS Bajaj, eds, Applied and Fundamental Aspects of Plant Cell Tissue and Organ Culture. Springer/Verlag, Berlin, pp 179-204
40. NAWASHIN SG 1898 Resultate einer Revision der Befruchtungs-Vorgange bei Lilium martagon und Fritillaria tenella. 12V Imp Akad Nauk 9: 377-382
41. NORSTOG K 1956 Growth of rye grass endosperm in vitro. Bot Gaz 117: 253-259
42. POLLARD PC, PW WAY, GB FINCHER 1981 Uptake and metabolism of hydroxyproline in endosperm cells of Lolium multiflorum (rye grass). Aust J Plant Physiol 5: 35-46
43. RAGHAVAN V 1976 Experimental Embryogenesis in Vascular Plants. Acad Press, London
44. RADHAKRISHNAN AN, KV GIRI 1954 The isolation of hydroxy-l-proline from sandal (Santalum album). Biochem J 58: 57-61
45. RANGASWAMY NS 1961 Experimental studies on female reproductive structures of Citrus microcarpa Bunge. Phytomorphology 11: 109-127
46. RANGASWAMY NS, PS RAO 1963 Experimental studies on Santalum album: Establishment of tissue culture of endosperm. Phytomorphology 13: 450-454
47. REINERT Y, YPS BAJAJ, B ZBELL 1977 Aspects of organization, organogenesis, embryogenesis, cytodifferentiation. In HE Street, ed, Plant Tissue and Cell Structure. Blackwell Sci, Oxford, pp 389-427
48. SATSANGI ASHA, HY MOHANRAM 1965 Continuously growing tissue cultures from mature endosperm of Ricinis communis L. Phytomorphology 15: 20-30
49. SRIVASTAVA PS 1973 Formation of triploid plantlets in endosperm cultures Putranjiva roxburghii. Z Pflanzenphysiol 69: 270-273
50. SRIVASTAVA PS, BM JOHRI 1974 Morphogenesis in mature endosperm cultures of Jatropha panduraefolia. Beitr Biol Pflanzen 50: 255-268
51. SRIVASTAVA PS 1983 Endosperm culture. In BM Johri, ed, Experimental Embryology of Vascular Plants. Springer/Verlag, Berlin, pp 174-193
52. STANLEY RG 1970 Biochemical approaches to forest genetics. In JA Romberger, P Mikola, eds, Int Rev For Res. Acad Press, pp 253-309
53. STRAUS J 1954 Maize endosperm tissue grown in vitro. II Morphology and cytology. Am J Bot 41: 833-839
54. STRAUS J 1960 Maize endosperm tissue grown in vitro. III Development of a synthetic medium. Am J Bot 47: 641-647
55. STRAUS J, CD LA RUE 1954 Maize endosperm tissue grown in vitro. I Culture requirement. Am J Bot 41: 687-694
56. STRENHEIMER EP 1954 Methods of culture and growth of maize endosperm in vitro. Bull Torrey Bot Club 81: 111-113
57. TAMAOKI T, AJ ULLSTRUP 1958 Cultivation in vitro of excised endosperm and meristem tissues of corn. Bull Torrey Bot Club 85: 260-272

58. WANG T, C CHANG 1981 Triploid *Citrus* plantlet from endosperm culture. *In* Pitman International Series in Applied Biology, Plant Tissue Culture, pp 463-468
59. WANG L, S CHEN, J ZUN, D WANG 1984 Induction of endosperm plants from *Lycium barbarum* and its ploidy level. Abstr Proc Genetic Manipulation in Crop Plants, pp 82
60. WINTON LL 1968 Plantlets from aspen tissue cultures. Science 160: 1234-1235
61. WINTON LL 1970 Shoot and tree production from aspen tissue cultures. Am J Bot 57: 904-909
62. WINTON LL 1978 Morphogenesis in clonal propagation of woody plants. *In* TA Thorpe, ed, Frontiers of Plant Tissue Culture. Int Assoc Plant Tissue Cult, Univ Calgary, Canada, pp 419-426
63. WOLTER K 1968 Root and shoot initiation in aspen callus culture. Nature 219: 509-510
64. WOLTER K, F SKOOG 1966 Nutritional requirement of *Fraxinus* callus cultures. Am J Bot 53: 263-269
65. WETHERELL DF, DK DOUGALL 1976 Sources of nitrogen supporting growth and embryogenesis in cultured wild carrot tissue. Physiol Plant 37: 97-103

15. COLD STORAGE OF TISSUE CULTURES

J. AITKEN-CHRISTIE and A.P. SINGH

Forest Research Institute, Private Bag, Rotorua, New Zealand

1. INTRODUCTION

Collections of seed and clonal material are traditional ways of storing genetic resources, while tissue cultures provide a complementary way. Freeze-preservation of cultured cells and meristems has become popular for long-term germplasm preservation (7,28,53). However, freeze-preservation requires the use of sophisticated techniques and expensive equipment, and often leads to tissue damage during freezing and thawing cycles. Another way of storing germplasm which does not require transfers is to limit growth by modification of the medium, e.g., by adding osmotically active substances. Cold storage of cultures at 0 to 15°C is a third method.

This last method, which can be as simple and as cheap as placing a container of sterile shoots in the refrigerator, is the subject of this chapter. Results of experimentation with woody species other than forest trees are included because (1) the techniques are widely applicable, and (2) published information on forest species is scarce. Results for non-woody species are also quoted where these illustrate important points. Published information on cold storage of herbaceous species has been reviewed by Bhojwani and Radzan (10).

2. METHODOLOGY OF COLD STORAGE

The following discussion of general methodology for cold storage of sterile cultures is based on procedures which have been developed for woody plants, drawing particularly on work with radiata pine, <u>Pinus radiata</u> D. Don, at the Forest Research Institute (FRI), Rotorua, New Zealand. Some problems encountered with the application of cold storage techniques are highlighted.

2.1. Materials

2.1.1. Type of culture. Table 1 lists references to tissue cultures and plant species which have been successfully cold-stored at -3 to 15°C. In addition, shoot cultures of apple, *Malus domestica* var. Northern Spy, (J.F. Hutchinson, pers. comm.), *Rhododendron* and *Azalea* spp. (B. Briggs, pers. comm.), and *Eucalyptus* spp. (e.g., *E. camaldulensis*, *E. blakelyi*, *E. regnans*, *E. spathulata*, *E. radiata*, *E. globulus* subsp. *bicostata*) (Hartney, pers. comm.) have been cold stored, all in the temperature range 4 to 6°C. Cold storage of shoot and plantlet cultures has been widely used, to provide a ready source of material for further multiplication and rooting *in vitro*, or for direct planting into soil. However, there are few reports of cold storage of callus and cell suspension cultures and none of cold-stored protoplasts. This may be because meristem, callus, cell, and protoplast cultures can be preserved by freezing (28,53).

2.1.2. Physiological state. The physiological state of shoots to a great extent determines the success or failure of a cold-storage program. Shoot health, size, and degree of vitrification are all important factors influencing tissue hardiness and response. In general, with juvenile *Pinus radiata* shoots, we have found that healthy vigorous shoot clumps, containing between three and six shoots, cold store well, but clones which are initially unhealthy continue to decline. Larger tissues appear to store better than smaller tissues. Barlass and Skene (9) found that single-rooted shoots of grapes, *Vitis* spp. and hybrids, survived better at 9.5°C than smaller proliferating shoot cultures without roots. Similarly, large 10-mm *Eucalyptus* shoots have been found to have better survival than small shoots under 5 mm (Hartney, pers. comm.). Larger, fully elongated *Pinus radiata* shoots over 20 mm are additionally to be preferred to smaller shoots (5 mm or less) because they are less prone to becoming "wet" or "vitrified". Normal "waxy" shoots store better than "vitrified" or "wet" shoots (see 2.2.4.; also Aitken-Christie *et al* (4) for a description of these three shoot types). Clonal differences also appear to be important. From a group of mature *Pinus radiata* clones, those which grew best under standard culture conditions for 6 months were also the healthiest in the cold store over the same time period (K. Horgan, pers. comm.).

TABLE 1. List of woody plant shoot, plantlet, cell and callus cultures cold stored at -3 to 15°C

Species	Reference	Cold storage temperature (°C)
A. SHOOT/PLANTLET CULTURES		
Cinchona ledgeriana	25	12
Deciduous trees	43	2
Eucalyptus dalrympleane	19	0-1
Eucalyptus gunnii	19	0-1
Eucalyptus spp.	6	0-1
Malus domestica	13,30,49	1-4
Malus spp.	12,50	2-4
Pinus radiata	1,2,3,41,47	4-5
Prunus amygdalus	39	4
Prunus avium	15	4
Prunus spp.	12,13,50	2-4
Prunus spp.	32	(-3)-8
Punica granatum	50	10
Pyrus communis	48	4
Pyrus pashia	50	4-10
Salix babylonica	20	5
Sequoia sempervirens	36	4
Vitis rupestris	21	9
Vitis spp. and hybrids	9	9.5
B. CELL/CALLUS CULTURES		
Citrus spp.	5	4-5
Coffea spp.	5	15
Mallotus japonicus	23	4
Pinus sylvestris	22	2

2.1.3. Substrate. Shoot cultures are most conveniently cold stored on the same substrates and nutrient medium used for normal culture conditions. Agar-gelled media and liquid with filter paper bridges have been the predominant substrates used (9,49). A conditioned medium which has had shoots on it for no longer than 2 to 3 weeks appears to be better than spent medium (16). The growth stage of the culture is an important factor in acclimatization to the cold. Actively growing younger callus and explants of *Chrysanthemum* stored at the time of maximum growth (20 days in culture) survived better than old callus or rooted shoots cold stored after 30 days in culture (8). At the FRI laboratory shoot cultures are routinely placed in the cold store 1 to 2 weeks after their last transfer. Hartney places all shoots under ambient conditions for 4 to 7 days before placing them in cold storage (pers. comm.).

The addition of 1-4% w/v mannitol to the culture medium enhanced the survival of shoots of quinine tree, *Cinchona ledgeriana* (semi-tropical species), cold stored at 12°C (25). However, mannitol added to the medium at normal growth temperatures (26 to 28°C) was inhibitory (25,48). Mannitol may be acting in a similar way to the cryoprotectants which protect cells during freeze preservation. It is usually necessary to protect shoots of cold-sensitive and semi-tropical species during cold storage and use of cryoprotectants, such as dimethylsulphoxide (DMSO), polyethyleneglycol (PEG) and sorbitol, may be worthwhile.

2.1.4. Containers. A variety of containers have been used to cold-store cultures, e.g., glass test-tubes (25 x 150 mm) (Fig. 2b), Erlenmoyer flasks (100 to 125 ml), polycarbonate tubes (25 x 80 mm), screw-cap polycarbonate vials, glass jars (600 ml) (Fig. 2a), and several commercial food containers. In our laboratory 600-ml jars are routinely used for *Pinus radiata* under normal and cold storage conditions. For long-term cultures, test-tubes (25 x 150 mm) are used because more replicates can be stored in the same space used by jars, reducing losses if contamination occurs.

Moisture loss and contamination are minimized by sealing the vessels and, as an extra precaution, containers are placed in a plastic bag.

2.2. Conditions for cold storage

2.2.1. Equipment and design. At FRI we use a walk-in cold storage room with the dimensions, 3.5 m (length) x 1.7 m (width) x 2.3 m (height). The room is fitted with 34 rust-proofed aluminium shelves and each shelf (0.36 m^2) holds thirty six 600-ml jars or 210 test-tubes of 25 x 150 mm each (Fig. 1). Each shelf has a small fluorescent white light 1.5 cm x 52 cm (Thorn 13 watts) enclosed in a plastic fitting (Fig. 1). (In some other applications light ballasts have been placed outside the room or light is "piped" in from fixtures outside the room - D.J. Durzan, pers. comm.) A 3/4-1 hp refrigeration unit cools the room to 4-5°C and a forced draught evaporator (blower) circulates chilled air around the room. The unit has electric automatic defrosting. Temperature is controlled within 1°C and there are two alarms, one to warn of increasing temperatures and one to prevent freezing. At FRI, clones are divided equally between two cold rooms to further reduce the risk of accidental loss. In normal circumstances the worst that can happen is that the refrigeration unit fails and the cultures come to room temperature and may need a transfer to fresh medium. Little harm is done to the cultures in this event; they will not die. This is a distinct advantage of cold storage methods compared with freeze preservation where cells and tissues could die if the freeze systems broke down and freezing and thawing protocols were not adhered to. The cold room is lined throughout with 10-cm polystyrene insulation and metal covering. The room should be kept clean and free of rotting materials; also wooden shelves should be avoided as these may harbour contaminants.

2.2.2. Temperature. Shoot and plantlet cultures are usually stored between 0 and 12°C depending on species requirements (Table 1). Generally cultures will previously have been growing at temperatures in the range 20 to 30°C. It is not possible to recommend a universal storage temperature applicable to all plant species. As a guide, the natural cold tolerance of a species will probably determine the range of temperatures over which material can be stored. Sub-tropical, cold-sensitive woody tissue will store between 9 and 12°C, whereas hardier conifers and other woody species require lower temperatures between 0 and 4°C (cf Chalupa, Volume 1).

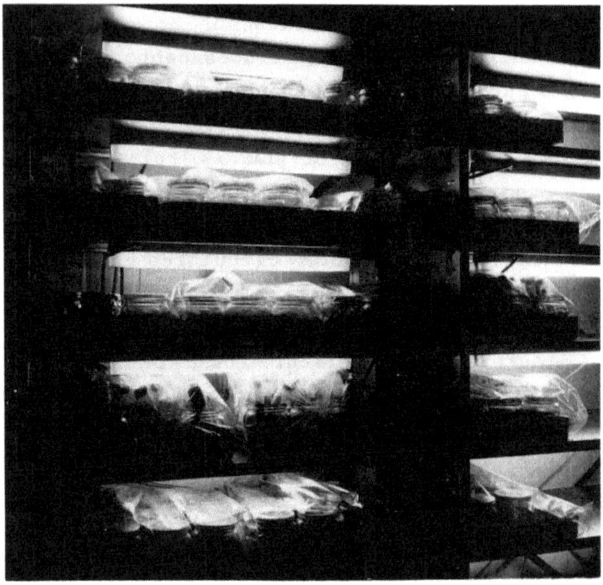

FIGURE 1. FRI cold storage facility showing design, cultures, shelving, and lighting.

2.2.3. Light. The significance of light in maintaining 'normal' state (non-etiolated) cultures is not clear. Cultures have been stored successfully in both light and dark. Some shoot cultures have been cold stored using a 12 to 16 hr photoperiod under a low light intensity of 8-34 $\mu Em^{-2}s^{-1}$ (3,9,24,25,32, Hutchinson, pers. comm.). Others have been stored in the dark for up to 6 months (15,43) and 12 months (30). Eucalyptus cultures have been stored in the dark for 9-29 months (Hartney, pers. comm.). Most of the Eucalyptus cultures did not grow or etiolate under these conditions but some developed roots. The temperature of storage may affect the response to light. Marino et al. (32) found that Prunus rootstocks (P. domestica 'D1869', P. persica x P. amygdalus 'GF 677' and P. cerasus 'CAB 11E') stored for 10 months at -3°C stored better in the dark than under a 16 hr photoperiod. However at 4°C and 8°C a 16 hr photoperiod was better than the dark. A natural light/cold treatment has been used as a rooting pretreatment (36). Light increases survival during cold storage of poplar, Populus sp., and artichoke, Cynara scolymus, callus (40,42). In one FRI experiment, green Pinus radiata shoots which had been cold stored for 1 year in the dark turned white or pale yellow. Such shoots became green again when placed in the normal culture room (2) but we do not know if these shoots are genetically stable. We use a low light level and a long day (16 hr) photoperiod to prevent discolouration and possibly genetic instability, and to avoid dormancy and thus maturation of the shoots.

2.2.4. Humidity and free water. Tissue cultures in cold storage suffer few or no humidity problems; a high humidity (95-100%) is achieved in a well sealed container. Gladwrap (thin plastic film) or Parafilm can be used as an additional seal.

Excess water in cold-stored containers of shoots may become a problem. "Wet" Pinus radiata shoots have been observed to become "translucent" or "vitrified" in the cold store, particularly where they touch the sides of the jar, whereas "waxy" shoots have remained normal. Wet shoots have been found to have less wax (4) and contain more intercellular spaces and free water than waxy shoots (C. Webb, pers. comm.). This may be why waxy shoots appear to cold store better. There is other evidence that reduced water content of plant tissues increases cold resistance (29). A cold storage treatment has also been used to change vitreous shoots to normal shoots (13,39) (cf Debergh et al, Volume 3).

2.2.5. Space. Cold storage uses less space than traditional methods of clonal storage. Lundergan et al. (30) were able to store about 2000 culture tubes of Malus domestica shoots in a 0.28 m^3 (10 cu. ft) refrigerator but calculated that it would require 5.7 ha (14 acres) to accommodate the same number of apple trees in a clonal bank, at a spacing of 4.6 x 6.1 m (12 x 20 ft). Using today's techniques about half this area would now be required in a clonal apple orchard in New Zealand (A. White, pers. comm.). At FRI, 2000 test-tubes take up 3.43 m^2 of cold store shelf space (or 3.83 m^3 room space) whereas in an outdoors clonal bank 2000 hedged ramets would require 4 ha at a spacing of 5.0 m x 4.0 m for a 12 to 15 year life. Wilkins (50) used 3.2 m^2 of shelf space for cold storing 2000 test tubes with Malus, Prunus, Pyrus, Morus spp., cultivars, while Hartney (pers. comm.) used approximately 1.0 m^2 of shelf space for 1000 polycarbonate vials, each with 3 to 4 shoots, giving a capacity of about 3000 shoots per square metre.

3. FACTORS AFFECTING COLD STORAGE

3.1. Requirement for subculture

Generally, cold-stored shoots have been subcultured annually to fresh medium (21,50,52). Some bud and shoot growth has been observed during the first year of cold storage (30,50). To date Pinus radiata shoots have been cold-stored at FRI for 5.5 years with only one subculture, necessitated by malfunction of the cold room. For very long term cold storage a method which does not require subculturing is superior to one using annual subculture. The addition of small volumes of liquid to replenish the media has been advocated by some researchers (34, Preil pers. comm.).

3.2. Length of time in cold storage

Shoots of many species have been successfully cold stored for up to 2 years (30,48,50). Pinus radiata shoots stored for 5.5 years have resumed growth in culture after return to high temperature. However, only shoots stored for up to 17 months were rooted (1). Figure 2 shows Pinus radiata cultures stored for 5.5 years and Prunus insititia stored for 18 months. Both cultures appear normal. Suttle (43) claims that stock cultures of deciduous trees can be maintained indefinitely if subcultured at least once a year. Based on work with strawberry, Fragaria spp., where 200 clones have been cold stored at 4°C for 10 years (12), it seems likely that many species presently being tissue cultured, including forest trees, will be able to be cold stored for long periods of time.

3.3. Survival

Survival of cold stored cultures can be assessed as percent surviving cultures after a return to normal conditions. Results have been generally good but variable between species. The first report (21) showed that 100% survival of grape, Vitis rupestris, plantlets after 9.5 months at 9°C was possible. Hutchinson (pers. comm.) found that Northern Spy apple, Malus domestica, shoots gave 95 to 100% survival after 1 to 6 months in cold storage (4°C) and

a b

FIGURE 2. a) Juvenile Pinus radiata shoots after 5.5 years storage at 4°C. b) Plum rootstock 'pixy', Prunus insititia, plantlets after 18 months storage at 4°C (courtesy of C. Wilkins, England).

55% after 12 months of storage. On the other hand Wanas (48) achieved 95% survival with pear, Pyrus communis var. caucasica, shoots which had been cold stored (4°C) for more than 2 years. Shoots recovered well and were transferrable to soil. High survival and increased proliferation rates have also been found in cold stored cultures of Golden Delicious apple, Malus domestica Borkh (30) and wild cherry, Prunus avium (15). We have found with Pinus radiata that 95 to 100% of shoots survived and grew upon transfer to normal conditions. By contrast rooting varied, depending on clone, from 37 to 85%. This variation is also found in non cold-stored clones.

3.4. Stability

Barlass and Skene (9) state that plants regenerated from cold stored grape, Vitis spp., shoots were phenotypically indistinguishable from unstored plants. Chromosome numbers were consistent with the diploid condition. It is important to examine the original and recovered cultures for any genotypic and phenotypic differences and for differences in survival, growth, and rooting. Pinus radiata plants produced from cold-stored shoots have appeared normal. However, the ability of shoots to grow or turn green does not necessarily mean that they are free of genetic instability. Problems of genetic aberration are likely to be worse if single cells or calluses are stored than with shoot cultures.

Cold resistance could be enhanced by cold storage of tissue cultures. Pinus sylvestris suspension cultures showed better cold acclimation after a short day 2°C treatment (22), and cold resistance in cell lines of tobacco, Nicotiana sylvestris, and capsicum, Capsicum annum was enhanced by cold storage between -3 and +5°C (17).

4. ULTRASTRUCTURE OF COLD-STORED CELLS

There are as yet no reports on the ultrastructure of tissue-cultured shoots of woody plants which have been cold stored, although ultrastructural studies have been done on freeze-preserved cells (51, 54). However, reports of biochemical changes in plant cells which have been subjected to cold temperatures to induce cold hardiness are numerous (14,27). Cold-induced cytological changes in plant cells include vacuolation of the cytoplasm, disruption of plastid and mitochondrial matrices, disorganisation of the cell wall and accumulation of lipids (26,31,35). Other studies, mainly concerned with the plasmalemma, have found either higher aggregation of protein particles (45) or no significant change in the distribution of protein particles (38). Despite these differences it is generally agreed that a cold-induced change in the lipids of plasmalemma does occur, albeit only a small proportion of the polar lipids may be involved in this change (37).

Studies with Pinus radiata shoots cold-stored at 4 ± 1°C for 1 year under a light intensity of 8 $\mu Em^{-2}s^{-1}$ and a 16 hr photoperiod showed striking changes in the fine structure and morphology of cell components of the apical meristem (compare Figs 3 and 4). Ribosomes (R), endoplasmic reticulum (ER), plastids (P), mitochondria (M), dictyosomes (D), nuclear envelope (NE), and plasmalemma (P) of cold stored cells all changed to varying degrees. Free cytoplasmic ribosomes were abundant in the cold-stored tissue imparting a high density to the ground cytoplasm (C). The endoplasmic reticulum and nuclear envelope appeared swollen in places. However, on account of the high density of cytoplasmic ribosomes it was difficult to know whether the endoplasmic

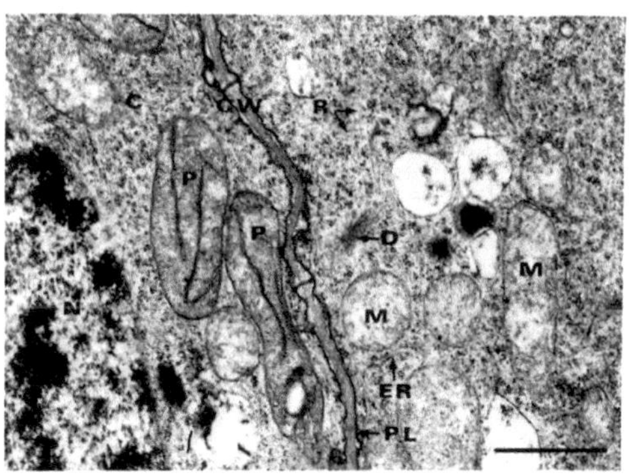

FIGURE 3. Transmission electron micrograph of Pinus radiata shoot tip cells (control at 25°C) showing the normal appearance of various cell structures. Bar = 1 μm.

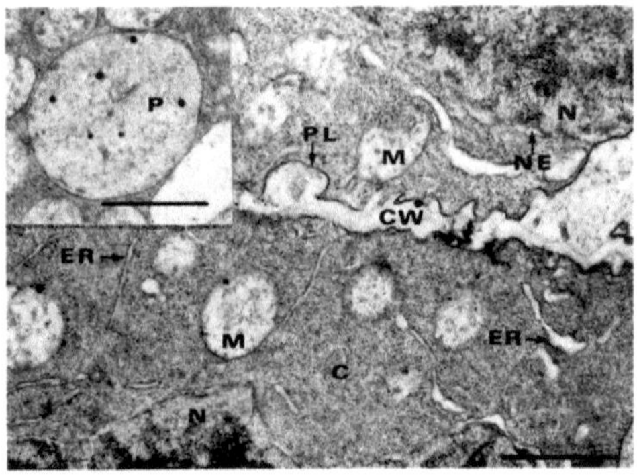

FIGURE 4. Transmission electron micrograph of Pinus radiata shoot tip cells (cold stored at 4°C for 1 year) showing structural changes (for details see text) in various cell components. Bar = 1 µm.

reticulum was granular. Plastids and mitochondria had fewer internal membranes and a more transparent interior. Dictysomes were somewhat disorganised. The plasmalemma appeared intact but invaginated in places and not as closely pressed to the cell wall as in the control. Increased vacuolation of the cytoplasm was observed at low magnification.

Collectively, the changes observed in cold stored Pinus radiata shoots suggest that a shift in metabolism occurs when cells survive prolonged periods of cold temperatures. Pinus radiata seedlings are known to tolerate -5 to -12°C frosts depending on the season and origin of the planting stock (33). Clearly, the ultrastructural changes are not detrimental because cold stored shoots resume growth when returned to normal culture conditions.

5. USES FOR COLD STORAGE

5.1. Short-term uses (less than 1 year)

Cold storage can be particularly useful when space is limiting in the culture room, when labour is in short supply for transfers or during holiday periods. It allows re-arrangement of transfer schedules and makes laboratory management an easier task. In a commercial tissue culture laboratory cold storage aids in smoothing out production peaks and, most importantly, allows the grower more flexibility as to when to plant in the greenhouse (43). While it is possible to produce micropropagated material on a year-round basis, the high cost of heating and lighting greenhouses during the winter months has also led to the use of cold storage during these months (43).

Some other short-term uses of cold storage for tissue cultures of woody plants are: to reduce vitrification (13,39), as a cold pretreatment of explants before sterilization (11,46), as a cold treatment of shoots in culture (20), a pretreatment for rooting (36) and as a treatment with gibberellic acid (GA_3) before potting up plantlets for transfer to the field (24).

5.2. Long-term uses (more than 1 year)

The long-term cold storage of juvenile clones in culture could facilitate clonal forestry on a large scale (2,41) as illustrated in Fig. 5. This figure shows a proposed scheme for testing and breeding superior clones being followed at FRI for *Pinus radiata*. Once the merits of cold-stored cultures and clonal fidelity have been tested superior clones can be multiplied from juvenile shoots (Fig. 5). Freeze preservation is another method suggested to store juvenile conifer tissue samples until the germplasm is proven in a progeny test (18).

In 1983 a trial was initiated at the FRI to cold store and field test 200 *Pinus radiata* clones from 19 of the best FRI control-pollinated families plus one unselected control family as shown in Fig. 5. Juvenile shoots obtained from *in vitro* induced axillary buds on fascicles of each clone are being held in cold storage at 4°C while the clonal test is being evaluated in the field over a 5 to 8 year period. This

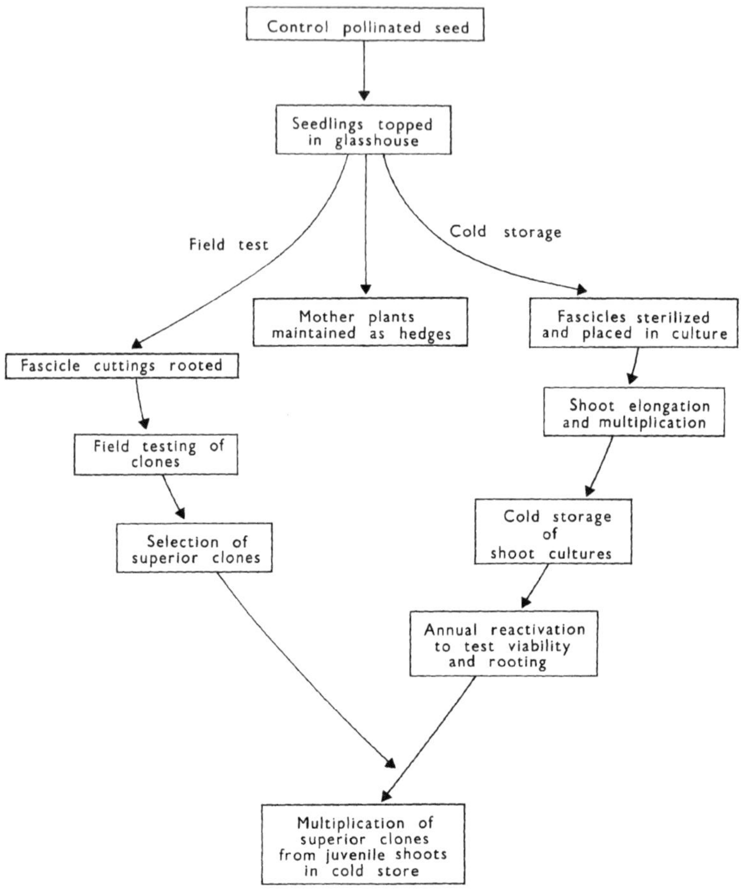

FIGURE 5. Proposed general scheme for cold storage and field testing of juvenile clones (fascicle option). This scheme is being used at FRI for Pinus radiata.

is a novel approach to clonal forestry which may minimize the maturation problems associated with the propagation of trees 6 to 8 years old. Slow initial growth rate, and difficulties in obtaining many cuttings and in rooting are problems associated with the propagation of material from 6 to 8 year old trees (44). At the end of the clonal test superior clones could be multiplied from the juvenile shoots in the cold store.

During our trials a low level of contamination (4%) occurred when 3,300 fascicle explants were sterilized from glasshouse-grown clones. When outside-grown material was sterilized, 90% of the explants were found to be contaminated. Clonal differences in requirement for charcoal or extra cytokinin in the medium, frequency of spontaneous rooting, vigour, and multiplication rate are also evident. These traits can be compared in cultured and field-grown clones to see if any correlations exist.

An advantage of this approach (Fig. 5) is that the original seedling is retained. Clonal mother plants are maintained as hedges to provide annual control cuttings (non tissue-cultured) for comparison with plantlets from reactivated cold-stored clones. This enables maturation state and genetic stability of the cold-stored clones to be assessed. The hedges may also provide cuttings for clonal multiplication if required. If seed instead of needle fascicles is tissue-cultured to produce planting stock and shoots for the cold store then there is no non tissue-cultured control for comparison with annually reactivated cold-stored material. Seed explants also produce predominantly "wet" shoots (4) which may not be suitable for long-term storage (see 2.1.1. and 2.2.4.) while needle fascicle explants produce waxy ones which store much better.

Use of the seed option (Fig. 6) would be feasible in a commercial laboratory provided that the benefits of cold storage had been proven, it had been confirmed that no genetic changes or maturation would occur and vitrification problems had been overcome. In this case adventitious shoots produced directly from embryos would be cold stored or rooted instead of fascicles.

Cold storage of clones in culture also offers an alternative or additional method to hedging of pine grafts or cuttings for germplasm storage (2). Cold storage of tissue cultures uses less space (2.2.5.), avoids losses from

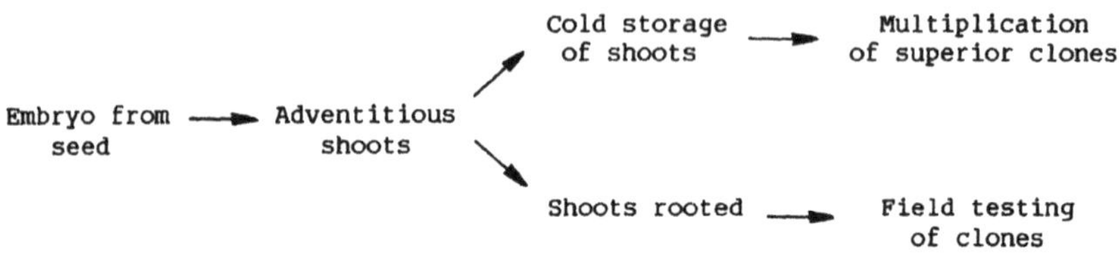

FIGURE 6. Proposed scheme for cold storage and field testing of juvenile clones that would be taken in an applied or commercial situation (seed option).

contamination and natural disasters and may have a lower annual maintenance cost than hedging. However, use of this option should be deferred until the results of long-term cold storage experiments, such as are being conducted at FRI, are available. It will be important to determine if cultures are genetically stable and remain juvenile during cold storage.

Cold stored germplasm has been successfully exported. Several fruit tree cultivars, Malus domestica 'Greensleeves' semi-dwarfing apple scion, Prunus insititia 'Pixy' plum rootstock, and Prunus domestica 'Jefferson's Gage' plum scion, have been transported from England to Mexico and then propagated (50).

6. CONCLUSIONS

Cold storage of tissue cultures (at 0 to 15°C) could prove to have considerable benefits in the conservation of genetic resources of forest trees. Necessary conditions for cold storage of woody cultures vary considerably from species to species. In many cases optimum conditions have not been well defined and the prescriptions given are coincident with environmental facilities available and media and containers in current use in a particular laboratory. A 95% survival of cultures cold-stored for 2 to 5.5 years has been achieved in some cases. However, further work is necessary to confirm the genetic stability of cold-stored tissues. Fine structural changes occur in cold-stored cells but these do not appear to be detrimental. Based on recent results it may be feasible for breeders to use cold storage as a tool for clonal forestry. It is already in use in commercial tissue culture laboratories. This simple system may also be used for the conservation of many other woody plants and forest trees.

ACKNOWLEDGEMENTS

The authors wish to thank Mr C. Wilkins for the use of Figure 2b and unpublished information, Dr J. Hutchinson, Dr V. Hartney, Mr C. Hunter, and W.N. Wanas for unpublished information and Dr L. Withers for a computer search.

8. REFERENCES

1. AITKEN-CHRISTIE J 1984 Micropropagation of Pinus radiata. The Plant Propagator 30: 9-11
2. AITKEN-CHRISTIE J, J GLEED 1985 Uses for micropropagation of juvenile radiata pine in New Zealand. In Proc Int Symp of Recent Advances in Forest Biotechnology, Traverse City, Michigan, pp 47-57
3. AITKEN-CHRISTIE J, KJ HORGAN, DR SMITH 1982 Microformering af skovtraeer. Ugeskrift for Jordbrug 19: 375-382
4. AITKEN-CHRISTIE J, C JONES, S BOND 1985 Wet and waxy shoots in radiata pine micropropagation. Acta Hort 166: 93-100
5. ANON 1982 Institutes working on tissue culture for genetic conservation. In L Withers, ed, International Board for Plant Genetic Resources Report. IBPGR Secretariat Rome, pp 64-69
6. ANON 1982 Les tailles d' Eucalyptus. In Culture de Biomasse Ligneuse, Taillis à Courte Rotation. AFOCEL-Armef, Nangis, pp 96-145
7. BAJAJ YPS 1979 Establishment of germplasm banks through freeze-storage of plant tissue culture and their implications in agriculture. In WR Sharp, PO Larsen, EF Paddock, V Raghavan, eds, Plant Cell and Tissue Culture, Principles and Applications. Ohio State Univ Press, Columbus, pp 745-774
8. BANNIER LJ, P STEPONKUS 1976 Cold acclimation of Chrysanthemum callus cultures. J Amer Soc Hort Sci 101: 409-412
9. BARLASS M, KGM SKENE 1983 Long-term storage in grape in vitro. Int Board Plant Genet Res Newslett 53: 19-21
10. BHOJWANI S, MK RADZAN 1983 Germplasm storage. In S Bhojwani, MK Radzan, authors, Plant Tissue Culture: Theory and Practice. Elsevier, Amsterdam, pp 373-385
11. BONGA JM 1981 Organogenesis in vitro of tissues from mature conifers. In Vitro 17: 511-518
12. BOXUS P, P DRUART 1980 Micropropagation, an industrial method of producing quality plants true to type and at a reasonable price. In F Sala, B Parisi, R Cella, O Ciferri, eds, Plant Cell Cultures: Results and Perspectives. Elsevier/North-Holland Biomed Press, pp 265-269

13. BOXUS P, P DRUART, E BRASSEUR 1979 Tissue culture of fruit species. Report for 1979 of the State Centre for Agricultural Research. Gembloux, Belgium, pp 116-117
14. BROWN GN 1978 Protein synthesis mechanisms relative to cold hardiness. In PH Li, A Sakai, eds, Plant Cold Hardiness and Freezing Stress. Academic Press, New York, pp 153-163
15. CORNU D, C CHAIX 1981 Multiplication par culture in vitro de merisiers adultes (Prunus avium). In Proc IUFRO Sect S2 01 5. Int Workshop "In Vitro" Cultivation for Tree Species, Fontainbleu, France, pp 71-79
16. DAMIANO C 1979 Cold storage of in vitro strawberry cultures and the resumption of multiplication. Ann Istit Sper Frutticolt 10: 53-58
17. DIX PJ, HE STREET 1976 Selection of plant cell lines with enhanced chilling resistance. Ann Bot 40: 903-910
18. FARNUM P, R TIMMIS, JL KULP 1983 Biotechnology of forest yield. Science 219: 694-702
19. FRANCLET A, M BOULAY 1982 Micropropagation of frost resistant Eucalypt clones. Aust For Res 13: 83-89
20. FEBVRE E 1981 Froid et croissance de boutures de saules (Salix babylonica L.) cultivées in vitro. In Proc IUFRO Sect S2 01 5. Int Workshop "In Vitro" Cultivation for Tree Species, Fontainbleu, France, pp 115-118
21. GALZY R 1969 Recherches sur la croissance de Vitis rupestris (Scheele) sain et court noué cultivé in vitro à different temperatures. Ann Phytopathol 1: 149-166
22. HELLERGREN J 1983 Cold acclimation of suspension cultures of Pinus sylvestris in response to light and temperature treatments. Plant Physiol 72: 992-995
23. HIRAOKA N, T KODAMA 1984 Effects of non-frozen cold storage on the growth, organogenesis and secondary metabolism of callus cultures. Plant Cell Tiss Org Cult 3: 349-357
24. HOWARD BH, HV OEHL 1981 Improved establishment of in vitro-propagated plum micropropagules following treatment with GA_3 or prior chilling. J Hortic Sci 56: 1-7
25. HUNTER CS 1986 In vitro propagation and germplasm storage of Cinchona. In LA Withers, PG Alderton, eds, Plant Tissue Culture and its Agricultural Applications. Butterworths, England, pp 291-300
26. ILKER R, AJ WARING, JM LYONS, RW BREIDENBACH 1976 The cytological responses of tomato-seedling cotyledons to chilling and the influence of membrance modifications upon these responses. Protoplasma 90: 229-252
27. KACPERSKA-PALACZ A 1978 Mechanism of cold acclimation in herbaceous plants. In PH Li, A SAKAI, eds, Plant Cold Hardiness and Freezing Stress. Academic Press, New York, pp 139-152
28. KARTHA KK 1981 Meristem culture and cryopreservation. In TA Thorpe, ed, Methods and Applications in Agriculture. Academic Press, New York, pp 181-211
29. LI PH, CJ WEISER 1971 Increasing cold resistance of stem sections of Cornus stolonifera by artificial dehydration. Cryobiology 8: 108

30. LUNDERGAN C, J JANICK 1979 Low temperature storage of in vitro apple shoots. Hortic Sci 14: 514
31. LYONS JM 1973 Chilling injury in plants. Ann Rev Plant Physiol 24: 445-466
32. MARINO G, P ROSATI, F SAGRATI 1985 Storage of in vitro cultures of Prunus rootstocks. Plant Cell Tiss Org Cult 5: 73-78
33. MENZIES MI, DG HOLDEN, LM GREEN, DA ROOK 1981 Seasonal changes in frost tolerance of Pinus radiata seedlings raised in different nurseries. NZ J For Sci 11: 100-111
34. MULLIN RH, DE SCHLEGEL 1976 Cold storage maintenance of strawberry meristem plantlets. Hortscience 11: 100-101
35. PLATT-ALOIA KA, WW THOMSON 1976 An ultrastructural study of two forms of chilling-induced injury to the rind of grapefruit (Citrus paradisi, Mac. Fed.). Cryobiol 13: 95-106
36. POISSONIER M, A FRANCLET, MJ DUMANT, JY GAUTRY 1980 Enracinement de tigelles in vitro de Sequoia sempervirens. Ann AFOCEL 1980: 231-253
37. RAISON JK, LC WRIGHT 1983 Thermal phase transitions in the polar lipids of plant membranes. Their induction by disaturated phospholipids and their possible relation to chilling injury. Biochem Biophys Acta 731: 69-78
38. ROBARDS, AW, DT CLARKSON 1984 Effects of chilling temperatures on root cell membrances as viewed by freeze-fracture electron microscopy. Protoplasma 122: 75-85
39. RUGINI E, DC VERMA 1982 Micropropagation of difficult to propagate almond (Prunus amygdalus) cultivar. Plant Sci Lett 28: 273-281
40. SAKAI A, Y SUGAWARA 1973 Survival of poplar callus at super-low temperatures after cold acclimation. Plant Cell Physiol 14: 1201-1204
41. SMITH DRS, KJ HORGAN, J AITKEN-CHRISTIE 1982 Micropropagation of Pinus radiata for afforestation. In A Fujiwara, ed, Plant Tissue Culture 1982. Japanese Assn for Plant Tissue Culture, Tokyo, pp 723-724
42. SUGAWARA Y, A SAKAI 1978 Cold acclimation of callus cultures of jerusalem artichoke. In PH Li, A Sakai, eds, Plant Cold Hardiness and Freezing Stress. Academic Press, New York, pp 197-210
43. SUTTLE GRL 1983 Micropropagation of deciduous trees. In TH Hartmann, AT Leiser, W Roberts, eds, Int Plant Propagators Soc Combined Proc Vol 33. IPPS, Boulder, pp 46-49
44. SWEET GB, LG WELLS 1974 Comparison of the growth of vegetative propagules and seedlings of Pinus radiata. NZ J For Sci 4: 399-409
45. TOIVIO-KINNUCAN MA, HH CHEN, PH LI, C STUSHNOFF 1981 Plasma membrane alternations in callus tissues of tuber-bearing Solanum species during cold acclimation. Plant Physiol 67: 478-483
46. VIEITEZ AM, A BALLESTER, ML VIEITEZ, E VIEITEZ 1983 In vitro plantlet regeneration of mature chestnut. J Hortic Sci 58: 457-463
47. VILLALOBOS VM 1983 The early events associated with organogenesis in cultured radiata pine cotyledons. PhD Thesis, Dep Biology, Univ Calgary, Canada, pp 69-71

48. WANAS WH, JA CALLOW, L WITHERS 1986 Growth limitation for the conservation of pear genotypes. In LA Withers, PG Alderton, eds, Plant Tissue Culture and its Agricultural Applications. Butterworths, England, pp 285-290
49. WILKINS CP, JH DODDS 1983 Tissue culture conservation of woody species. In JH Dodds, ed, Tissue Culture of Trees. AVI Publishing Co Inc, Connecticut, pp 113-136
50. WILKINS CP, JH DODDS 1985 Storage in vitro of temperate fruit trees. Int Board Plant Genet Res Newslett (in press)
51. WITHERS LA 1978 A fine-structural study of the freeze-preservation of plant tissue cultures. II. The thawed state. Protoplasma 94: 235-247
52. WITHERS LA 1982 Storage of plant tissue cultures. In LA Withers, JT Williams, eds, Crop Genetic Resources - The Conservation of Difficult Material. IUBS/IBPGR, Paris, pp 49-82
53. WITHERS LA 1984 Freeze preservation of cells. In IK Vasil, ed, Cell Culture and Somatic Cell Genetics of Plants Vol 1. Academic Press, Florida, pp 608-620
54. WITHERS LA, MR DAVEY 1978 A fine-structural study of the freeze preservation of plant tissue cultures. I. The frozen state. Protoplasma 94: 207-209

16. CRYOPRESERVATION OF WOODY SPECIES

THH CHEN and KK KARTHA

1. INTRODUCTION

The irretrievable loss of the naturally occurring germplasm is a problem of great concern to tree breeders. Many countries are facing serious gene pool losses in populations of some valuable tree species (4). With increasing world demand for forest products and continuing deforestation, the conservation of genetic resources has become an urgent issue. The preservation of available germplasm for use in future tree improvement is imperative.

Seed storage has been the most common practice for maintaining plant germplasm throughout our agricultural history. For plant species in which germplasm is maintained in the form of seeds, cold storage at temperatures below -18 C and moisture contents of 5-7% can greatly extend the seed viability. For long-term seed storage, seed are stored in liquid nitrogen (27). However, seed storage cannot apply to the preservation of some plant species. First, some species produce seeds that cannot be dried without immediate injury and viability loss (i.e., recalcitrant seeds) (11,27). Second, most long-lived forest trees, including angiosperms and gymnosperms, do not produce seed for several years. Third, plants which do not produce seed naturally must be stored in a vegetative form as tubers, roots, cuttings, etc.

Maintaining vegetatively propagated plants in the field is labor-intensive and expensive, and also often results in the loss of valuable genetic material due to disease and pest problems. Other problems are associated specifically with the conservation of forest trees by conventional means. Among them, the most obvious one is the immense land area required. As most forest trees are genetically heterozygous, it is necessary to conserve many trees at the same time to maintain as much as possible of the genetic variation within a population. In view of the problems associated with conventional preservation of woody species, alternative strategies are needed. Recently, there has been increasing interest in the potential for application of *in vitro* techniques to plant genetic conservation.

In applying tissue culture techniques to the special problems posed by plant genetic conservation, some techniques are more readily applicable than others. The primary reason for this is the observation of genetic variation among cells or plants derived from tissue cultures (17). The ultimate goal of any tissue culture system employed for genetic conservation is to preserve indefinitely specific and unique genotypes. Among the available tissue culture techniques, meristem culture seems to be most suitable for the purpose of germplasm preservation, since progeny populations regenerated from in vitro culture of meristems contain very little genetic variation (11).

With increasing interest in genetic engineering of forest tree species (4,20,29) and in vitro selection, the preservation of cultured cells with unique characteristics is becoming more important. Cryopreservation (i.e., storage at the temperature of liquid nitrogen) of ornamental, forest and fruit tree germplasm in vitro is a potentially reliable means of eliminating the high cost of conventional labor-intensive nursery practices and the problem of mutation over cell generations in tissue cultures.

2. GENERAL PROCEDURE

2.1. Background

Cryogenic storage of biological material involves reduction of metabolic activities by lowering the sample temperature to approach that of liquid nitrogen (-196°C) while retaining viability. When plant cells are exposed to ultra-low temperature of -196°C, cellular metabolic activities arrest due to unavailability of liquid water. Consequently, genetic changes are minimized. Unfortunately, only a few plant materials can easily be frozen to such a low temperature without adversely affecting viability. Very few examples of successful cryopreservation of plant material involved simple transfer from room temperature to liquid nitrogen and back. Very winter hardy woody plant materials, when sufficiently dehydrated by extracellular freezing, were found to survive upon exposure to liquid nitrogen temperature. Winter vegetative buds of several fruit trees prefrozen to temperatures ranging from -30°C to -50°C, have survived immersion in liquid nitrogen (16, 23). To our knowledge, there is no example of actively growing plant cells surviving liquid nitrogen immersion without cold preconditioning or the application of cryoprotective compounds. It is normally necessary to carry out several specific operations prior to, during and after, freezing and thawing.

The discovery of cryoprotective compounds has allowed significant advances in the development of cryopreservation techniques for the long term storage of living samples (8). The majority of our current techniques for cryopreservation of plant materials have been derived by modification of those applied to animal materials. The mechanisms of freezing injury and cryoprotection are still poorly understood.

The procedures provided below are generalized protocols used in the cryopreservation of plant materials, not all of which have been adequately tested on woody species.

2.2. Plant materials

As mentioned earlier, meristems and cell cultures are the candidates for cryopreservation. Various considerations regarding cryopreservation of meristems have been noted by Kartha (10,11,12, 13,15). For cell cultures, high viability was obtained by various workers when cells in a late lag or early exponential growth division phases were used for cryopreservation (24,28,34). The sensitivity of cells in early lag phase or stationary phase to freeze-thaw stress is attributable to the greater cell size, vacuolar volume and water content. In the case of alkaloid producing and non-producing cell cultures of *Catharanthus roseus*, cells sampled 3-4 days after subculture to fresh medium were more resistant to freeze and thaw injury than those sampled several days later (5, 14). The enhanced survival of the earlier sample may be attributable to the dense cytoplasm present in the predominantly small sized cells.

2.3. Pregrowth conditions

Preconditioning of plant material prior to cryopreservation has been beneficial in increasing the viability in certain species. Cold treatment (4°C for 3 days) of clonal carnation plants dramatically increased the survival of shoot apices after recovering from liquid nitrogen storage (26). Supplementation of the culture medium with osmotically active compounds such as mannitol, sorbitol, sucrose and proline has resulted in an increase in freezing resistance. For example, preculture of Acer pseudoplatanus and Capsicum annuum cells in mannitol enriched medium reduced cell volume and increased post-freezing survival (4) . However, this treatment did not enhance the survival of Nicotiana sylvestris cells (18), or alkaloid-producing cell cultures of Catharanthus roseus (5).

Preculturing cells or meristems in a DMSO-supplemented medium for a period of 24 to 48 hours prior to freezing has resulted in increased survival (12,13,14). However, DMSO also exhibits cytotoxicity in the preculture medium at concentrations exceeding 5%. For example, preculturing the alkaloid non-producing C. roseus cells for 24 hours in medium supplemented with 5% DMSO resulted in 34% viability loss prior to freezing. However, survival of the remaining cells after cryopreservation has been very high compared to those which were not precultured (14). On the other hand, a similar method was not applicable to the cryopreservation of 3 cell lines of the alkaloid producing C. roseus (5).

Sorbitol has also been used either as an osmotic agent in the preculture medium or as the sole cryoprotectant. Weber et al. (33) successfully cryopreserved cell cultures of Glycine max, Datura innoxia, Brassica napus and Daucus carota and cell culture derived protoplasts of G. max. Alkaloid producing C. roseus cell cultures could not be cryopreserved using sorbitol as the only cryoprotectant (5). They could be successfully cryopreserved using a regime involving preculturing the cells in nutrient medium containing 1 M sorbitol for 6-20 hours as well as using 1 M sorbitol + 5% DMSO as cryoprotectants. Although there was considerable survival subsequent to preculturing the cells in medium supplemented with 1 M concentrations of glucose, trehalose, and sucrose, the overall survival and recovery was much superior when sorbitol was used as the preculturing agent (5).

The sensitivity of cultured plant cells to sorbitol differs markedly between species. However, it has been observed that many types of cultures can be successfully cryopreserved by paying careful attention to sorbitol concentration and preculture duration in the sorbitol-enriched medium, in addition to strictly controlling the cooling program (Kartha, unpublished results).

2.4. Cryoprotectant treatment

The application of cryoprotectant, singly or in combination, is essential for cryopreservation of plant materials except for very cold hardy samples (8). Cryoprotectants protect living cells against the deleterious effects of freezing and thawing. The cryoprotectants generally used for freezing plant materials fall into two categories, namely, permeating and non-permeating, with respect to the plasma membrane. The most commonly used permeating additives are DMSO and glycerol. The non-permeating compounds include sugars, sugar alcohols and high molecular weight additives, such as polyvinylpyrolidone, polyethylene glycol, dextran, and hydroxyethyl starch. Most of them exhibit cytotoxicity at relatively higher concentrations.

In instances where application of a single cryoprotectant does not result in high survival, mixtures of cryoprotectants have been beneficial (5,6,8,39). For example, Ulrich et al. (31) demonstrated the beneficial effects of using a mixture of cryoprotectants involving PEG, glucose, and DMSO (10-8-10 v/v) in the cryopreservation of sugarcane callus. In subsequent studies several other cell or callus lines have been cryopreserved using the same mixture (8). Withers and associates (34-40) have applied a cryoprotectant mixture of 0.5M DMSO + 0.5M glycerol + 1M sucrose or 0.5M DMSO + 0.5M glycerol + 1M proline, to a wide variety of cell cultures.

2.5. Freezing

2.5.1. Slow freezing. When plant cells are exposed to progressive temperature reduction, ice formation is usually initiated extracellularly. It is generally believed that the plasma membrane acts as a barrier that prevents the ice from seeding the cytoplasm. Because there is no ice nucleator in the cytoplasm, it becomes supercooled. As the temperature is further lowered, the concentration of extracellular liquid increases as more liquid water is converted into ice. Since the vapor pressure of the frozen solution is lower than the same concentration in the supercooled liquid, the vapor pressure of slowly cooled cells reach equilibrium with the external ice by the efflux of water. Slow freezing is thought to prevent the intracellular ice formation, thus preventing freezing injury. It is generally agreed that upon extracellular freezing the cytoplasm will be effectively concentrated and plant cells survive better when adequately dehydrated. There are different ways of achieving protective dehydration, including slow cooling at constant or varying cooling rates, or holding the samples at one or more intermediate subzero temperatures.

The success of slow freezing depends on a number of factors including cooling rates, pretreatments and cyroprotection, type and physiological states of the material, and the temperature prior to immersion in liquid nitrogen. The most commonly used methods for the cryopreservation of plant cells generally involve the slow cooling at a constant rate of 0.5 to 2°C/min to terminal temperatures between -30°C to -40°C followed by storage in liquid nitrogen. The optimal cooling rates depend on the type of cells, the cryoprotectant used and preculturing methods. Each plant specimen is characterized by its unique narrow range of optimal cooling rates. Once the cells are frozen to the terminal temperature, allowing sufficient dehydration, they may be immersed in liquid nitrogen and stored indefinitely.

Another way to achieve adequate protective dehydration is sequential slow cooling to one or more intermediate sub-zero temperatures, prior to plunge into liquid nitrogen. At any given freezing temperature, the degree of dehydration is increased with time of holding. Therefore, holding samples at a sub-zero temperature will increase dehydration. The most appropriate holding temperatures will be determined by the physiological states of the specimen, the pre-growth conditions and the cryoprotectant in use, but are generally in the region of -30° to -40°C (35,39).

2.5.2. Rapid freezing. Rapid freezing is generally accomplished by direct immersion of the cryoprotectant-treated specimens in liquid nitrogen. The cooling rate is in the order of several hundred degrees per minute. At such rapid cooling rates, the intracellular fluids do not have sufficient time to equilibrate with the external ice. The ultra-rapid cooling may prevent the growth of intracellular ice crystals by rapidly passing the cells through the temperature zone in which lethal ice crystal growth occurs. The result of the presence of many small ice crystals in the cytoplasm is believed to be less harmful than ice crystals which may be induced by slow freezing. Rapid freezing method have been used to cryopreserve shoot-tips of carnation (25), potato (9), Brassica napus (36), and peanut and chickpea (1), although the level retention of viability varies among species. Most of the studies of cryopreservation using rapid freezing have been done with cryoprotectant-treated meristems or shoot -tips. Rapid freezing has been generally unsuitable for the cryopreservation of cell cultures.

2.5.3. Droplet freezing. The droplet freezing method was originally developed by Kartha *et al.* (15) for the preservation of cassava meristems. The basis of the droplet freezing method is essentially the same as that of slow freezing. However, the cryoprotectant solution is dispensed in droplets of 2-3 μl on an aluminum foil contained in a petri dish. Each meristem is then transferred to a droplet and frozen by slow cooling (0.5°C/min) to a sub-zero temperature between -20°C and -40°C prior to immersion in liquid nitrogen. Probably, freezing plant specimens in droplets on aluminum foil, which has an efficient thermal conductivity, results in more homogeneous cooling of the sample. Also, since a very small amount of cryoprotectant solution is used for each meristem, a uniform freezing of cell occurs and lesser amount of extracellular ice forms, thus avoiding the excess mechanical pressure exerted on cell walls as the ice propagates. The droplet freezing technique has been applied to the cryopreservation of both cassava meristems (15) and cultured cells (Chen and Kartha, unpublished results) with better results than using conventional slow freezing in glass or plastic cryogenic ampules.

2.6. Storage

One of the important criteria for storage of frozen plant samples is the maintenance of a sufficiently low temperature, free of temperature fluctuation. Plant cells cooled below their eutectic point are subject to possible ice crystal growth or recrystallization during storage. Small ice crystals have higher surface free energies than large ice crystals and so tend to reduce their surface energies by growing or by fusing with other small crystals (19). In pure water, recrystallization occurs at temperatures as low as -100°C. In gels and concentrated aqueous solutions, it is expected to occur at higher temperatures (19). The rate of ice crystal growth depends on both the size of the crystals formed during cooling and the subsequent temperatures to which they are exposed. The rate is higher in smaller crystals and at higher temperatures. Little information is available on the effect of higher storage temperatures on the retention of viability. Results from pea meristems indicated a gradual loss of viability at a storage temperature of -86°C as opposed to no such viability loss at -196°C (Kartha, unpublished). Generally, frozen plant samples are stored in liquid nitrogen (-196°C), which is cold enough to prevent growth of ice crystals during storage.

Most reports on cryopreservation of plant specimens in liquid nitrogen deal only with storage for a short period of time. The frequency of plant regeneration after various periods of storage in liquid nitrogen have been reported only for pea and strawberry meristems (12,13). Most people believe that if the specimen can withstand liquid nitrogen temperature for the first few minutes, it should be able to survive prolonged storage.

2.7. Thawing

Rapid thawing has been adopted in most cryopreservation protocols. The basis of applying rapid thawing is to avoid the damaging ice recrystallization which may occur during slow warming. In general, thawing is carried out by removing the sample from liquid nitrogen storage and immersing in a 35-40°C water bath. A complete thawing (the disappearance of ice crystals) is generally achieved within 1-2 minutes.

2.8. Viability testing

In the case of meristems, the tests for the retention of viability is the development of the cryopreserved meristems into shoot or plantlets. However, other forms of regrowth are often observed in the forms of callus or leaves (11,15). Similarly, the most accurate test for assessing viability of cryopreserved cells is simply regrowth, although the data may require days or months to obtain.

Rapid viability testing using vital stains, such as the triphenyltetrazolium chloride (TTC), fluorescein diacetate (FDA), and Evan's blue, may be used to estimate viability within hours. Each rapid viability test method has shortcomings, however, and thus a combination of a number of assays would be a safer approach. In some cases high TTC or FDA readings have been observed immediately after thawing while no regrowth could be obtained from the same samples. Conversely, samples with very low TTC readings have produced vigorous regrowth. These differences between viability tests and regrowth may be related to the varying ability of the cultures to repair damage encountered during freezing and thawing.

2.9. Post-thaw regrowth

In most cases, the objective of the cryopreservation operation is to reestablish the culture of interest from the thawed sample. The factors which affect post-thaw regrowth include removal of cryoprotectant, mode of culture, and culture medium composition (39).

Conventionally, prior to assessing viability, thawed cell cultures are washed with chilled nutrient medium and returned to culture either in liquid or semi-solid medium. This technique has proven to be satisfactory for a number of cell cultures. However, it has been reported that viability of suspension cultured sugarcane cells protected with a mixture of cryoprotectants (PEG + glucose + DMSO) was poorer when they were washed with a cold solution than with a 22°C wash (7). The washing step has been found to be deleterious to a few cell cultures such as the alkaloid-producing _C. roseus_ (5), _Zea mays_, Paul's Scarlet rose and _Acer pseudoplatanus_ (37,38). According to Withers (39), washing of post-thawed samples caused two harmful effects: 1) removal of a chemical factor or factors which are supportive of recovery and growth; and 2) physical injury, perhaps related to deplasmolysis. Therefore, she suggested that unless absolutely essential, the washing step should be avoided.

While a special post-thaw culturing technique had to be developed to enhance regrowth of the alkaloid-producing cultures of _C. roseus_ (5), the alkaloid non-producing cell culture of the same species could be successfully returned to either liquid or semi-solid medium immediately following post-thaw wash (14). Other researchers have also reported special regrowth techniques to obtain reasonable regrowth (39).

3. EXAMPLES OF CRYOPRESERVATION OF WOODY PLANT MATERIAL

Table 1 summarizes reported examples of cryogenic storage of woody plant materials. It is obvious that efforts in this direction have not been very extensive. However, the list of successful examples does include suspension cultures, callus cultures and meristematic tissues. Although meristems seem to be the best candidate for germplasm preservation, only two reports exist detailing successful cryopreservation of such tissues (16,23). Since plant regeneration from suspension and callus cultures (4,29) is possible, in some woody species, other tissue types may also be useful in terms of germplasm preservation if genetic stability could be ensured in the regenerants.

In most cases of cryopreservation of woody plant samples, plant regeneration from cultured cells recovered from cryopreserved samples was not reported. There has been only one report of successful regeneration of whole plants after liquid nitrogen storage (32). This report, however, notes the importance of using morphogenetically competent cultures for the long-term storage of tree germplasm. As noted earlier, variation arising from cell culture is a widespread phenomenon in tissue culture systems, and some off-type plants will probably be found in the regenerated plant populations. However, in most cases, most of the regenerants should be identical to the parent genotype. Thus, while tissue culture-induced variation is undesirable for germplasm preservation, the effort required to eliminate deviant regenerants may not preclude the use of tissue and cell cultures for that purpose.

The regenetation of plants in woody species generally occurs via the induction of multiple bud formation, followed by shoot elongation and rooting (4,29). Thus, the buds induced _in vitro_ or meristems derived from those buds are also potential candidates for cryopreservation. Research exploring the use of _in vitro_ buds and/or meristems from conifers for cryopreservation is currently underway at the Plant Biotechnology Institute, National Research Council of Canada, Saskatoon. Another candidate for cryopreservation of tree germplasm is zygotic embryos, especially hybrid embryos. Preliminary results to store conifer zygotic embryos in liquid nitrogen indicate that the zygotic embryos of white spruce (_Picea glauca_) can be viably stored in liquid nitrogen.

4. POTENTIAL APPLICATION OF CRYOPRESERVATION IN TREE IMPROVEMENT

The potential applications of tissue culture techniques to tree improvement or the enhancement of forest production have been reviewed

(4,20,29). Figure 1 represents an over-simplified scheme of the possible application of cryopreservation in forestry.

One of the problems in the tissue cultures of woody plants is the poor plant regeneration capacity of cultures derived from explants of mature trees. By regenerating plants from mature tissues, we might be able to select truely superior trees from forests or orchards. These could be used to exert a stronger and tighter selection pressure in breeding programs, which could result in much more rapid breeding progress. However, technology for routine regeneration of plants from mature tissue is currently not well developed.

The cryogenic storage of juvenile tissues would bypass the difficulties of regeneration of plants from mature tissues. Long-term storage of juvenile clones could allow the researchers to perform field evaluation of the same clones. At the end of the test period (while the clones in the field have passed their juvenile stage and lost their potential for in vitro cloning), the investigators could pull the juvenile clones of interest out of cryogenic storage and multiply them to provide seedlings for nurseries. The cold storage technique described in Chapter 15 is another method to achieve this objective in the interim.

Although plants that originated from hybrid seeds generally perform better than the progeny of open-pollinated seeds, such breeding is very costly and it often takes many years to collect enough seeds to generate enough seedlings for a plantation (20). An alternative may be to produce clones from limited supplies of hybrid seeds. Cryogenic storage of some of this material either as zygotic embryos, seeds, or in vitro induced buds or meristems, could provide a continuous supply of superior clones of desirable genotypes. On the other hand, cryopreservation could also be employed as a versatile tool not only for preserving the existing natural variability in the germplasm but also for conserving the variability induced in vitro through biotechnological manipulations.

5. REFERENCES

1. BAJAJ YPS 1983 Production of normal seeds from plants regenerated from the meristems of Arachis hypogaea and Cicer arietinum cryopreserved for 20 months. Euphytica 32: 425-430

2. BINDER WD, B ZAERR 1980 Freeze preservation of suspension cultured cells of a gymnosperm, Douglas-fir. Cryobiology 17:624 (Abst)
3. BINDER WD, B ZAERR 1980 Freeze preservation of suspension cultured cells of a hardwood, poplar. Cryobiology 17: 624-625 (Abst)
4. BONGA JM 1977 Application of tissue culture in forestry. In J Reinert, YPS Bajaj, eds, Applied and Fundamental Aspects of Plant Cell, Tissue, and Organ Culture. Springer-Verlag, Berlin, pp 93-108
5. CHEN, THH, KK KARTHA, NL LEUNG, GWG KURZ, KB CHATSON, F CONSTABEL 1984 Cryopreservation of alkaloid-producing cell cultures of periwinkle (Catharanthus roseus). Plant Physiol 75: 726-731
6. FINKLE BJ, J ULRICH 1979 Effects of cryoprotectants in combination on the survival of frozen sugarcane cells. Plant Physiol 63: 598-604
7. FINKLE BJ, J ULRICH 1982 Cryoprotectant removal as a factor in the survival of frozen rice and sugarcane cells. Cryobiology 19: 329-335
8. FINKLE BJ, ME ZAVALA, J. ULRICH 1985 Cryoprotective compounds in the viable freezing of plant tissues. In KK Kartha, ed, Cryopreservation of Plant Cells and Organs, CRC Press, pp 75-113
9. GROUT BWW, GG HENSHAW 1978 Freeze preservation of potato shoot-tip cultures. Ann Bot 42: 1227-1229
10. KARTHA KK 1982 Cryopreservation of germplasm using meristem and tissue culture. In DT Tomes, BE Ellis, PM Harney, KJ Kasha and RL Peterson, eds, Application of Plant Cell and Tissue Culture to Agriculture and Industry, Univ Guelph, Guelph, Ontario, pp 136-161
11. KARTHA KK 1985 Meristem culture and germplasm preservation. In KK Kartha, ed, Cryopreservation of Plant Cells and Organs, CRC Press, pp 115-134
12. KARTHA KK, NL LEUNG, OL GAMBORG 1979 Freeze-preservation of pea meristems in liquid nitrogen and subsequent plant regeneration. Plant Sci Lett 15:7-15
13. KARTHA KK, NL LEUNG, K. PAHL 1980 Cryopreservation of strawberry meristems and mass propagation of plantlets. J Am Soc Hortic Sci 105:481-484
14. KARTHA KK, NL LEUNG, P GAUDET-LAPRAIRIE, F CONSTABEL 1982 Cryopreservation of periwinkle Catharanthus roseus cells cultured in vitro. Plant Cell Rep 1: 135-138
15. KARTHA KK, NL LEUNG, LA MROGINSKI 1982 In vitro growth responses and plant regeneration from cryopreserved meristems of cassava (Manihot esculenta Crantz). Z Pflanzenphysiol 107:133-140
16. KATANO M, A ISHIHARA, A SAKAI 1983 Survival of dormant apple shoot tips after immersion in liquid nitrogen. HortScience 18:707-708
17. LARKIN PJ, WR SCOWCROFT 1981 Somaclonal variation - a noval source of variability from cell cultures. Theor Appl Genet 60: 197-214
18. MADDOX AD, FG GONSALVES, R. SHIELDS 1983 Successful cryopreservation of suspension cultures of three Nicotiana species at the temperature of liquid nitrogen. Plant Sci Lett 28:157-162
19. MAZUR P 1970 Cryobiology: The freezing of biological systems. Science 168:939-949
20. POWLEDGE TM 1984 Biotechnology touches the forest. Biotechnology 2:763-772
21. NAG KK, HE STREET 1975 Freeze-preservation of cultured plant cells. I. The pretreatment phase. Physiol Plant 34:254-260

22. SAKAI A, Y SUGAWARA 1973 Survival of poplar callus at super-low temperatures after cold acclimation. Plant Cell Physiol 14: 1201-1204
23. SAKAI A, Y NISHIYAMA 1978 Cryopreservation of winter vegetative buds of hardy fruit trees in liquid nitrogen. HortScience 13:225-227
24. SALA F, R CELLA, F ROLLO 1979 Freeze-preservation of rice cells. Physiol Plant 45:170-176
25. SEIBERT M 1976 Shoot initiation from carnation shoot apices frozen to -196°C. Science 191:1178-1179
26. SEIBERT M, PG WETHERBEE 1977 Increased survival and differentiation of frozen herbaceous plant organ cultures through cold treatment. Plant Physiol 59: 1043-1046
27. STANWOOD PC 1985 Cryopreservation of seed germplasm for genetic conservation. In KK Kartha, ed, Cryopreservation of Plant Cells and Organs, CRC Press, pp 200-226
28. SUGAWARA Y, A SAKAI 1974 Survival of suspension cultured sycamore cells cooled to the temperature of liquid nitrogen. Plant Physiol 52:722-724
29. THORPE TA, S BIONDI 1984 Conifers. In WR Sharp, DA Evans, PV Ammirato, Y Yamada, eds, Handbook of Plant Cell Culture, Vol 2, Macmillian Pub Co, pp 435-470
30. TOWILL LE, P MAZUR 1974 Effects of cooling and warming rates on survival of frozen Acer saccarum tissue cultures. Plant Physiol 54 (Suppl):10
31. ULRICH JM, BJ FINKLE, PH MOORE, H GINOZA 1979 Effect of a mixture of cryoprotectants in attaining liquid nitrogen survival of callus cultures of a tropical plant. Cryobiology 16:550-556
32. ULRICH JM, RA MICKLER, BJ FINKLE, DF KARNOSKY 1984 Survival and regeneration of American elm callus cultures after being frozen in liquid nitrogen. Can J For Res 14:750-753
33. WEBER G, EJ ROTH, HG SCHWEIGER 1983 Storage of cell suspensions and protoplasts of Glycine max (L.) Merr, Brassica napus (L.), Datura innoxia (Mill.) and Daucus carota (L.) by freezing. Z Pflanzenphysiol 109: 29-39
34. WITHERS LA 1978 The freeze-preservation of synchronously dividing cultured cells of Acer pseudoplatanus. Cryobiology 15:87-92
35. WITHERS LA 1980 Tissue Culture Storage for Genetic Conservation. Tech Report, Int Board Plant Genetic Resources, FAO, United Nations, Rome
36. WITHERS LA 1982 The development of cryopreservation techniques for plant cell, tissue and organ cultures. In Fujiwara, ed, Proc Fifth Int Congr Plant Tissue Culture, Tokyo, pp 793-794
37. WITHERS LA, PJ KING 1979 Proline- A novel cryoprotectant for the freeze-preservation of Zea mays. Plant Physiol 64:675-678
38. WITHERS LA, PJ KING 1980 A simple freezing unit and routine cryopreservation method for plant cell cultures. Cryo Lett 1:213-220
39. WITHERS LA 1985 Cryopreservation of cultured plant cells and protoplasts, In KK Kartha, ed, Cryopreservation of Plant Cells and Organs. CRC Press, pp 243-267
40. WITHERS LA, HE STREET 1977 The Freeze-preservation of cultured plant cells: III The pregrowth phase, Physiol Plant 39:171-178

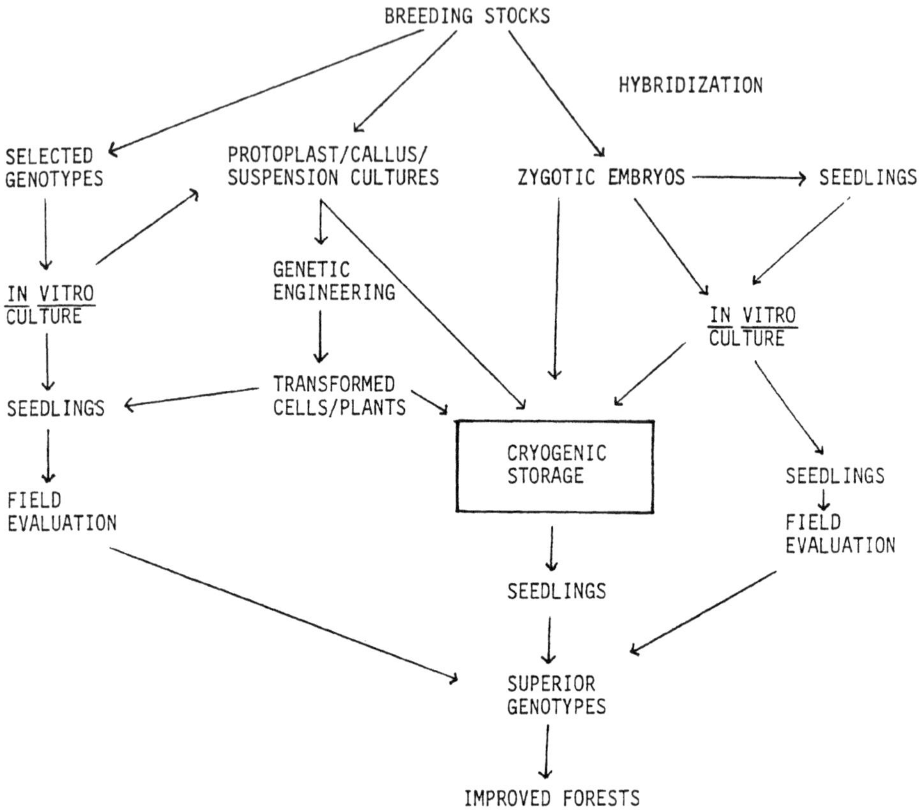

Figure 1. Possible role of cryopreservation in tree improvement.

Table 1. A Brief Summary of Cryopreservation protocol for Woody Material

Species	Culture type	Cryoprotectant	Freezing regime and storage	Thawing	Results Viability	Results Regrowth	References
<u>Acer</u> <u>pseudoplatanus</u>	Suspension	12% DMSO +5% glycerol	Prefreezing to -30/-50 C, LN[1]	+40 C	25% TTC	+	28
"	"	10% glycerol	2 C/min to -100 C, LN	+37 C	25% FDA	?[2]	21
"	"	10% glycerol +5% DMSO	"	"	22% FDA	?	"
"	"	"	1 C/min to -100 C, LN	+40 C	30% FDA	?	40
"	Synchronized suspension	"	"	"	25% FDA 45% TTC	?	34
"	Suspension pregrowth in Medium plus 3.3% Mannitol	"	"	"	40% FDA	+	40
"	Same as above but with 6% Mannitol	0.5M DMSO +0.5M glycerol +1 M Sucrose or proline	1 C/min to -35 C, held for 40 min, LN		75-90% FDA	+	"
<u>Acer</u> <u>saccharum</u>	Suspension	DMSO?	Less than 3 C/min to -40 C	Rapid ?	20% TTC	?	30
<u>Populus</u> <u>euramericana</u>	Callus (cold hardened)	none	Prefreezing to -30, -70, -120 C, then LN	Air at RT	?	+	22
"	Callus	5% glycerol	1 C/min to -40, LN	+40C	?	+	3
<u>Populus</u> <u>euramericana</u>	Suspension	5% DMSO +5% glycerol +4% sucrose	1 C/min to -40 C, LN	"	40% TTC	?	"
<u>Ulmus</u> <u>americana</u>	Callus	10% PEG + 8% glycerol +10% DMSO	1 C/min to -30 C, LN	+45 C	?	Callus regrowth and plant re-generation	32
<u>Pseudotsuga</u> <u>menziesii</u>	Suspension	5% DMSO + 1% glycerol +4% sucrose	1 C/min to -40 C, LN	+40 C	65% TTC	-	2
<u>Malus</u> <u>spp</u>	Winter bud	none	0.5 C/min to -15 C, -70 C, LN	+38 C		70 - 100% survival	16
<u>Malus</u> <u>domestica</u>	Winter bud	none	Cooling rate unknown, Cooled to -40 C, LN	0 C	?	77 - 100% survival	23
<u>Pear</u> (<u>pyrus</u>) <u>spp</u>[3]	"	"	Cooled to -40, -70% C, LN	0 C	?	13 - 100% survival	"

1. LN = liquid nitrogen.
2. ? = no data given in the original publication.
3. No species name was given.

17. NURSERY HANDLING OF PROPAGULES

J.A. DRIVER and G.R.L. SUTTLE

1. INTRODUCTION

Propagation of trees by conventional means tends to be highly geared to local requirements, with commercial firms often specializing on certain species. This is especially exemplified by the horticultural tree nursery industry. With the development and refinement of micropropagation techniques, a new means of propagation is available, especially for species that are difficult to propagate by conventional methods.

Our experience is chiefly with horticultural tree crops. It is believed that these crops will lead the way in defining commercial micropropagation practices, and the experience gained will be applicable to forestry. Also horticultural plant systems may be understood in greater detail because of substantial conventional asexual propagation experience.

From a conventional nursery perspective micropropagation should be viewed as another propagation tool. It is with this perspective that specific aspects of laboratory practice that affect commercial production are addressed.

2. COMMERCIAL NURSERY NEEDS VS. LABORATORY PRACTICE

The product from micropropagation must meet three criteria: 1. It must be needed. 2. It must be cost effective. 3. The micropropagation system must be defined adequately to deliver the product in the quality and quantity desired. Given these criteria, the number of tree species that 'should' be micropropagated is small. If dependable conventional methods exist for asexual propagation of the species (ie. air layering, mist propagation, hardwood cuttings, etc.), and if the end products are saleable units, conventional methods should be the methods of choice. However, it may be advantageous to establish mother stock from tissue culture for specific objectives (virus elimination, rejuvenation, rapid increase of a new cultivar) and subsequently use conventional methods to mass propagate the improved product. The greatest promise for forestry is in the propagation of elite genotypes from species where conventional asexual propagation is difficult or impossible (24).

Thus, if the objective for micropropagation is production of large quantities of a particular species, and it is cost effective, tissue culture propagation will gain in importance. With some forest species, there is progress towards this objective. This is due in large part to a more complete definition of the micropropagation system for each particular species. An example of this is the optimization of such media components as the hormones and mineral elements (51). It must be noted that many species demonstrate a high degree of specificity with regard to elemental concentration and the resulting morphological and physiological responses (69).

3. SEASONALITY OF GROWTH AND PRODUCTION CYCLES

For each particular crop, the commercial nursery has production cycles that correspond to seasonal climatic changes. This is especially true for temperate tree crops. Dormancy requirements or rest periods must be satisfied for each crop to allow successful propagation and growth (36). The tissue culture laboratory usually does not have these constraints during multiplication. Once the propagule is placed in the greenhouse or field, the plant becomes subject to seasonal factors such as daylength and temperature that control subsequent growth and development.

The effort to control output from the lab is thus complicated by these factors. Production should be constant throughout the year, since facilities are utilized most efficiently that way. Bulking up of plant material at various stages in vitro with cold temperature storage, is an aid in the control of output (1, 56).

4. MICROPROPAGATION OPTIONS

4.1. Trends in commercial micropropagation

It may be presumptive to assume trends in commercial micropropagation when so few firms are actually doing truely commercial production. One trend that is identifiable is specialization. Difficult to propagate tree species require a great research effort, and generalizations about propagation systems are difficult to make. The high degree of specificity requires that each production system be optimized. For recalcitrant species of high value, many years of effort may be required to adequately define a micropropagation scheme.

4.1.1. Contract micropropagation. If the quantities of plant material needed are not great, micropropagation is often contracted with existing firms that specialize in that crop, or with a firm whose business is providing contract research. For those nurseries that have large and narrow needs, the choice often is to conduct the required research in house and structure output around ones' own needs.

5. FACTORS AFFECTING SURVIVAL AND GROWTH

Several reviews of acclimatization and transplanting (12,20,23,37, 49,51, 57), suggest that considerable attention needs to be focused in this area. From a commercial nurseries' perspective acclimatization of some tree species has proved problematic. Procedures that have been described, may not be applicable on a commercial scale. Fortunately, progress is being made as more workers are turning their attention to this crucial area. Some species root and survive with ease, while other remain recalcitrant. *In vivo* rooting has been suggested as a means of avoiding acclimatization problems, and of lowering cost (40,48). While *in vivo* rooting has been successful for some species (ie., *Acer rubrum*, Suttle unpublished), other species respond better under an *in vitro* rooting scheme.

The current approach to rooting and acclimatization of micropropagated plantlets may be summarized as follows:

1. Root induction *in vitro*. After roots are induced, plantlet is transferred to the greenhouse, and placed in potting mix or soil medium; beginning with very high humidity (sweat tents, fog, mist), and then reducing the humidity over time, the plantlet is acclimated to the general greenhouse environment. The plant may be transferred to a screenhouse before field establishment.
2. Root induction *in vivo*. Tissue culture derived shoots are treated by a dip in root inducing compounds (if necessary) and placed in the greenhouse with very high humidity (sweat tents, fog, mist), reducing humidity over time till the plantlet is acclimated to the greenhouse environment.
3. Combinations of the above treatments. For example, with some crops the root inducing treatment takes place *in vitro*, but the cutting is transferred to the greenhouse before visible roots appear. Which of these three methods is used is determined by the specific needs of the crop. See (7,12,23,43), for reviews.

5.1. Hardening of propagules *in vitro*.

Many woody tree species do not acclimatize easily. *In vitro* treatments to prepare the propagule for this transition include adjustments of the medium, temperature and light. These adjustments often help to overcome morphological problems. Morphological 'abnormalities' from plants grown *in vitro* have been well documented (5,6,17,18,26,29,30,31,32,54,55,64,65,67,73). Absence of a waxy cuticle, or open stomates, are examples of the challenges faced when introducing the plant to a new environment. Donnelly *et al.* (19), describes the structural changes during the acclimatization process for red raspberry (*Rubus idaeus* L.), and how light and other factors affect survival. Studies of this type are greatly needed for tree species.

5.2. Greenhouse considerations.

Perhaps the greatest challenge to the use of micropropagation as an alternate means of production of tree species is the melding of the laboratory product, be it a cutting or a rooted plantlet, with the climatic phase of growth in the field. This means getting a plant the right size, in the right phase of growth, during the right time of year to be planted in the field. Most horticultural trees are valued in the marketplace by size of caliper or height of tree. Thus the overall control of growth must begin in the laboratory to optimize these attributes.

Commercial labs today use the greenhouse as an intermediary step between the lab and field planting. Primary factors promoting optimum growth are temperature (minimal basal heat 15C, maximum air temp. 38C), adequate daylength and light intensity, high relative humidity during early stages of growth, and adequate nutrition and pest/disease control (23).

Experience has shown that with *Pyrus*, *Prunus*, *Malus*, (Suttle, unpublished), and with *Juglans*, *Prunus* species (Driver, unpublished), very little top growth is achieved during the months of September through February. Rooting does proceed, albeit more slowly than between March and August. Top growth is maximum during the warm summer months of June - August. Suttle (unpublished) has found that exposing *in vitro* rooted plantlets to low temperature (2-4C) for at least 1000 hours in darkness, results in significantly greater growth and uniformity of some *Pyrus* and *Malus* species when subsequently planted under normal springtime greenhouse conditions. Others (3,37,57) also have discussed a chilling benefit. This benefit is similar to that found when dormant material and actively growing material (conventionally propagated) are compared after being planted side by side in the field (29).

5.3. Field planting.

Most horticultural tree species are grown under field conditions for one to three years before being sold as finished trees. Common nursery practice is to plant young 'liners' as early in the spring as possible or to sow seed during the fall or winter months. This takes advantage of as much of the growing season as possible. With micropropagated derived plantlets, the season of choice for field planting is also the spring. Some planting is done in the fall, but more often the plants are overwintered in cold storage (or ambient conditions) to be planted dormant in the spring. Increasing use of drip irrigation and advanced irrigation systems are allowing field planting into hot summer months.

5.4. New approaches: Direct field rooting

In addition to the described shoot acclimatization problems there may be additional root acclimatization problems (14,16,20). This can often compound losses to noneconomic levels.

Plant Research Laboratory (Driver), has tested a new process (patents pending), that allows successful rooting and acclimatization in the field. It is unique in that the greenhouse is completely avoided. The general steps are: pretreatment of tissue culture shoots *in vitro*: induction of roots in high density flats in the growth room; and before root emergence, placement of induced shoots under protective coverings in the field.

5.4.1. Pretreatment in vitro. It is necessary to encourage lignification-woodiness of the cutting (28,35,51,52,59,68). With walnut rootstock, 'Paradox,' (*Juglans regia* x *J. hindsii*), a species that suffers extensive root shock, cuttings 3-10 cm are cut from multiplication cultures and placed for a minimum of 1 week on a basal medium [DKW-C, (21) less cytokinins] with the gelrite content increased from 2.1g/L to 2.4g/L. (also see chapter on Juglans in Volume 3 for corrected media). The vessels with their cuttings are then placed under different culture conditions (17,20,42,46). The light intensity is increased from 55 μE m-2 s-1 to 66 μE m-2 s-1, and the day length is changed from 24 to 17 hrs. Temperature is lowered from 28 to 19 C. The results reported here will only be for Paradox walnut. All data were obtained from plants in high density trays (Castle & Cook Techniculture) in the growth room, not from field plants.

The pretreatment step encouraged stem lignification and had a profound effect upon subsequent rooting and acclimatization in the field. In addition, the organogenic capacity of the cuttings is influenced. For example, a preliminary experiment, in which the length of pretreatment varied from 5 to 50 days, maximum rooting (75%) was obtained with a 15 day exposure (Table 1). Without a pretreatment step, direct rooting and acclimatization percentage ranged from 0 to 10.2%. The propagules should root uniformly and rapidly in the field. Thus any adjustment of the media that caused rapid rooting is of benefit. Others have noted that balances of growth regulators, nutrients, carbohydrates, and light/temperature, are necessary to be effective in organ initiation (4,9,10,11,27,28,33,34,39,41,44,50,52,53,59,62,63,66,69,70). Our approach was to direct these developmental processes to a new commercial process in which the propagule would be able to avoid root shock that some recalcitrant species demonstrate under conventional micropropagation rooting and acclimatization programs.

By optimization of: sucrose concentration (Table 2); pretreatment exposure length (Table 2); auxin concentration (Tables 4 and 5); nitrogen concentration (Table 6); and length of exposure to darkness (etiolation) (Table 7); rooting and acclimatization events were influenced. Where applicable, optimum levels of medium components were suggested by solving the 1st order derivative of the best-fit curve for dy = 0 (15). The sequential experiments included the suggested optimum concentration of the compound of the prior experiment(s) when testing the effect of a different component.

Within all these parameters, rooting improved from 55.5% to 100%, and the mean number of roots/treatment rose from 1.75 to 3.85. However, the mean length of the longest root per treatment was not always correlated positively with percent rooting.

Increases in sucrose concentration caused the greatest organogenic response (Table 2). This is in contrast to many reports (4,9,11,25,47,48,51,52,66,70,71,72), stating that a reduction in sucrose increased rooting. Only a few references (8,28,41,45) reported improved rooting in response to increases in sucrose. The elevated sucrose concentration treatments showed a marked increase in stem hardening (lignification) of the cutting, and higher field acclimatization success. Thorpe (58,59), reported a positive relationship between lignification and sucrose in regard to organ initiation. Haissig (33), and Thorpe (60,61,62,63), has found that root initiation is a high energy process, requiring sucrose as a carbon source and as an osmotic agent. As these reports suggest, auxin and sucrose interactions are likely. Our data show that an exposure to a high concentration of sucrose for a short time gave a large response. The shorter the length of exposure, the greater the concentration needed for optimum response.

The results of a concentration/duration experiment (Tables 2 and 3), suggest that a 7 day exposure or 14 day exposure to high sucrose are probably best for good root formation. The 21 day exposure treatments did not indicate a trend in regard to sucrose concentration, and although rooting percentages were high, the propagules showed senescence and poor vigor. Therefore, the 7 and 14 day treatments were the best ones, and they were used in all subsequent experiments.

The concentration of auxin (IBA) in the pretreatment medium also demonstrated considerable effect (Tables 4 and 5). In the 7 day experiment (Table 4), the concentration/response did not resemble the 'normal' curve relationship that is expected in an equal increment dose-response experiment (15). Therefore, an experiment with a wider range of concentrations and a greater number of replicates was tested (Table 5).

The effect of concentration of nitrogen (DKW group stock 'A') showed that a lowering of levels was beneficial (table 6). This is generally in agreement with other reported results from _in vitro_ rooting (4,7,9,11,13,34,38,39,47,48, 52,53,66,70,71,72). In these studies most mineral salts were reduced in concentration, but we found that equal results could be achieved by lowering nitrogen alone as did Hyndman _et al_. (38).

An experiment to test the effect of etiolation shows (Table 7) no benefit to dark treatment. At longer exposure to darkness, shoots showed pronounced leaf burning. Also, the shoots did not harden as desired, but remained succulent. This contrasts to other reports where etiolation was beneficial (2,10,22,70,71).

The media and exposure time (21) for pretreatment of paradox walnut was thus modified: Exposure-7 day, sucrose-52.64g/L, NH4 NO3-456.2mg/L, Ca(NO3)2 -634.0mg/L, IBA-0.15mg/L.

5.4.2. _Root induction_. The pretreatment cutting is taken from the vessel and handled two ways. The first is to dip the cutting in a root induction compound (powder or liquid), and then place the cutting in a high density flat (ie. Castle and Cook Techniculture or other). This flat is covered with plastic (P.I. chamber, Knight Hollow Nursery, Inc. Madison Wis.) to maintain humidity and placed back in the culture room for 1 to 2 weeks till roots become visible. The cutting at this point is placed in the field under protective coverings (see 5.4.3). Secondly, it is possible to induce the cutting to root without a dip. The bypass of this step for Paradox walnut however, causes uneven time of rooting, and rooting percentages of the cuttings are reduced 10 to 30 percent. Because of the uneven rooting time, subsequent growth of the cuttings is also uneven.

5.4.3. Field placement. The pretreated and induced cutting is taken from the laboratory environment and stuck directly into the soil in the field. It is immediately protected by a translucent covering(s) that will: a) maintain adequate humidity, b) maintain a certain temperature range, c) allow proper light intensity.

This has been achieved by using two to three cups, one over the other pressed into the soil. When using three cups, a 16 oz. styrofoam cup is placed over a 10 oz. one, which in turn is placed over a 9 oz. translucent plastic cup. The foam cup(s) maintains proper light and temperature, the air space between the cups helps temperature control also; and the plastic cup maintains proper humidity. Constant moisture is supplied to the plant either through short duration irrigations (sprinkler), or through a drip irrigation system. Soil moisture is monitored by observing the moisture on the surface of the soil surrounding the cup(s) and occasional checking of the humidity within the coverings. Moisture from the soil will adequately condense on the inner surfaces of the plastic cup to maintain proper atmospheric humidity in the cup and thus protect the cutting from wilting. While this procedure requires frequent attention, it is quite manageable, even under intense horticultural nursery conditions. Juglans paradox, Juglans regia, and Prunus persica, cuttings have been rooted/acclimatized under this regime with ambient field temperatures of 5C to 40C.

Roots then grow into the soil, not being disturbed in any way. The top also begins growth. When the root system is adequate to support removal of the cups (8 to 10 weeks), the cups may be removed and reused. Coverings other than cups, for example commercial hotcaps, have been used also, but these do not provide as much environmental control as cups.

Thus this new process exhibits several unique features that are advantageous. It is cost effective, because a greenhouse is not required. A better product is produced with less loss. The root system is better developed as damage due to plant movement does not occur. It should be noted that a given species must already have an optimized micropropagation scheme (media, etc.) to take advantage of this direct process. With clonal peach, (3 varieties) or nectarine (1 variety) some of the aspects of this protocol were changed (different optimized media) the general direction (increased sucrose, reduced nitrogen in pretreatment) and steps are the same. It is expected that a number of tree and plant species could benefit from such a process. This may bring the cost of micropropagation within reach of new crops in forestry.

6. CONCLUSION

Continued improvement in micropropagation techniques will allow the commercial exploitation of more forest species. Mature clones of high values species are expected to yield to the considerable effort being focused upon their propagation.

7. REFERENCES

1. AITKEN-CHRISTIE J, TA THORPE 1984 Clonal Propagation: Gymnosperms, In IK Vasil, ed, Cell Culture and Somatic Cell Genetics of Plants. Vol I, Academic Press, New York, pp 82-95
2. ANDERSON WC 1981 Etiolation as an aid to rooting. Proc Int Plant Propagators Soc 31:138-141
3. BONGA JM 1984 Tissue culture techniques. In JM Bonga and DJ Durzan, eds, Tissue Culture in Forestry. Martinus Nijhoff, The Hague, pp 4-35
4. BOULAY M, A FRANCLET 1977 Recherches sur la propagation vegetative du Douglas: Pseudotsuna menziesii (Mirb.) Franco, CR Acad Sci (Paris) 284:1405-1407
5. BRAINERD KE, LH FUCHIGAMI 1981 Acclimatization of aseptically cultured apple plants to low relative humidity. J Am Soc Hortic Sci 106:525-518
6. BRAINERD KE, LH FUCHIGAMI 1982 Stomatal functioning of in vitro and greenhouse apple leaves in darkness. J Exp Bot 33:388-392
7. BROOME OC, RH ZIMMERMAN 1984 Culture of shoot meristems: fruit plants. In Vasil IK, ed, Cell Culture and Somatic Cell Genetics of Plants, vol 1, Academic Press, New York, pp 111-122
8. BUTCHER DN, HE STREET 1960 The effects of gibberelins on the growth of excised tomato roots. J Exp Bot 11:206-216
9. CHALUPA V 1977 Organogenesis in Norway spruce and Douglas fir tissue cultures. Commun Inst For Cech 10:79-87
10. CHANCEL LM, JJ MACHEIX, R JONARD 1980 Les conditions du microbouturage in vitro du Pecher (Prunus persica Batch): influence combinees des substances de croissance et de divers composes phenoliques. Physiol Veg 18:597-608
11. CHENG TY, TH VOQUI 1977 Regeneration of Douglas-fir plantlets through tissue culture, Science 198:306-307
12. CONNER AJ, MP THOMAS 1981 Re-establishing plantlets from tissue culture: A review. Proc Int Plant Propagators Soc 31:342-357
13. DAVID A 1982 In vitro propagation of gymnosperms. In JM Bonga and DJ Durzan, eds, Tissue Culture in Forestry, Martinus Nijhoff, The Hague, pp 72-108
14. DAVIS MJ, R BAKER, JJ HANAN 1977 Clonal multiplication of Carnation by

micropropagation. J Am Soc Hortic Sci 102:48-53
15. DE FOSSARD RA 1976 Records, experimental designs, statistical analysis of data, and the broad spectrum tissue culture experiment. In Tissue culture for plant propagators, Univ of New England Printery, Armidale, NSW, Aust, pp 95-152
16. DEBERGH PC, LJ MAENE 1981 A scheme for commercial propagation of ornamental plants by tissue culture. Sci Hortic 14:335-345
17. DONNELLY DJ, WE VIDAVER 1984 Leaf anatomy of red raspberry transferred from culture to soil. J Am Soc Hortic Sci 109:172-176
18. DONNELLY DJ, WE VIDAVER 1984 Pigment content and gas exchange of red raspberry in vitro and ex vitro. J Am Soc Hortic Sci 109:177-181
19. DONNELLY DJ, WE VIDAVER, K COLBOW 1984 Fixation of 14 CO2 in tissue cultured red raspberry prior to and after transfer to soil, Plant Cell Tissue Organ Cult 3:313-317
20. DONNELLY DJ, WE VIDAVER, KY LEE 1985 The anatomy of tissue cultured red raspberry prior to and after transfer to soil. Plant Cell Tissue Organ Cult 4:43-50
21. DRIVER JA, AH KUNIYUKI 1984 In vitro propagation of paradox walnut rootstock. HortScience 19:507-509
22. DRUART P, C KEVERS, P BOXUS, T GASPAR 1982 In vitro promotion of root formation by apple shoots through darkness effect on endogenous phenols and perodidases. Z Pflanzenphysiol 108:429-436
23. DUNSTAN DI, KE TURNER 1984 The acclimatization of micropropagated plants. In Vasil IK, ed, Cell Culture and Somatic Cell Genetics of Plants. Vol 1, Academic Press, New York, pp 123-129
24. DURZAN DJ 1982 Cell and tissue culture in forest industry. In JM Bonga and DJ Durzan, eds, Tissue Culture in Forestry. Martinus Nifhoff, The Hague, pp 36-71
25. FELICIANO AJ, M de ASSIS 1983 In vitro rooting of shoots from embryo-cultured peach seedlings. HortScience 18:705-706
26. FUCHIGAMI LH, TY CHENG, A SOELDNER 1981 Abaxial transpiration and water loss in aseptically cultured plum. J Am Soc Hortic Sci 106:519-522
27. GASPAR T, D SMITH, T THORPE 1977. Arguments supplementaires en faveur d'une variation inverse du niveau auxinique endogene au cours des des deux premieres phases de la rhizogenese. CR Acad Sci (Paris) 285:327-330
28. GREENWOOD MS, GP BERLYN 1973 Sucrose-indole-3-acetic acid interactions on root regeneration by Pinus lambertiana embryo cuttings. Am J Bot 69(1): 42-47
29. GROUT B, MJ ASTON 1977 Transplanting of cauliflower plants regenerated from meristem culture, I Water loss and water transfer related to changes in leaf wax and to xylem regeneration. Hortic Res 17:1-7
30. GROUT B, MJ ASTON 1977 Transplanting of cauliflower plants regenerated from meristem culture, II Carbon dioxide fixation and the development of photosynthetic ability. Hortic Res 17:65-71
31. GROUT B, MJ ASTON 1978 Modified leaf anatomy of cauliflower plantlets regenerated from meristem culture. Ann Bot 42:993-995
32. GROUT BWW 1975 Wax development on leaf surfaces of Brassica olerace var. Currawona regenerated from meristem culture. Plant Sci Lett 5:401-405
33. HAISSIG BE 1974 Metabolism during adventitious root primordium initiation and development. N Z J For Sci 4:324-327
34. HAMMERSCHLAG F 1982 Factors influencing in vitro multiplication and rooting of the plum rootstock myrobalan (Prunus cerasifera Ehrh.). J Am Soc Hortic Sci 107:44-47
35. HARKIN JM, JR OBST 1973 Lignification in trees: Indication of exclusive peroxidase participation. Science 180:296-298
36. HARTMANN HT, DE KESTER 1983 Plant propagation, Principles and Practices.

Prentice-Hall, Inc, Englewood, NJ 4th Ed
37. HOWARD, BH, VH OEHL 1981 Improved establishment of in vitro propagated plum micropropagules following treatment with GA3 or prior chilling. J Hortic Sci 56:1-7
38. HYNDMAN SE, PM HASEGAWA, RA BRESSAN 1982 Stimulation of root initiation from cultured rose shoots through the use of reduced concentrations of mineral salts. HortScience 17:82-83
39. JAMES DJ 1983 Adventitious root formation 'in vitro' in apple rootstocks (Malus pumila) 1 Factors affecting the length of auxin sensitive phase in M9. Physiol Plant 57:149-153
40. LLOYD G, B MCCOWN 1980 Commercially-feasible micropropagation of mountain laurel, Kalmia latifolia, by use of shoot tip culture. Proc Int Plant Propagators Soc 30:421-427
41. MARETZKI A, P HIRAKI 1980 Sucrose promotion of root formation in plantlets regenerated from callus of sacharum spp. Phyton 38:85-88
42. MCCOMB, JA 1978, Clonal propagation of woody plants using tissue culture, with special reference to apples. Proc Int Plant Propagators Soc 28:413
43. MCGUIRE JJ 1980 Root initiation: A survey of current literature, Proc Int Plant Propagators Soc 30:282-288
44. MEHRA-PALTA A, RL MOTT, RH SMELTZER 1977 Hormonal control of induced organogenesis from excised plant parts of Loblolly pine (Pinus taeda L.). TAPPI Conf Pap; For Biol Wood Chem Conf 1977, pp 15-20
45. MINOCHA SC 1980 Callus and adventitious shoot formation in excised embryos of white pine (Pinus strobus) Can J Bot 58:366-370
46. MURASHIGE T 1977. Clonal crops through tissue culture, In W Barz, E Reinhard, and MJ Zenk, eds, Plant Tissue Culture and its Bio-technological Application. Springer Verlag, Berlin, pp 392-403
47. SIMMONDS J 1984 Induction, growth and direct rooting of adventitious shoots of Begonia x hiemalis. Plant Cell Tissue Organ Cult 3:283-289
48. SIMMONDS J 1983 Direct rooting of micropropagated M26 apple rootstocks. Scientia Hortic 21:223-241
49. SKOLMEN RG, MO MAPES 1978. Aftercare procedures required for field survival of tissue culture propagated Acacia koa. Proc Int Plant Propagators Soc 28:156-164
50. SKOOG F, CO MILLER 1957 Chemical regeneration of growth and organ formation in plant tissues cultured in vitro. Symp Soc Exp Biol 11:118-131
51. SOMMER HE, LS CALDAS 1981 In vitro methods applied to forest trees. IN TA Thorpe, ed, Plant Tissue Culture, Methods and Applications in Agriculture. Academic Press, New York, pp 349-358
52. SRISKANDARAJAH S, MG MULLINS 1981 Micropropagation of Granny Smith apple: factors affecting root formation in vitro. J Hortic Sci 56:71-76
53. SRISKANDARAJAH S, MG MULLINS, Y NAIR 1982 Induction of adventitious rooting in vitro in difficult-to-propagate cultivars of apple. Plant Sci Lett 24:1-9
54. SUTTER E, RW LANGHANS 1982 Formation of epicuticlar wax and its affect on water loss in cabbage plants regenerated from shoot-tip culture. Can J Bot 60:2896-2902
55. SUTTER E, RW LANGHANS 1979 Epicuticular wax formation on carnation plantlets regenerated from shoot tip culture J Am Soc Hortic Sci 104:493-496
56. SUTTLE GRL 1983 Micropropagation of deciduous trees. Proc Int Plant Propagators Soc 33:46-49
57. THOMPSON DG, JC GORDON 1977 Propagation of poplars by shoot apex cultures and nutrient film technique. TAPPI Conf Pap; For Biol Wood Chem Conf 1977, pp 77-82
58. THORPE TA 1978 Physiological and biochemical aspects of organogenesis in

vitro. In: TA Thorpe, ed, Frontiers of Plant Tissue Culture 1978, Int Assoc Plant Tissue Cult, Univ Calgary, Alberta, pp 49-58
59. THORPE TA, S BIONDI 1981 Regulation of plant organogenesis, In K Maramorosch, ed, Advances in Cell Culture. Academic Press, New York, pp 213-239
60. THORPE TA, T MURASHIGE 1970 Some histochemical changes underlying shoot initiation in tobacco callus cultures. Can J Bot 48:277-285
61. THORPE TA, T MURASHIGE 1974 Carbohydrate availability and shoot formation in tobacco callus cultures. Physiol Plant 30:77-81
62. THORPE TA, T MURASHIGE 1968 Starch accumulation in shoot-forming tobacco callus cultures. Science 260:421-422
63. THORPE TA 1982 Carbohydrate utilization and metabolism. In JM Bonga and DJ Durzan, eds, Tissue Culture in Forestry, Martinus Nijhoff, The Hague, pp 325-368
64. WARDLE K, A QUINLAN, I SIMPKINS 1979 Abscisic acid and regulation of water loss in plantlets of Brassica oleracea var. botrytis regenerated through apical meristem culture. Ann Bot 43:745-752
65. WARDLE K, EB DOBBS, KC SHORT 1983. In vitro acclimatization of aseptically cultured plantlets to humidity. J Am Soc Hortic Sci 108:386-389
66. WELANDER T 1979 Influence of medium composition on organ formation in explants of Begonia x hiemalis in vitro. Swed J Agric Res 9:163-168
67. WETZSTEIN HY, HE SOMMER 1982 Leaf anatomy of tissue-cultured Liquidambar stvraciflua (Hammelidaceae) during acclimatization. Am J Bot 69:1579-1586
68. WOLTER KE, JC GORDON 1975 Peroxidases as indicators of growth and differentiation in aspen callus cultures. Physiol Plant 33:219-223
69. ZAERR JB, MO MAPES 1982 Action of growth regulators. In JM Bonga and DJ Durzan, eds, Tissue Culture in Forestry, Martinus Nijhoff, The Hague, pp 231-255
70. ZIMMERMAN, RH 1985 Rooting apple cultivars in vitro: interactions among light, temperature, phloroglucinol and auxin. Plant Cell Tissue Organ Cult 3:301-311
71. ZIMMERMAN RH, I FORDHAM 1985 Simplified method for rooting apple cultivars in vitro. J Am Soc Hortic Sci 110:34-38
72. ZIMMERMAN RH, OC BROOME 1980 Apple cultivar micropropagation. Proc Conf Nursery Prod Fruit Plants through Tissue Cult-Appl, USDA, Sci Educ Admin Agric Res Results ARR-NE-11, pp 54-58
73. ZIV M, G MEIR, AH HALEVY 1983 Factors influencing the production of hardened glaucus carnation plantlets in vitro. Plant Cell Tissue Organ Cult 2:55-65

Table 1. Effect of pretreatments on adventitious root development in tissue culture derived cuttings of paradox walnut. Data are means of 144 values from two replications expressed on a per cutting basis. Root induction by powder dip method (IBA- 2g/100g talc), following pretreatment.

Pretreatment length (days)	No. roots	Length (cm) longest root	% Rooted
5	1.75	3.31	53
15	2.69	3.49	75
20	2.14	2.72	61
35	2.15	2.74	59.7
50	1.60	2.42	45.1

Results were read 21 days after dip.

Table 2. Effect of concentration of sucrose during pretreatment on adventitious root development in tissue culture derived cuttings of paradox walnut. Data are means of 72 values from one replication expressed on a per cutting basis. Root induction by powder dip method (IBA - 2g/100g talc), following pretreatment.

Pretreatment sucrose conc. (g/L)	No. roots	Length (cm) longest root	% Rooted
7 Day	Optimum concentration = 52.6 g/L		
20	2.15	3.57	55.5
30	2.31	3.63	80.5
40	2.88	2.77	86.9
50	3.05	3.04	93.0
60	2.59	3.79	93.3
14 Day	Optimum concentration = 40.2 g/L		
20	2.52	3.29	81.9
30	2.77	3.28	91.7
40	2.48	3.65	93.1
50	2.23	3.73	90.3
60	2.07	3.74	83.3
21 Day	Optimum concentration = N/A		
20	3.04	2.18	86.7
30	3.01	2.70	81.5
40	3.19	2.16	91.2
50	3.09	2.78	84.7
60	3.85	3.05	91.9

Results were read 21 days after dip.

Table 3. Effect of pretreatments on adventitious root development in tissue culture derived cuttings of paradox walnut with different concentrations of sucrose. Data are means of 72 values, expressed on a per cutting basis. Root induction by powder dip method (IBA - 2g/100g talc), following pretreatment.

Pretreatment length (days)	No. roots	Length (cm) longest root	% Rooted
20g/L Sucrose		Optimum # of Days = 19	
7	2.15	3.57	55.5
14	2.52	3.29	81.9
21	3.04	2.18	86.7
30g/L Sucrose		Optimum # of Days = 14	
7	2.31	3.63	80.5
14	2.77	3.28	91.7
21	3.01	2.70	81.5
40g/L Sucrose		Optimum # of Days = 15	
7	2.88	2.77	86.9
14	2.48	3.65	93.1
21	3.19	2.16	91.2
50g/L Sucrose		Optimum # of Days = 4	
7	3.05	3.04	93.0
14	2.23	3.73	90.3
21	3.09	2.78	84.6
60g/L Sucrose		Optimum # of Days = N/A	
7	2.59	3.79	93.3
14	2.07	3.74	83.3
21	3.85	3.05	91.9

Results were read 21 days after dip.

Table 4. Effect of concentration of IBA during pretreatment on adventitious root development in tissue culture derived cuttings of paradox walnut. Data are means of 36 values from two replications, expressed on a per cutting basis. Root induction by powder dip method (IBA - 2g/100g talc), following pretreatment.

Pretreatment IBA conc. (mg/L)	No. roots	Length (cm) longest root	% Rooted
7 Day	Sucrose concentration = 52.6 g/L		
0	2.25	3.69	90.5
.001	2.45	3.63	94.4
.003162	2.59	3.51	87.5
.01	2.31	3.79	87.5
.03162	3.40	3.97	95.8
.1	3.22	4.05	100.00
14 Day	Sucrose concentration = 40.2 g/L		
0	2.91	3.43	88.7
.001	2.99	3.17	83.3
.003162	2.91	3.32	88.9
.01	2.65	3.35	95.9
.03162	2.85	3.91	94.4
.1	3.65	3.55	95.8

Results were read 21 days after dip.

Table 5. Effect of concentration of IBA during Pretreatment on adventitious root development in tissue culture derived cuttings of paradox walnut. Data are means of 72 values from one replication, expressed on a per cutting basis. Root induction by powder dip method (IBA - 2g/100g talc), following pretreatment.

Pretreatment IBA conc. (mg/L)	No. roots	Length (cm) longest root	% Rooted
7 Day	Sucrose concentration = 52.6 g/L		
0	2.47	3.15	81.4
.01	2.60	3.40	90.3
.03162	2.63	3.22	93.1
.1	3.09	3.50	93.1
3.162	3.64	3.55	93.1
10.	3.14	3.30	90.3

Results were read 21 days after dip.

Table 6. Effect of concentration of nitrogen (DKW 'A' stock) during Pretreatment on adventitious root development in tissue culture derived cuttings of paradox walnut. Data are means of 36 values from two replications expressed on a per cutting basis. Root induction by powder dip method (IBA - 2g/100c talc), following pretreatment. "A" stock concentration (100X) (NH_4NO_3 = 141.6g/L; $Ca(NO_3)_2$ = 196.8g/L). Log value/L = the log of the volume in ml from the group stock 'A' (100X) (2.0 = 100ml/L; 1.5 = 31.62ml/L; 1.0 = 10ml/L; 0.5 = 3.162ml/L; 0.1 = 0.1ml/L; 0 = none).

Pretreatment 'A' conc. (log value/L)	No. roots	Length (cm) longest root	% Rooted
7 Day		Sucrose concentration = 52.6 g/L	
0	2.67	3.31	84.4
0.1	2.69	3.65	91.7
0.5	3.03	3.45	94.4
1.0	2.77	4.09	86.1
1.5	2.60	4.22	55.6
2.0	1.00	0.70	5.6
14 Day		Sucrose concentration = 40.2 g/L	
0	2.11	2.37	75.0
0.1	2.56	2.75	75.0
0.5	2.88	3.20	91.7
1.0	2.86	4.26	97.2
1.5	3.08	3.93	33.3
2.0	0	0	0

Results were read 21 days after dip.

Table 7. Effect of length of exposure to etiolation after 7 days Pretreatment on adventitious root development in tissue culture derived cuttings of paradox walnut. Exposure to dark began upon pretreatment. Data are means of 48 values, expressed on a per cutting basis. Root induction by powder dip method (IBA - 2g/100g talc), following pretreatment.

Etiolation length (days)	No. roots	Length (cm) longest root	% rooted
0	2.15	3.37	95.8
1	2.35	3.63	95.9
2	2.11	3.89	95.9
3	2.44	3.41	93.8
4	2.46	3.31	85.4
5	3.15	3.60	97.3
6	1.00	5.53	6.3
7	2.87	4.03	66.7

Results were read 21 days after dip.

18. MYCORRHIZAE

ROBERT K. DIXON and DONALD H. MARX

1. INTRODUCTION

Forest trees are obligately dependent upon mycorrhizal symbiosis for normal growth and development. Because the benefits derived from different species and isolates of mycorrhizal fungi vary for a given tree or site, manipulation of these fungi should improve the uniformity and yield of artificially regenerated forests. Experimental inoculation of hardwood and conifer seedlings with specific fungi in nurseries and greenhouses has dramatically improved survival and growth following outplanting on routine and adverse reforestation sites in the U.S. and abroad. The development of mycorrhizal inoculation procedures offers a biotechnological opportunity for foresters to increase the yield and quality of forest stands.

2. ROLE OF MYCORRHIZAE IN TREE GROWTH AND DEVELOPMENT

The feeder roots of forest trees are associated with specific root-inhabiting fungi. The infected feeder roots are called mycorrhizae (fungus-roots). The relationship between roots and these fungi is symbiotic since both partners benefit from the association (25). The mycorrhizal structure is involved in nutrient uptake (7,26), water absorption (18,80), protection against feeder root pathogens (37), and extremes of soil temperature (51), pH (53), and soil toxins (87).

There are three basic types of mycorrhizae: ectomycorrhizae, endomycorrhizae, and ectendomycorrhizae. The types of greatest importance to foresters are ectomycorrhizae and endomycorrhizae. Ectomycorrhizae occur naturally as feeder roots on most conifers including all species of the family Pinaceae, as well as some angiosperms, such as members of the Fagaceae, Betulaceae, Juglandaceae, and Myrtaceae (66). The classes of fungi that form ectomycorrhizae are Basidiomycotina and Ascomycotina.

Endomycorrhizae, the most prevalent type of mycorrhizae, occur on about 90% of all herbaceous and woody plants. Many of our economically important hardwood genera such as Liquidambar, Ulmus, Juglans, Fraxinus, and Liriodendron form endomycorrhizae. Fungi which form endomycorrhizae are mainly Phycomycetes in the Endogonaceae (95).

The dependence of forest trees on mycorrhizal symbiosis has long been recognized (25,27). Certain trees, such as pines (Pinus spp.), will not develop normally without ectomycorrhizae. Pine species establishment trials in parts of the world where pine are not native inevitably fail in the absence of suitable ectomycorrhizal fungus inoculum (97,98). In the past three decades, the role of mycorrhizal inoculation in promoting survival and growth of planted seedlings has also been clearly illustrated. This paper discusses the potential value of forming mycorrhizae on tree seedlings or plantlets in nurseries with specific fungi to improve field performance. Inoculum production methods,seedling inoculation procedures, and field performance of mycorrhizal planting stock will be briefly reviewed.

3. PRODUCTION AND APPLICATION OF ECTOMYCORRHIZAL FUNGUS INOCULUM

The concept of improving field performance of seedlings by forming ectomycorrhizae on them with specific fungi ecologically adapted to the planting site was originally developed by Moser (73) in Austria. This subject has been reviewed comprehensively by Bowen (6), Mikola (67), Trappe (94), Marx (46), and Molina and Trappe (70). Using various modifications of Moser's techniques, Takacs (90) in Argentina, Theodorou and Bowen (92) in Australia, and Vozzo and Hacskaylo (98) in the United States showed that field survival and growth of tree seedlings with specific ectomycorrhizae exceeded the performance of seedlings that lacked or had few native ectomycorrhizae at planting. Since 1966, research has been done by the USDA Forest Service in the southern United States to determine the significance of ectomycorrhizae formed by Pisolithus tinctorius (Pers.) Coker and Couch to survival and growth of tree seedlings on a variety of sites. This interest was prompted by the wide geographic distribution, range of environmental tolerance, and ease of pure culture propagation of P. tinctorius (41,71,72,87). Marx (46) has discussed the early testing and development of vegetative inoculum of P. tinctorius for research.

3.1. Bareroot Stock

Vermiculite-based vegetative inoculum of P. tinctorius has formed ectomycorrhizae on pine, oak (Quercus sp.), pecan (Carya illinoensis Wangenh), and cedar (Cedrus atlantica Manetti) seedlings in microplots (42,43,44,52,56,85) and on pine, fir (Abies spp.) and oak seedlings in conventional bare-root nurseries throughout the United States (60). In most of these nursery tests and in others (46,47,54,57,58), seedlings with abundant P. tinctorius ectomycorrhizae grew larger than control seedlings with fewer, naturally occurring ectomycorrhizae. Fumigation of nursery soil prior to inoculation improves ectomycorrhizal development (60) because it lowers populations of 1) soil microorganisms that can colonize introduced inocula (8,46); 2) feeder root pathogens that damage roots and thus reduce ectomycorrhizal development (54,82); and 3) indigenous competing ectomycorrhizal fungi such as Thelephora terrestris (Ehrh.) Fr. from previous tree crops (54,58). Vermiculite-based inoculum broadcast at a concentration of about 1 liter/m^2 of soil surface and mixed into soil has been as effective as higher rates in forming abundant P. tintorius ectomycorrhizae on pine in southern nurseries (46,60).

Basic techniques have also been developed to use basidiospores as inoculum to form specific ectomycorrhizae. Spores have been used to coat (91,93) or encapsulate (60) pine seeds, pelletized for broadcast application (49), mixed with hydromulch to cover seed, added directly into soil, and dusted onto soil or seedlings (2,39,52,57,84). Puffball-producing fungi, such as Rhizopogon and Pisolithus, produce large quantities of easily obtainable spores and, thus, have been used in most basidiospore inoculation tests.

3.2. Container-grown stock

Vermiculite-based vegetative inoculum and spore inoculum also have been used to form abundant P. tinctorius ectomycorrhizae on tree seedlings grown in various media and containers (14,15,16,17,24,35,36,38,59,68,69,76,81,83,84).

4. FIELD TRIALS WITH ECTOMYCORRHIZAL PLANTING STOCK

The value of P. tinctorius ectomycorrhizae to forest regeneration has been shown by the performance of seedlings under diverse field conditions. Significant, and often dramatic, enhancement of survival and growth of pine seedlings with abundant P. tinctorius ectomycorrhizae, in comparison to seedlings with natural ectomycorrhizae produced in containers or bare-root nurseries, has been reported from studies on acid coal spoils in Appalachia (4,48,99), kaolin spoils in Georgia (40,75), severely eroded sites of the Copper Basin in Tennessee (5), borrow pits in South Carolina (83) and North Carolina (22), and prairie soil (3). Similar improvements in pine seedling performance were reported on routine reforestation sites in Florida and North Carolina (55), Oklahoma and Arkansas (65,85,86), Georgia (45), Mississippi (30), and Louisiana (96). Dramatic improvements in performance of planted P. caribaea Mor. seedlings were observed on afforestation sites in Nigeria (72), the Congo (13), and Liberia (60). Growth of various species of native pine in Mexico was also significantly stimulated by P. tinctorius ectomycorrhizae on severely eroded hillsides near Mexico City (60). Recently,, Dixon et al (15) reported that container-grown oak seedlings with P. tinctorius ectomycorrhizae survived and produced more shoot and root growth than noninoculated oak seedlings after one year on a reforestation site in Missouri.

On poor soils amended with fertilizer or sewage sludge, pine seedlings with T. terrestris ectomycorrhizae have survived and grown as well as those with abundant P. tinctorius (5,55). Inoculation with P. tinctorius, however, did not stimulate growth of pine and spruce (Picea spp.) on a high-elevation, molybdenum tailing pond (pH 8.0) in Colorado (24), pine on a grassland (pH 8.4) in North Dakota (81), pine on a fertile abandoned field in Georgia (78), or bare-root oak seedlings on a reforestation site in Missouri (15).

To obtain significant survival or growth increases of southern pine seedlings by P. tinctorius on reforestation sites, a threshold level of at least half of all ectomycorrhizae on seedling roots must be those of P. tinctorius at planting time (30,55,86). Frequently, pine seedlings with less than half of all ectomycorrhizae formed by

P. tinctorius survive and grow at the same rate as seedlings with abundant T. terrestris ectomycorrhizae on routine reforestation sites.

Operational use of P. tinctorius in forest regeneration programs has been prevented by the lack of sufficient quantities of effective inoculum. In 1976, the Institute for Mycorrhizal Research and Development (IMRD) and State and Private Forestry, USDA Forest Service, and Abbott Laboratories, North Chicago, Illinois, began a cooperative program of research and nursery evaluation to develop commercial methods of producing vegetative inoculum of P. tinctorius. They also tested the effectiveness of different inoculum formulations under diverse nursery cultural regimes on different tree species throughout the United States. A comprehensive report was published (59) on the testing of various formulations of the commercial inoculum for ectomycorrhizal development on container-grown tree seedlings of various species. Inoculum from IMRD and Abbott formed P. tinctorius ectomycorrhizae on 10 pine species, Douglas-fir (Pseudotsuga menziessii Mirb. and Franco.), western hemlock (Tsuga heterophylla Raf. and Sarg.), and bur oak (Q. macrocarpa Michx.) in tests using different containers and cultural practices at six locations in the United States and one in Canada. From the Canadian test, Navratil et al (74) reported that the container-grown pine seedlings with P. tinctorius ectomycorrhizae formed by 1978 formulations of IMRD and Abbott inocula grew significantly more than initially larger, noninoculated seedlings after one growing season on a boreal forest site in Ontario.

Tests of IMRD and Abbott inoculum formulations in bare-root seedling nurseries were completed in 1981 (60). This research, done in 33 different tree nurseries in 25 states on 11 pine species and varieties, Douglas-fir and oak, showed that inoculum formulations that were effective in the containerized seedling tests were also effective in most bare-root nursery tests. Over 70 outplanting tests in 23 states are currently being monitored yearly to determine the field significance of seedling inoculation with P. tinctorius in bare-root nurseries. A nursery seedbed inoculator has also been designed and tested (11). This machine is currently being marketed for operational inoculation of nursery soil with commercial vegetative inoculum formulations. Machine inoculation of nursery soil uses two-thirds less inoculum, by banding the inoculum in soil, than does broadcast application; the results of both methods are comparable.

Recently, IMRD and State and Private Forestry also initiated a program of research and nursery evaluation of various types of vegetative inoculum of different ectomycorrhizal fungi produced by Sylvan Spawn, Worthington, Pennsylvania, USA.

5. PRODUCTION AND APPLICATION OF ENDOMYCORRHIZAL INOCULUM

Many commercially important hardwood trees are hosts for fungi that form endomycorrhizae (63). Unfortunately, development of hardwood seedlings that normally form endomycorrhizae has not been satisfactory in most forest tree nurseries (88) and, consequently, seedling quality remains a major obstacle in artificial regeneration of hardwood plantations (34). This problem is due in part to a lack of endomycorrhizal development (31). Until recently, little research has focused upon the use and manipulation of endomycorrhizal fungi to improve hardwood seedling quality.

Studies conducted in forest tree nurseries indicate that endomycorrhizal fungi are a major factor in production of quality hardwood seedlings (9,89). Kormanik *et al* (32) reported large increases in the growth of endomycorrhizal sweetgum (*Liquidambar styraciflua* L.) inoculated with various *Glomus* spp. when compared to noninoculated control seedlings. Significant growth differences among half-sib progeny inoculated with *G. mosseae* (Nicol. and Gerd.) Gerdemann and Trappe, and *G. etunicatum* Becker and Gerdemann have also been observed (31). Inoculation of red maple (*Acer rubrum* L.), black walnut (*Juglans nigra* L.), green ash (*Fraxinus pennsylvanica* Marsh.), boxelder (*Acer negundo* L.), sycamore *Platanus occidentalis* L.), and black cherry (*Prunus serotina* Ehrh.) with the fungal symbionts *G. fasciculatum* (Thaxt.) Gerdemann and Trappe, *G. mosseae*, and *G. etunicatum* in nursery microplots resulted in 2- to 80-fold increases in shoot dry weight over noninoculated control seedlings (34). *Citrus* and yellow poplar (*Liriodendron tulipifera* L.) expressed similar dramatic growth responses following inoculation with *Glomus* and *Gigaspora* species (63,77).

Commercial production of endomycorrhizal inoculum for introduction into forest tree nurseries, greenhouses, or on adverse sites is presently being attempted at several locations worldwide (63). Currently, the only method to produce suitable quantities of endomycorrhizal inoculum is on the roots of "trap plants" growing in sterilized soil under controlled conditions (19) or by the timing of soil fumigation and cover cropping in tree nurseries (33). The possibility of pathogenic organisms contaminating endomycorrhizal inoculum is a serious problem when the fungal symbiont is grown in semisterile cultures. For this reason, many foresters and horticulturists favor mass production of endomycorrhizal fungi under axenic conditions. Production of large quantities of endomycorrhizal spores in pure culture which will infect tree seedlings and survive under field conditions has remained an elusive goal.

The key to economically successful inoculation with endomycorrhizal fungi in forest tree nurseries is the efficient introduction of superior fungal symbionts, which are capable of promoting better seedling growth than indigenous fungi. Menge et al (64) found that layering inoculum below the seed was superior to seed inoculation. Commercial applications of endomycorrhizal fungus inoculum using banding and seeding machinery were successfully carried out in Citrus nurseries in California. Hattingh and Gerdemann (28) reported significant growth responses of sour orange (Citrus aurantium L.) following inoculation of fumigated nursery soil.

The selection of efficient isolates of endomycorrhizal fungi is critical to the development of commercial inoculum for use in forestry. Factors that contribute to symbiont efficiency include inoculum density, storage, placement of inoculum in soil, dependable spore germination, rapid growth in soil, competitive capabilities, survival mechanisms, and rapid infection rates (63). However, evidence is mounting that the amount and location of external hyphae, which absorb soil nutrients, as well as the relative carbon demand, may also be critical factors governing isolate efficiency (26).

6. FIELD TRIALS WITH ENDOMYCORRHIZAL PLANTING STOCK

Although inoculation of tree seedlings with endomycorrhizal fungi has resulted in significant growth responses, few large-scale tests of seedling outplanting performance have been published. Recently, Dixon (unpublished data) found that survival and growth of black walnut inoculated with G. fasciculatum were significantly greater than in nonmycorrhizal control seedlings in southern Minnesota. Endomycorrhizal fungi have been conclusively shown to improve efforts to revegetate coal spoils, strip mine overburden, and other disturbed areas (1,12,79). On these disturbed sites, populations of endomycorrhizal fungi are usually reduced or absent. Adding endomycorrhizal fungi to improve endomycorrhizal development provides a nutritional advantage to tree seedlings, in addition to increasing tolerance to environmental extremes (29).

7. RESEARCH OPPORTUNITIES

Research on mycorrhizal associations has increased dramatically during the past three decades, particularly in applied areas. Currently, there are opportunities for foresters to apply mycorrhizal technology in seedling production and reforestation programs. Nonetheless, wide gaps remain in the knowledge of mycorrhizal symbiosis and the cultural techniques suited to different tree and fungal species and different sites. Fundamental physiological and genetic questions should also receive the attention of "mycorrhizasts."

Growth of tissue culture plantlets of trees in the greenhouse or nursery is extremely slow when compared to seedlings (61,62). Generally, plantlets have few lateral roots compared to seedlings of the same age. Growth regulators and nutrients supplied by mycorrhizae may promote root and shoot growth of plantlets (23). This situation suggests the need for more research on the influence of mycorhizae on the growth and development of tissue culture plantlets during the production phase and following outplanting.

The influence of genetic composition of trees on mycorrhizae formation has not been well defined. Wright and Ching (100) demonstrated that significant differences exist in the ability of Douglas-fir seedlings from different seed sources to form ectomycorrhizae during the first growing season. Marx and Bryan (50) and Cline and Reid (10) observed that genotypes of slash (P. elliotti Engelm.), lodgepole (P. contorta Dougl.), and ponderosa pine (P. ponderosa Laws.) influenced the degree of ectomycorrhizae formation with the fungus P. tinctorius, as well as the response of the host seedling to ectomycorrhizal colonization. Specificity in mycorrhizal relationships (26,70), including the identification of dominant genes for fungal infectivity potential and host receptivity, deserve further investigation (21,23).

Only limited efforts have been made to modify the genetic constitution of mycorrhizal fungi (20). Impediments to genetic manipulation of fungal symbionts include the inability to culture the fungi free of their hosts, unreliable methods of germinating spores of ectomycorrhizal symbionts, the multinucleate nature of fungi, and a weak understanding of fungal genomes. In the future, the knowledge and techniques will exist to alter the genetic constitution of fungi. Possible goals include nitrogen-fixing capabilities, tolerance to extremes of temperature, soil pH, moisture levels, and other environmental factors, as well as greater efficiency in uptake of nutrients, wide ranges of host x fungus compatibility, and reduced carbon drain from the host.

8. SUMMARY

The use and manipulation of mycorrhizal symbiosis has enormous potential benefits in tree nurseries and forest regeneration programs. Mycorrhizae benefit trees by enhancing nutrient and water uptake and increasing tolerance to adverse environmental extremes. Several-fold increases in the growth of hardwood and conifer seedlings in nursery and field conditions have been observed after manipulation of fungal symbionts. Progress toward commercial production of ectomycorrhizal and endomycorrhizal inoculum is rapid. Despite these advances, many areas of applied and basic research on mycorrhizal symbiosis remain fertile for exploration. To maximize the vast benefits from mycorrhizal inoculation in field applications and to achieve genetic alterations to improve desired characteristics, the fundamental principles and processes of mycorrhizal symbiosis and testing procedures need to be thoroughly examined.

9. REFERENCES

1. ALDON EF 1975 Endomycorrhizae enhance survival and growth of four-wing saltbush on coal mine spoils. USDA For Serv Res Note RM-294, 7 pp
2. ALVAREZ IF, JM TRAPPE 1983 Effects of application rate and cold soaking treatment of Pisolithus spores on effectiveness as nursery inoculum on western conifers. Can J For Res 13:533-537
3. BAER NW, JD OTTA 1981 Outplanting survival and growth of ponderosa pine seedlings inoculated with Pisolithus tinctorius in South Dakota. For Sci 27:277-280
4. BERRY CR 1982 Survival and growth of pine hybrid seedlings with Pisolithus ectomycorrhizae on coal spoils in Alabama and Tennessee. J Environ Qual 11:709-715
5. BERRY CR, DH MARX 1980 Significance of various soil amendments to borrow pit reclamation with loblolly pine and fescue. Reclam Rev 3:87-94
6. BOWEN GD 1965 Mycorrhiza inoculations in forestry practice. Aust For 29:231-237
7. BOWEN GD 1973 Mineral nutrition of ectomycorrhizae. In GC Marks and TT Kozlowski, eds, Ectomycorrhizae: Their Ecology and Physiology. Academic Press, New York, pp 151-205
8. BOWEN GD, C THEODOROU 1979 Interactions between bacteria and ectomycorrhizal fungi. Soil Biol Biochem 11:119-126
9. BRYAN WC, PP KORMANIK 1977 Mycorrhizae benefit survival and growth of sweetgum seedlings in the nursery. South J Appl For 1:21-23
10. CLINE ML, CPP REID 1982 Seed source and mycorrhizal fungus effects on growth of containerized Pinus contorta and Pinus ponderosa seedlings. For Sci 28:237-250
11. CORDELL CE, DH MARX 1983 The operational application of Pisolithus tinctorius ectomycorrhizae in forest tree nurseries for custom seedling production. In Proc 1982 South Nursery Conf, Tech Publ R8-TP4, pp 117-120
12. DAFT MJ, E HACSKAYLO 1977 Growth of endomycorrhizal and nonmycorrhizal red maple seedlings in sand and anthracite spoil. For Sci 23:207-216
13. DEWAULLE JC, J GARBAYE, GO OKOMBI 1982 Stimulation de la croissance initiale de Pinus caribaea Morelet dans une plantation du Congo par controle de la mycorrhization. Bois For Trop 196:25-323
14. DIXON RK, HE GARRETT, GS COX 1979 Containerized shortleaf pine seedlings show superior growth and ectomycorrhizal development with mist foliar fertilization. South J Appl For 3:154-157
15. DIXON RK, HE GARRETT, GS COX, PS JOHNSON, IL SANDER 1981 Container- and nursery-grown black oak seedlings inoculated with Pisolithus tinctorius: growth and ectomycorrhizal development following outplanting on an Ozark clear-cut. Can J For Res 11:492-296

16. DIXON RK, GM WRIGHT, HE GARRETT, GS COX, PS JOHNSON, IL SANDER 1981 Container- and nursery-grown black oak seedlings inoculated with Pisolithus tinctorius: growth and ectomycorrhizal development during seedlings production period. Can J For Res 11:487-491
17. DIXON RK, HE GARRETT, GS COX, DH MARX, IL SANDER 1984 Inoculation of three Quercus species with eleven isolates of ecotmycorrhizal fungi. I. Inoculation success and seedling growth relationships. For Sci 30:364-372
18. DUDDRIDGE JA, A MALIBARI, DJ READ 1980 Structure and function of mycorrhizal rhizomorphs with special reference to their role in water transport. Nature 287:834-836
19. FERGUSON JJ, SH WOODHEAD 1982 Increase and maintenance of vesicular-arbuscular mycorrhizal fungi. In NC Schenek, ed, Methods and Principles of Mycorrhizal Research, Am Phytopathol Soc, St Paul, pp 47-54
20. GILES KL, H WHITEHEAD 1976 Uptake and continued metabolic activity of Azotobacter within fungal protoplasts. Science 193:1125-1126
21. GODBOUT C, JA FORTIN 1985 Synthesized ectomycorrhizae of aspen: fungal genus level of structural characterization. Can J Bot 63:252-262
22. GOODWIN OC 1982 Four year research of loblolly and Virginia pine containerized seedlings colonized by Pisolithus tinctorius ectomycorrhizae. NC For Serv, Div For Res, Raleigh For Note 57, 3 pp
23. GRELLIER B, R LETOUZE, DG Strullu 1984 Micropropagation of birch and mycorrhizal formation in vitro. New Phytol 97:591-599
24. GROSSNICKLE SC, CPP REID 1982 The use of ectomycorrhizal conifer seedlings in the revegetation of a high-elevation mine site. Can J For Res 12:354-361
25. HARLEY JL 1969 The Biology of Mycorrhiza. Leonard Hill, London, 344 pp
26. HARLEY JL, SE SMITH 1983 Mycorrhizal Symbiosis. Academic Press, NY, 483 pp
27. HATCH AB 1937 The physical basis of mycotrophy in Pinus. Black Rock For Bull No 6, 168 pp
28. HATTINGH MJ, JW GERDEMANN 1975 Inoculation of Brazilian sour orange seed with an endomycorrhizal fungus. Phytopathol 65:1013-1016
29. HAYMAN DS 1983 The physiology of vesicular-arbuscular endomycorrhizal symbiosis. Can J Bot 61:944-963
30. KAIS AG, GA SNOW, DH MARX 1981 The effects of benomyl and Pisolithus tinctorius ectomycorrhizae on survival and growth of longleaf pine seedlings. South J Appl For 5:189-195
31. KORMANIK PP, WC BRYAN, RC SCHULTZ 1977 Influence of endomycorrhizae on growth of sweetgum seedlings from eight mother trees. For Sci 23:500-506
32. KORMANIK PP, WC BRYAN, RC SCHULTZ 1977 Quality hardwood seedlings require early mycorrhizal development in nursery beds. In Proc South Forest Tree Improv Conf 14:289-293
33. KORMANIK PP, WC BRYAN, RC SCHULTZ 1980 Increasing endomycorrhizal fungus inoculum in forest nursery soil with cover crops. South J Appl For 4:151-153

34. KORMANIK PP, RC SCHULTZ, WC BRYAN 1982 The influence of vesicular-arbuscular mycorrhizae on the growth and development of eight hardwood tree species. For Sci 28:531-539
35. MARONEK DM, JW HENDRIX, PL CORNELIUS 1982 Slow-release fertilizers optimize mycorrhizal development in container-grown pine seedlings inoculated with Pisolithus tinctorius. J Am Soc Hortic Sci 107: 1104-1110
36. MARONEK DM, JW HENDRIX, CD STEVENS 1981 Fertility mycorrhizal-isolate interactions in production of containerized pin oak seedlings. Scientia Hortic 15:283-289
37. MARX DH 1973 Growth of ectomycorrhizal and nonmycorrhizal shortleaf pine seedlings in soil with Phytophthora cinnamomi. Phytopathol 68:18-23
38. MARX DH 1975 Mycorrhizae of exotic trees in the Peruvian Andes and synthesis of ectomycorrhizae on Mexican pines. For Sci 21:353-358
39. MARX DH 1976 Synthesis of ectomycorrhizae on loblolly pine seedlings with basidiospores of Pisolithus tinctorius. For Sci 22:13-20
40. MARX DH 1977 The role of mycorrhizae in forest production. Tech Advisors Pulp Pap Ind Conf Pap, Annu Meet, Atlanta GA, pp 151-161
41. MARX DH 1977 Tree host range and world distribution of the ectomycorrhizal fungus Pisolithus tinctorius. Can J Microbiol 23:217-223
42. MARX DH 1979 Synthesis of Pisolithus ectomycorrhizae on white oak seedlings in fumigated nursery soil. USDA For Serv Res Note SE-280, 4 pp
43. MARX DH 1979 Synthesis of ectomycorrhizae by different fungi on northern red oak seedlings. USDA For Serv Res Note SE-282, 8 pp
44. MARX DH 1979 Synthesis of Pisolithus ectomycorrhizae on pecan seedlings in fumigated soil. USDA For Serv Res Note SE-283, 4 pp
45. MARX DH 1979 Pisolithus ectomycorrhizae survive cold storage on shortleaf pine seedlings. USDA For Serv Res Note SE-281, 4 pp
46. MARX DH 1980 Ectomycorrhizal fungus inoculations: a tool for improving forestation practices. In P Mikola, ed, Tropical Mycorrhiza Research, Oxford Univ Press, London, pp 13-71
47. MARX DH, JD ARTMAN 1978 Growth and ectomycorrhizal development of loblolly pine seedlings in nursery soil infested with Pisolithus tinctorius and Thelephora terrestris in Virginia. USDA For Serv Res Note SE-256, 4 pp
48. MARX DH, JD ARTMAN 1979 Pisolithus tinctorius ectomycorrhizae improve survival and growth of pine seedlings on acid coal spoils in Kentucky. Reclam Rev 2:23-31
49. MARX DH, W BELL 1985 Formation of Pisolithus ectomycorrhizae on loblolly pine seedlings with spore pellet inoculum applied at different times. USDA For Serv Res Paper, SE-249, 6 pp
50. MARX DH, WC BRYAN 1971 Formation of ectomycorrhizae on half-sib progenies of slash pine in aseptic culture. For Sci 17:488-492
51. MARX DH, WC BRYAN 1971 Influence of ectomycorrhizae on survival and growth of aseptic seedlings of loblolly pine at high temperature. For Sci 17:37-41

52. MARX DH, WC BRYAN 1975 Growth and ectomycorrhizal development of loblolly pine seedlings in fumigated soil infested with the fungal symbiont Pisolithus tinctorius. For Sci 21:245-254
53. MARX DH, B ZAK 1965 Effect of pH on mycorrhizal formation of slash pine in aseptic culture. For Sci 11:66-75
54. MARX DH, WC BRYAN, CE CORDELL 1976 Growth and ectomycorrhizal development of pine seedlings in nursery soils infested with the fungal symbiont Pisolithus tinctorius. For Sci 22:91-100
55. MARX DH, WC BRYAN, CE CORDELL 1977 Survival and growth of pine seedlings with Pisolithus ectomycorrhizae after two years on reforestation sites in North Carolina and Florida. For Sci 23:363-373
56. MARX DH, K JARL, JL RUEHLE, W BELL 1984 Development of Pisolithus tinctorius ectomycorrhizae on pine seedlings using basidispore-encapsulated seeds. For Sci 30:897-907
57. MARX DH, JG MEXAL, WG MORRIS 1979 Inoculation of nursery seedbeds with Pisolithus tinctorius spores mixed with hydromulch increases ectomycorrhizae and growth of loblolly pines. South J Appl For 3:175-178
58. MARX DH, WG MORRIS, JG MEXAL 1978 Growth and ectomycorrhizal development of loblolly pine seedlings in fumigated and nonfumigated soil infested with different fungal symbionts. For Sci 24:193-203
59. MARX DH, JL RUEHLE, DS KENNEY, CE CORDELL, JW RIFFLE, RJ MOLINA, WH PAWUK, S NAVRATIL, RW TINUS, OC GOODWIN 1982 Commercial vegetative inoculum of Pisolithus tinctorius and inoculation techniques for development of ectomycorrhizae on container-grown tree seedlings. For Sci 28:373-400
60. MARX DH, CE CORDELL, DS KENNEY, JG MEXAL, JD ARTMAN, JW RIFFLE, RJ MOLINA 1984 Commercial vegetative inoculum of Pisolithus tinctorius and inoculation techniques for development of ectomycorrhizae on bare-root tree seedlings. For Sci Monograph 25, 101 pp
61. MCCOWN BH, R AMOS 1979 Initial trials with commercial micropropagation of birch selections. Proc Int Plant Propagators Soc 29:387-393
62. MCKEAND SE, HL ALLEN 1984 Nutritional and root development factors affecting growth of tissue culture plantlets of loblolly pine. Physiol Plant 61: 523-528
63. MENGE JA 1983 Utilization of vesicular-arbuscular mycorrhizal fungi in agriculture. Can J Bot 61:1015-1024
64. MENGE JA, H LEMBRIGHT, ELV JOHNSON 1977 Utilization of mycorrhizal fungi in citrus nurseries. Proc Int Soc Citriculture 1:129-137
65. MEXAL JG 1980 Aspects of mycorrhizal inoculation in relation to reforestation. N Z J For Sci 10:208-217
66. MEYER FH 1973 Distribution of ectomycorrhizae in native and man-made forests. In GC Marks and TT Kozlowski, eds, Ectomycorrhizae: Their Ecology and Physiology. Academic Press, NY, pp 79-106
67. MIKOLA P 1973 Application of mycorrhizal symbiosis in forestry practice. In GC Marks and TT Kozlowski, eds, Ectomycorrhizae: Their Ecology and Physiology. Academic Press, NY, pp 383-411

68. MOLINA R 1979 Ectomycorrhizal inoculation of containerized Douglas-fir and lodgepole pine seedlings with six isolates of Pisolithus tinctorius. For Sci 25:585-590
69. MOLINA R 1980 Ectomycorrhizal inoculation of containerized western conifer seedlings. USDA For Serv Res Note PNW-357, 10 pp
70. MOLINA R, JM TRAPPE 1982 Applied aspects of ectomycorrhizae. In NS Subba Rao, ed, Advances in Agricultural Microbiology, Oxford and IBH Publ Co, New Delhi, pp 305-324
71. MOLINA R, JM TRAPPE 1982 Patterns of ectomycorrhizal host specificity and potential among Pacific Northwest conifers. For Sci 28:423-458
72. MOMOH ZO, RA GBADEGESIN 1980 Field performance of Pisolithus tinctorius as a mycorrhizal fungus of pines in Nigeria. In P Mikola, ed, Tropical Mycorrhiza Research. Oxford Univ Press, London, pp 72-79
73. MOSER M 1958 Die kunstliche Mykorrhizaimpfung an Forestpflanzen. I. Erfahrungen bei der Reinkulture von Mykorrhizapilzen. Forstwiss Centrabl 77:32-40
74. NAVRATIL S, NJ PHILLIPS, A WYNIA 1981 Jack pine seedling performance improved by Pisolithus tinctorius. For Chron 57:212-217
75. OTROSINA WJ 1977 Microbiological and ectomycorrhizal aspects of kaolin spoil. PhD thesis, Univ GA, Athens, 78 pp
76. PAWUK WH, JL RUEHLE, DH MARX 1980 Fungicide drenches affect ectomycorrhizal development of container-grown Pinus palustris seedlings. Can J For Res 10:61-64
77. POPE PE, WR CHANEY, JD RHODES, SH WOODHEAD 1983 The mycorrhizal dependency of four hardwood tree species. Can J Bot 61:412-417
78. POWERS HR Jr, SJ ROWAN 1983 Influence of fertilization and ectomycorrhizae on loblolly pine growth and susceptibility to fusiform rust. South J Appl For 7:101-103
79. REEVES FB, D WAGNER, T MOORMAN, J KIEL 1979 The role of endomycorrhiza in revegetation practices in the semi-arid west. I. A comparison of incidence of mycorrhizae in severely disturbed vs. natural environments. Am J Bot 66:6-13
80. REID CPP 1979 Mycorrhizae and water stress. In A Reidacker, J Gagnaire-Michard, eds, Proc IUFRO Symp Root Physiol and Symbiosis, Nancy, France, pp 392-408
81. RIFFLE JW, RW TINUS 1982 Ectomycorrhizal characteristics, growth and survival of artificially inoculated ponderosa and Scots pine in a greenhouse and plantation. For Sci 28:646-660
82. RUEHLE JL 1973 Nematodes and forest trees -- Types of damage to tree roots. Annu Rev Phytopathol 11:99-108
83. RUEHLE JL 1980 Growth of containerized loblolly pine with specific ectomycorrhizae after 2 years on an amended borrow pit. Reclam Rev 3:95-101
84. RUEHLE JL, DH MARX 1977 Developing ectomycorrhizae on containerized pine seedlings. USDA For Serv Res Note SE-242, 8 pp
85. RUEHLE JL, DH MARX, M ABOUROUH 1981 Development of Pisolithus tinctorius and Thelephora terrestris ectomycorrhizae on seedlings of coniferous trees important to Morocco. Ann Rech For Maroc 21:181-296

86. RUEHLE JL, DH MARX, JP BARNETT, WH PAWUK 1981 Survival and growth of container-grown and bare-root shortleaf pine seedlings with Pisolithus and Thelephora ectomycorrhizae. South J Appl For 5:20-24
87. SCHRAMM JR 1966 Plant colonization studies on black wastes from anthracite mining in Pennsylvania. Trans Am Phil Soc 56:1-194
88. SCHULTZ RC 1979 Tree growth response to changes in soil moisture, fertility, and microorganisms on difficult sites. In Proc 5th North Am Forest Biol Workshop, Gainesville, FL, pp 2-18
89. SOUTH D 1977 Artificial inoculation of fumigated nursery beds with endomycorrhizae. Tree Plant Note 31:3-5
90. TAKACS EA 1967 Produccion de cultivos puros de hongos micorrizogenos en el Centrol Nacional de Investigaciones Agropecuarias, Castelar. India Supplemento Forestal 4:83-87
91. THEODOROU C 1971 Introduction of mycorrhizal fungi into soil by spore inoculation of seed. Aust For 35:23-26
92. THEODOROU C, GD BOWEN 1970 Mycorrhizal responses of radiata pine in experiments with different fungi. Aust For 34:183-191
93. THEODOROU C, GD BOWEN 1973 Inoculation of seeds and soil with basidiospores of mycorrhizal fungi. Soil Biol Biochem 5:765-771
94. TRAPPE JM 1977 Selection of fungi for ectomycorrhizal inoculation in nurseries. Annu Rev Phytopathol 15:203-222
95. TRAPPE JM, NC SCHENCK 1982 Taxonomy of the fungi forming endomycorrhiza. In NC Schenck, ed, Methods and Principles of Mycorrhizal Research, Am Phytopathol Soc, St. Paul, MN, pp 1-10
96. VERMILLION TI 1982 Loblolly seed encapsulated with Pt mycorrhizae shows early growth response. Res Note 82-1, Crown Zellerbach, Southern Timber Division, 5 pp
97. VOZZO JA 1971 Field inoculations with mycorrhizal fungi. In E Hacskaylo, ed, Mycorrhizae, USDA For Serv Misc Publ No. 1189, pp 187-196
98. VOZZO JA, E HACSKAYLO 1971 Inoculation of Pinus caribaea with ectomycorrhizal fungi in Puerto Rico. For Sci 17:239-245
99. WALKER FR, DC WEST, SB MCLAUGHLIN 1982 Pisolithus tinctorius ectomycorrhizae reduce moisture stress of Virginia pine on a southern Appalachian coal spoil. In BA Thielges, ed, Proc 7th N Am For Biol Workshop, Univ KY, pp 374-383.
100. WRIGHT E, KK CHING 1962 Effect of seed source on mycorrhizal formation on Douglas-fir seedlings. Northwest Sci 36:1-6

19. TISSUE CULTURE APPLICATION TO FOREST PATHOLOGY AND PEST CONTROL

A.M. DINER and D.F. KARNOSKY

1. INTRODUCTION

The use of tissue culture methodologies for studies of pathogen physiology, host vs. pathogen interrelationships and disease resistance has a short but productive history in agronomy and horticulture (23, 44, 45). The very recent applications of these procedures for forest trees show tremendous potential to circumvent some of the difficult logistical problems of studying disease *in vivo*. Technical and informational gains made in other plant sciences are now being examined by forest pathologists to supplement and accelerate traditional breeding programs (60, 66).

In contrast to most agronomic crops, trees require 10 to 40 years or more in place to harvest, while facing windborn or vector-inoculated microbial adversaries capable of undergoing innumerable genetic recombinations in a single season. Rapid gains in resistance may be made only through an understanding of the genetics of host and pathogen, and their physiological interrelationships.

In vitro techniques avoid many limitations implicit in field studies where large acreages and long time periods are required to grow, inoculate, characterize the pathology, and appraise the susceptibility or resistance of host plants. These studies are usually quantitative and relative, and are often confounded by site and microclimate variability, physiological age of tissues at the inoculation site, infection by opportunistic saprophytes, and other biotic and abiotic stresses which are generally beyond our control *in vivo* (7, 24, 30, 33, 46, 59, 63, 77, 81, 89, 101, 113, 129, 133). Further major impediments to *in vivo* resolution of a plant pest problem are the phenophases of pest and host, where virulent pest and susceptible host are simultaneously available for only a short period of the host's growing season (115). Assays of infection or resistance are also complicated by seasonal fluctuations in climate, which can render some infections temporarily latent (112).

The practical applicability of rapid *in vitro* expression of characteristic disease pathology and selection of resistant variants requires substantial background genetic information on both the tree host and its pathogens (45, 56). This information is currently scant, and must be evolved to define single genotypes (98) of these genetically flexible pathogens for which multigenic or horizontal (132) host resistance needs to be found or established. The job ahead is thus enormous, and *in vitro* procedures should be developed to assist traditional programs.

In vitro systems offer ready access to selected organs and tissues of the tree under controlled and aseptic conditions, permitting close scrutiny of inoculation events and mechanisms, and host and pathogen physiological interrelationships. The synthesis, regulation and activities of such chemical mediators as phytotoxins, phytoalexins, peroxidases, and polyphenols may be selectively explored *in vitro*. And, genetics of both pathogenesis and resistance may be examined "in the bottle" through use of cloned hosts and pathogens. Yet, behavior of host or pathogen in the laboratory can sometimes prove artifactual (34, 57) so that the researcher must appreciate both the advantages and limitations of *in vitro* culture and/or host challenge.

Numerous inroads in forest pathology are being made in the tissue culture laboratory by plant scientists studying an increasing variety of tree diseases. Research objectives include life cycles of the specific pathogens and their physiologies in monoculture, host vs. pathogen physical and physiological interrelationships, biochemical mechanisms of both disease specificity and host resistance exclusion of the pathogen, and the identification, modification, and selective transfer and expression of genetic determinants underlying those mechanisms.

Excellent reviews describe tissue culture applications in phytopathology (23, 44, 45) and forestry (85) in particular. However, a considerable amount of research has been done since their publication. This chapter will concentrate on *in vitro* studies of forest tree diseases, carried out in the past few years.

2. HOST AND PATHOGEN: CULTURE AND CHALLENGE

2.1. Organized tissues of the host

The intact plant is logically the best representative of the natural host, since mechanisms determining host/pathogen specificity and host resistance mechanisms are often expressed at or near the intact epidermis (49, 68, 116), and wounded or otherwise stressed plants may show anomalous reactions to inoculation with pathogen (80, 89). Young embryos, seedlings, isolated plant organs or organ parts, and even plantlets generated via tissue or cell culture have been inoculated with specific pathogens or their products.

Young embryos of sugar pine, *Pinus lambertiana* Dougl. were inoculated *in vitro* with vegetative hyphae of the blister rust fungus, *Cronartium ribicola* J.C. Fisch, ex. Rabenh., grown from basidiospores in axenic culture (19). Characteristic genetic resistance was expressed within 2 weeks by these succulent embryos which were commonly inoculated through and between epidermal cells (17). Stomatal penetration, the natural route of inoculation by basidiospore germ tubes, was rarely seen, and thus not required for resistance expression *in vitro*.

Basidiospores of the fusiform rust fungus *Cronartium quercuum* (Berk.) Miyabe ex Shirai f. sp. *fusiforme* were used to inoculate loblolly pine, *Pinus taeda* L. embryos (31, 32) and seedlings (57) *in vitro*. Embryos from seed lines showing field resistance to fusiform rust exhibited rapid (7-day) incompatible necrosis of fungal hyphae, and formation of wall appositions in penetrated cells of the host (31). Later work (28) assessed 3 different host resistance reactions to inoculation with this fungus. Disease resistance expressed by half-sib families in the field showed a 0.986 correlation coefficient with *in vitro* expression of these 3 reactions. Mechanisms of pest resistance expressed within days or weeks *in vitro* offer much hope in battles against blister and fusiform rust diseases, where field assays take several years. Basidiospores of eastern gall rust fungus, *Cronartium quercuum* (Berk.) Miyabe ex Shirai f. sp. *banksianae* Burdsall & Snow were used to inoculate tissue culture propagules of Jack pine, *Pinus banksiana* Lamb. *in vitro*, (Diner and Karnosky, unpublished data). Host propagules were generated from embryonic tissues (62). Intracellular haustoria were observed microscopically in needle epidermis and cortex within 7 days following application of spores.

Seedling stem wounds of *Pinus monticola* Dougl. have been inoculated with vegetative *C. ribicola* grown from infected stem callus (41). *P. monticola* propagules generated in cotyledon culture were routinely inoculated *in vitro* through stem tip wounds using *C. ribicola* hyphae grown in culture from basidiospores (21). Similar cultures also inoculated primary needles through stomates. Such inocula have not yet been used to challenge clones of tissue culture plantlets generated from resistant seed lines, although methods exist to clone both host (84) and pathogen (18) in culture.

Organs or parts of organs of trees grown under natural conditions have been isolated *in vitro* and inoculated with specific pests or phytotoxic pest products. Leaf discs from field-grown clones of eastern cottonwood, *Populus deltoides* Bartr., showing either resistance or susceptibility to *Melamspora medusae* Thum. were inoculated with urediospores (117). Resultant *in vitro* expression of resistance was histopathogically similar to that seen under conditions of natural inoculation of the tree. Further, clonal fidelity was displayed *in vitro*. In another study (128), pathotoxins of *Hypoxylon mammatum* (Wahl.) Miller were purified and used to reflect clonal, genetic resistance or susceptibility to the fungus itself, using excised leaves of quaking aspen, *Populus tremuloides* Michx. The assay required 48 hours for resolution. A method, using the purified mammatoxin, has also been described (135) to screen organ cultures of quaking aspen for resistance to the fungus, directly identifying and selecting resistant clones *in vitro*.

2.2. Cell cultures of the host

The use of callus or suspension cultures as host tissues for *in vitro* studies of pathology and resistance in trees is a tantalizing prospect that has had disappointing results to date. Under circumstances where characteristic tissue pathology or forms of resistance to disease are expressed by specific organized tissues of the mature host (8), attempts to employ cell cultures may be ill-considered. However, in at least one case, that of *P. lambertiana* versus *C. ribicola* (20), cell cultures of a tree host reliably displayed characteristic genetic resistance when inoculated with a vegetative culture of the pathogen. Resistance expression was rapid and similar to that seen in young embryos (19) and in basidiospore-inoculated primary and secondary needles and cotyledons (64). Unfortunately, immediate usefulness of this resistance mechanism has proven limited (65) since it is regulated by a single dominant allele, thus adding support to arguments which favor breeding for multigenic resistance (74).

Selective modifications of the physical or chemical environment may be used to alter *in vitro* growth of host, the pathogen or their co-cultures. Cell cultures derived from a few tree species have been manipulated in attempts to elucidate the physiology of infection processes. The list of hosts includes cell cultures of American elm, *Ulmus americana* L. (47), American chestnut, *Castanea dentata* (Marsh.) Borkh. (9, 43), Douglas fir, *Pseudotsuga menziesii* (Mirb.) Franco (40) and 3- (57, 88) and 5-needle pines (20, 39, 109, 110). However, cell cultures of the naturally susceptible host may sometimes not support inoculation. Basidiospores or vegetative hyphae of *C. fusiforme* did not inoculate *P. taeda* cell cultures; the callus produced a water-soluble, agar-diffuseable substance lethal to the pathogen (57). *Pinus banksiana* cell cultures were similarly lethal to vegetative hyphae of the Eastern gall rust fungus grown from basidiospores on cell-free medium (Diner and Karnosky, unpublished data). An understanding of how syntheses of these as yet unidentified toxic substances are regulated in culture, may provide insight into their possible regulation in whole tree physiology.

The use of genetic vectors and recombinant DNA techniques will likely be soon applied to breeding programs in forestry. Cell cultures may prove useful as hosts or as a rapid screening tool for host species susceptibility-to-inoculation with vectors such as the agrobacteria. Cell cultures generated from chopped embryos of European larch, Larix decidua were rapidly colonized and killed by Agrobacterium rhizogenes (Riker et al.) Conn; similarly inoculated cultures of Jack pine were essentially uncolonized during the same 2-week inoculation period (Diner and Karnosky, unpublished data). Such differential responses to this microorganism correlated positively with susceptibility to in vitro inoculation of young seedlings of these two conifers (see "Genetic Vectors", below).

2.3. Monocultures of the pest

Understanding pest physiology and mechanisms of its genetic regulation, as well as timing and development of pest virulence versus host susceptibility can enhance both management of the tree crop and the progress of in vivo and in vitro pest resistance screening. Numerous studies have been undertaken to explore pest physiology in monoculture, where effects of selected modifications in the in vitro environment can be studied.

The effects of light, gravity, temperature and medium composition on spore germination and axenic, in vitro hyphal development and growth of some Cronartia have been examined (2, 16, 31, 38, 41, 76). In vitro growth of specific (13, 42, 76) and opportunistic (82) fungal pathogens on media supplemented with selected carbon sources examined interactions between these pests and metabolism of host constituents. Colonial morphology of at least one important tree pathogen, Endothia parasitica (Murr.), P.J. & H.W., has been used to screen for virulence or hypovirulence (14, 70). Hypovirulent isolates grown in vitro were converted to virulence by cycloheximide elimination of double-stranded RNA (29). Subsequent virulence in vivo was correlated with colonial morphology of the cultured, altered isolates. This indicates that *in vitro* screening can be employed directly against genetic variants of a tree pathogen for study of the molecular regulation of virulence.

Numerous other tree-pathogenic fungi have been isolated and grown in cell-free culture, both for studies of pest life cycle and physiology and for continual availability of virulent inocula. Such pathogens include the butternut canker fungus, *Sirococcus clavigignenti-juglandacearum* Nair *et al.*, (94), elm black spot fungus, *Stegophora ulmea* (78), hypoxylon canker fungus, *Hypoxylon mammatum* (Wahl.) Miller (35), root rot fungus, *Heterobasidion annosum* (Fr.) Bref. (51), peach canker fungus, *Cytospora leucostoma* (Pers. ex Fr.)(7), cypress canker fungus, *Seirdium cardinale* (Wag.) Sutton *et al*. Gibson (106), poplar rust fungus, *Melampsora occidentalis* Jacks. (72), scleroderris canker fungus, *Gremmeniella abietina* (Lagerb.) Morelet (22), and the willow rust fungi *Melampsora capraearum* (DC.) Thumen, *M. epiphylla* Dietel, *M. epitea* (Kunze et Schm.) Thumen, and *M. humilis* Dietel (140).

Life cycles and nutritional requirements of insect pests of trees are resolved with great difficulty in the natural environment, if at all. Yet the specific physical and physiological requirements shown by the different developmental stages of many insects complicates their rearing *in vitro*. Among the few tree insect pests reported successfully reared *in vitro* are larvae of the southern pine beetle, *Dendroctonus frontalis* Zimmerman (88), eastern spruce budworm, *Choristoneura fumiferana* Clem.(79) and fall webworm, *Hyphantria cunea* Drury (131), thus offering promise for similar advances in studies of other parasitic insects. Growth and development of other parasitic insects *in vitro*, as well as responses of some to controlled application of certain host plant chemicals (130) offer further insight into physiological requirements of these pests.

Another potentially important tool in forest protection is the *in vitro* rearing of parasites of insect pests. Defoliating insects such as larvae of the gypsy moth, *Lymantria dispar* (Linnaeus) and other leaf eaters may be partially controlled by application of Bacillus thuringensis berliner spores, generated in batch culture in the laboratory. Insect parasites of Diptera spp. were mass-reared in vitro (137) and readily attacked representative host larvae. Larvae of the fall webworm, reared on several artificial diets were inoculated with nuclear and cytoplasmic polyhedrosis viruses, to determine susceptibility to infection (71).

3. SOMACLONAL VARIATION

Adventitious embryogeny from plant cell cultures frequently suffers from chromosomal aberrations and associated intra-clonal variability (5, 26). Changes in the DNA in cultured cells may involve changes in either the nuclear (4, 99) or cytoplasmic (103) fraction. These changes, which may be detectable in culture, are only one possible source of variation within clones of regenerant plants (somaclonal variation). Three other likely sources are induced mutagenesis (134), deregulation or decontrol of genome expression in the unnatural "microculture environment" (92), and differential responses by component cell types to plant growth regulators added to heterogenous cell cultures.

Whatever their causes, these high frequency variants (92) may be experimentally exposed under defined in vitro conditions of cell culture to such selection pressures as inoculation with a specific pathogen (43) or its toxins (51). Subsequent selection of disease-resistant variants may be made from the partially inhibited (118) or even killed (135) residual cell population.

Somaclonal variation in cell cultures derived from tissues of any multicellular organism should not be an unexpected phenomenon. Even in a genome so small as that found in the bacterium Escherichia coli (Migula) Castellani and Chalmers, the frequency of a spontaneous point mutation approximates 10^{-7} per cell per generation. Should any such mutation be both non-lethal and identifiably expressed, a significant population of mutant progeny may be selected for, within several hours. Greater opportunities for such mutational events exist in cell cultures of trees. The genomic DNA of a cultured cell of Pinus coulteri D. Don. shows a 2C value of 17 picograms (99) in contrast to the

0.02 picograms of E. coli DNA (136), thus offering a 850-fold increase in target size for any particular mutagen.

A further possible source of somaclonal variation is the cellular heterogeneity in the explant itself. Collections of tissues representing juvenile or mature explants commonly used to initiate cell cultures cannot be relied upon to generate a cytologically homogeneous cell population (1). Furthermore, application of natural or synthetic growth regulators may differentially stimulate the various cell types to give rise to phenotypically mixed populations of regenerants. Proportions of regenerant types may be further influenced by intrinsic selection pressures which may favor the growth of component fractions within mixed cell cultures (118), pressures exerted prior to medium supplementation with those higher levels of growth regulators effective for regeneration of plantlets.

The phenomenon of somaclonal variation may prove a mixed blessing. If universal, a genetically uniform cell population in culture may be an impossibility, thus dictating the perpetual need for progressively more refined selection devices for desired regenerants. However, genetic instability in culture may prove to be peculiar to certain species or to conditions of culture. Loblolly (108) and lodgepole pine, Pinus contorta Dougl ex Loud. (100) as well as Norway spruce, Picae abies (L.) Karst. (36) plantlets regenerated from embryonic tissues showed no apparent chromosomal alterations. Such stability was also expressed by cell cultures of several pine species (108).

Resistance to disease may not be quickly expressed by a somaclonal variant, similar to limitations shown by some seedlings. For example, blister rust-resistant western white pines show no genetic resistance in primary needles; seedlings are not experimentally inoculated until at least 2 years of age (97). Thus, the prospect of maintaining necessarily very large unselected host populations of regenerants in extensive ex vitro acreages, to an age when inoculation might resolve differences in resistance to pathogen, may be overwhelming.

Should selectively induced variation in plants offer economic usefulness, the most likely crop candidates would seem to be agronomic annuals such as alfalfa, tobacco, corn, tomatoes and some cereals (91, 102), for which extensive genetic information is available. Disease resistance expression in developing annuals would hopefully not suffer the long and vulnerable lag sometimes shown by young tree seedlings.

Obviously, resolution of the usefulness of variant annuals would be far more rapid than the decade or more required for forest trees. Still, even among agronomic annuals, selectable and economically important markers are few (91). Thus, rewards from attempts to capture somaclonal variation in tree tissues may be significantly less immediate.

Somaclonal variation in trees has been documented for _Populus_ spp. regenerated from callus culture (75). When examined, cultured tissues of other broadleaf trees may also show somaclonal variation which may then be captured in regenerants. For reasons expressed above, cell cultures likely generate variants more rapidly than might similarly treated embryonic tissues _in vitro_. However, embryonic tissues are those commonly used for conifer micropropagation (52, 62, 87). Conifers have not been reliably regenerated from cell cultures.

Ongoing attempts to apply somaclonal technology to forest trees (92) include use of both embryo and callus cultures of conifers as well as various tissues of broadleaf trees. Use of conifer cell cultures is not unrealistic. Obstacles to conifer regeneration from cell culture will likely soon be eliminated. And conifer calli have shown characteristic histopathology and genetic resistance to disease (20).

Somaclonal variation may prove to be the detectable consequence of genetic events common in all higher plants. Cell divisions in apical meristems _in vivo_ likely provide plants with mutant cells leading to chimeras and other genetically mixed tissue populations, though mutation frequency may be only a fraction of that seen from culture. A pertinent question is therefore "How is the genetic integrity of the plant body maintained against the constant pressure of somatic mutation?" (67). Possibly it is not. Clones of common cypress, _Cupressus sempervirens_ L., inoculated with the etiologic agent of cypress canker showed deviant responses among transplanted ramets (106). Susceptible and resistant ramets were observed within resistant and susceptible clones, respectively. Conversely, chromosomal instability or aberrations may be common in tree species showing little, if any, phenotypic variation. Porcelia goyazensis R.E.F.R. shows frequent structural heterozygosity in C-bands of the first chromosome pair (86). Yet, Porcelia has shown essential morphological stability throughout its extensive pantropical range.

The fact that most mutations are lethal, indicates that most are never fully expressed. Of the small fraction that survive, most are somehow deleterious. Thus, only an extremely small percentage of mutants which evolve either in vivo or in vitro might generate a population of usefully altered tissues or plantlets. We must somehow be able to identify these alterations. Plant tissue culture methodologies in controlled in vitro environments appear to offer the greatest opportunities to induce and identify mutations and then to select desired regenerants.

4. VIRUS AND MYCOPLASMA ELIMINATION

Probably the most significant current application of in vitro techniques to resolve disease problems involves the elimination of viruses, mycoplasmas and other microbial pathogens by meristem culture, either alone or in combination with heat or chemical treatments (23). Systemic pathogens are difficult, and in some cases impossible, to eliminate from stock plants through chemical sprays or treatments. Furthermore, these pests are transferred from generation to generation when vegetative propagation is used. Morel and Martin (83) first demonstrated that meristem culture could produce virus-free plants. Since then, meristem culture and/or heat treatments applied prior to meristem culture have been used to eliminate viruses from stock plants in horticultural species such as potatoes, strawberries, peaches, cherries and apples (58, 104). Transfer of germplasm between laboratories and between countries is also facilitated by isolation of disease-free plants via meristem culture (111).

Only a limited application has been made of virus elimination in forest trees. Berbee *et al.* (5, 6) demonstrated that it was possible to eliminate viruses from poplars, via meristem culture in combination with heat treatments in the range of 35 to 50C. The early field tests from this work indicated that the disease-free poplars grew at twice the rate of the same hybrid poplars that were not rendered disease-free (3). Whitehead and Giles (139) showed that rapid micropropagation of poplar was possible *in vitro*, and they discussed the advantages of this technique including the ability to exchange disease-free material internationally and to maintain stock plant material in a juvenile form for later propagation. As more emphasis is placed on clonal forestry in the future, it is likely that maintaining disease-free stock plants will become as important in forestry as it now is in horticulture.

5. GENETIC VECTORS

Although pathogen elimination remains possibly the most important current application of *in vitro* techniques related to diseases, the most exciting future application may be the use of bacterial vectors for transfer of DNA. By far the most promising vector system appears to be the tumor-inducing (Ti) and (Ri) plasmids of the soil bacteria *Agrobacterium tumefaciens* (Smith and Townsend) Conn. (10, 12, 50, 55, 107) and *A. rhizogenes* (11, 127, 138), respectively. *Agrobacterium* species can be modified so that the tumor-inducing genes are deleted and replaced with genes-of-choice for subsequent transfer to host plant cell . These systems have been demonstrated by the incorporation of a kanamycin-resistance gene into petunia, tobacco, and tomato plants (53), and by the transfer and expression of the phaseolin gene from bean to sunflower (90). In addition, the neomycin phosphotransferase gene from the bacterial transposon Tn5 and the yeast alcohol dehydrogenase gene have been introduced into tobacco, and the maize zein gene into sunflower (73). The stability of the transformed plants has been documented by Mendelian inheritance of the foreign genes by transformant offspring (54, 93, 95).

A wide range of susceptibility to agrobacteria which harbor Ti and Ri plasmids has been demonstrated (96). The total host range of *Agrobacterium* probably includes thousands of different plant species (15). The known hosts are predominantly dicotyledonous plants, but some species of gymnosperms have also been reported to be susceptible to *A. tumefaciens* (69, 120, 121, 122, 123, 124, 125, 126). None of the nine pine species or three larch species tested by De Cleene and De Ley (15) were susceptible to infection by *A. tumefaciens*.

The only reported successful infection of a pine species with *A. tumefaciens* was by Sederoff *et al.* (114). They described inoculation and infection in a few *Pinus taeda* seedlings. Use of only one of 19 different *A. tumefaciens* strains tested resulted in successful infection.

We have recently succeeded in infecting young *Larix decidua* seedlings with *A. rhizogenes*. Hypocotyl and shoot apex inoculations resulted in infection symptoms in 74 of 91 *L. decidua* seedlings inoculated. Symptoms included stem swelling, blue needles (short, wide, blue needles with dense aggregations of stomates), adventitious roots, and multiple buds.

Verification of genetic transformation demonstrable by opine production, and by establishing that the bacterial DNA has been incorporated into host tissue DNA yet remains to be done with both the *Pinus taeda* and *Larix decidua* systems. However, the preliminary results look encouraging.

Because trees are long-lived, slow to reach sexual maturity, large, and sporadic seed producers, traditional tree improvement via breeding and selection is a slow, laborious, and costly exercise (27, 119). The use of *Agrobacterium* vectors offers numerous opportunities for genetic engineering of trees. These opportunities include the ability to transfer:

1. genes from one species to another otherwise impossible through sexual crosses;
2. genes in a single step, rather than through repeated crosses or backcrosses;
3. only specific target genes exclusive of detrimental genes which may be linked to the target gene, and

4. genes rapidly (105).

The likely first targets-of-choice for gene transfer into trees will be those genes controlling resistance to commonly used herbicides. Resistance to S-triazine herbicides seems to be under control of single genes (25, 48) and thus may be opportune targets for genetic engineering. Conifers show varying degrees of sensitivity to S-triazine herbicides such as Velpar (37) and many such trees (including *Larix* species) could benefit from increased resistance to them. Poplars and other hardwood species could be improved by incorporating glyphosate-(Roundup) resistance genes.

The usefulness of genetic information transfer into some trees via vectors such as the *Agrobacterium* species is somewhat limited by our current inability to regenerate whole plants from single cells and/or protoplasts (61). To date, only a few hardwood trees and no conifers have been reliably regenerated from single cells, protoplasts or callus cultures. Plants regenerated after genetic manipulation will also have to be thoroughly field tested before being released for commercial production. In spite of these logistical difficulties, the future looks very bright for the use of nucleic acid vectors for genetic engineering of trees.

6 CONCLUSION

The successes with *in vitro* studies reported here suggest that many further such opportunities exist. Reliable techniques for selective genetic recombination and protoplast fusion, as well as plantlet regeneration from tissues-of-choice, will hasten our ability to develop genetically improved forest trees. Programs of intensive forest management, required for efficient growth of tree crops, will be enhanced by contributions made progressively more refined as we learn more of pest physiology, pest versus host physical and physiological relationships, and the molecular regulation of pest resistance. However, information generated in the laboratory must be verified *in vivo*; resistance expression under natural, field conditions must be the final arbiter.

7. REFERENCES

1. AHUJA MR 1983 Isolation and culture of mega and normal protoplasts in aspen. Silvae Genet 32: 5-6
2. AMERSON HV, RL MOTT 1978 Technique for axenic production and application of Cronartium fusiform basidiospores. Phytopathology 68: 673-675
3. ANONYMOUS 1974 Scientists claim to double growth rate for hybrid poplar. J For 72: 299
4. BAYLISS MW, AR Gould 1974 Studies on the growth in culture of plant cells. XVIII Nuclear cytology of Acer pseudoplatanus suspension cultures. J Exp Bot 25: 772-783
5. BERBEE FM, JG BERBEE, AC HILDEBRANT 1972 Induction of callus and trees from stem tip cultures of a hybrid poplar. In vitro 7: 269
6. BERBEE JG, JO OMUEMU, RR MARTIN, JP CASTELLO 1976 Detection and elimination of viruses in Poplulus. In Intensive Plant Culture. USFS-USDA Tech Rep NC21 N Cent For Exp Sta pp 85-91
7. BIGGS AR 1984 Boundary zone formation in peach bark in response to wounds and Cytospora leucostoma infection. Can J Bot 62: 2814-2821
8. BINGHAM RT 1983 Blister rust resistant western white pine for the inland empire: The story of the first 25 years of the research and development program. USDA Gen Tech Rep INT-146 pp 45
9. BORROD G 1974 Parasitic attack on chestnut by Phytophthora cinnamomi: Enzymatic activities of the fungus. Ann de Phytopathologie 6: 83-90
10. CAPLAN A, L HERRERA-ESTRELLA, D INZE, EV HAUTE, MV MONTAGU, J SCHELL, P ZAMBRYSKI 1983 Introduction of genetic material into plant cells. Science 222: 815-821
11. CHILTON M-D, DA TEPFER, A PETIT, C DAVID, F CASSE-DELBART, J TEMPE 1982 Agrobacterium rhizogenes inserts T-DNA into the genomes of the host plant root cells. Nature 295: 432-434
12. COCKING EC, MR DAVEY, D PENTAL, JB POWER 1981 Aspects of plant genetic manipulation. Nature 293: 265-270
13. CUTTER VM JR 1959 Studies on the isolation of plant rusts in host tissue cultures and upon synthetic media. I. Gymnosporangium. Mycologia 51: 248-295
14. DAY PR, JA DODDS, JE ELLISTON, RA JAYNES, SL ANAGNOSTAKIS 1977 Double-stranded RNA in Endothia parasitica. Phytopathology 67: 1393-1396
15. DE CLEENE M, J DE LEY 1979 The host range of crown gall. Bot Rev 42: 389-411
16. DINER AM, RL MOTT 1982a Axenic cultures from basidiospores of Cronartium ribicola. Can J Bot 60: 1950-1955
17. DINER AM, RL MOTT 1982b Direct inoculation of five-needle pines with Cronartium ribicola in axenic culture. Phytopathology 72: 1181-1184
18. DINER AM, RL MOTT 1982c A method of cloning the

blister rust fungus in axenic culture. Phytopathology 72: 982
19. DINER AM, RL MOTT 1982d A rapid axenic assay for hypersenstive resistance of Pinus lambertiana to Cronartium ribicola. Phytopathology 72: 864-865
20. DINER AM, RL MOTT and HV AMERSON 1984 Cultured cells of white pine show genetic resistance to axenic blister rust hyphae. Science 224: 407-408
21. DINER AM, RL MOTT 1985 In vitro inoculation of western white pine tissue culture propagules with vegetative hyphae of Cronartium ribicola. Phytopathology 75: 1130-1131
22. DORWORTH CE, J KRYWIENCZYK 1975 Comparisions among isolates of Gremmenilla abietina by means of growth rate, conidia measurement, and immunogenic reaction. Can J Bot 53: 2506-2525
23. EARLE ED 1982 Application of plant tissue culture in the study of host-pathogen interactions. In DT Tomes, DE Ellis, PM Harney, KJ Kasha, RL Peterson, eds, Application of Plant Cell and Tissue Culture to Agriculture and Industry. Univ Guelph, Guelph, Canada pp 45-62
24. EISENSMITH SP, TM SJULIN, AL JONES, CE CRESS 1982 Effects of leaf age and inoculum concentration on infection of sour cherry by Coccomyces hiemalis. Phytopathology 72: 574-577
25. ERICKSON JM, M RAHIRE, JD ROCHAIX 1985 Herbicide resistance and cross-resistance: Changes at three district sites in the herbicide-binding protein. Science 228: 204-207
26. EVANS DA, WR SHARP, HP MEDINA-FILHO 1984 Somaclonal and gametoclonal variation. Am J Bot 71: 759-774
27. FARNUM P, R TIMMIS, JL KULP 1983 Biotechnology of forest yield. Science 219: 694-702
28. FRAMPTON LJ, JR, HV AMERSON, RJ WEIR 1983 Potential of in vitro screening of loblolly pine for fusiform rust resistance. In Proc 17th South For Tree Impr Conf, pp 325-31
29. FULBRIGHT DW 1984 Effect of eliminating ds RNA in hypovirulent Endothia parasitica. Phytopathology 74: 722-724
30. GRANT BR, PN BYRT 1984 Root temperature effects on the growth of Phytophthora cinnamoni in the roots of Eucalyptus marginata and E. calophylla. Phytopathology 74: 179-184
31. GRAY DJ, HV AMERSON 1983 In vitro resistance of embryos of Pinus taeda to Cronartium quercuum f sp fusiforme: Ultrastructure and cytology. Phytopathology 73: 1492-1499
32. GRAY DJ, HV AMERSON, CG VAN DYKE 1983 Ultrastructure of the infection and early colonization of Pinus taeda by Cronartium quercuumm formae specialis fusiforme. Mycologia 75: 117-130
33. GREGORY SC 1984 Micro-organisms isolated from wounded stems of Picea sylvestris. Trans Br Mycol Soc 83: 683-725

34. GRIESBACH RJ 1984 An introduction to somatic cell genetics. HortScience 19: 367-371
35. GRIFFIN DH, PD MANION 1985 Host-pathogen interactions as measured by bioassay of metabolites as produced by _Hypoxylon mammatum_ with its host _Populus tremuloides_. Phytopathology 75: 674-678
36. HAKMAN IC, S VON ARNOLD, A BENGTSSON 1984 Cytofluorometic measurements of nuclear DNA in adventitious buds and shoots of _Picea abies_ regenerated _in vitro_. Physiol Plant 60: 321-325
37. HALLETT RD 1983 Seedling injury by simazine and other triazine herbicides. Can For Serv Inf Rep M-X-145 pp 1-3
38. HARE RC 1978 Axenic culture of _Cronartium fusiforme_ from three spore forms. Can J Bot 56: 2641-2647
39. HARVEY AE 1979 Pine tissue affected by blister rust despite dialysis barrier between the organisms in culture. USDA-USFS Res Note INT-265 pp 7
40. HARVEY AE, JL GRASHAM 1971 Inoculation of a non-host tissue culture with _Crontarium ribicola_. Can J Bot 49: 881-882
41. HARVEY AE, JL GRASHAM 1974 Axenic culture of the mononucleate stage of _Cronartium ribicola_. Phytopathology 64: 1028-1035
42. HEBARD FV, GJ GRIFFIN, JR ELKINS 1984 Developmental histopathology of cankers incited by hypovirulent and virulent isolates of _Endothia parasitica_ on susceptible and resistant chestnut trees. Phytopathology 74: 140-149
43. HEBARD FV, PB KAUFMAN 1978 Chestnut callus cultures: Tannin content and colonization by _Endothia parasitica_. Proc Am Chestnut Symp Morgantown WV West Virginia Univ Books pp 63-70
44. HELGESON JP, BJ DEVERALL, eds, 1983 Use of Tissue Culture and Protoplasts in Plant Pathology. Academic Press, New York
45. HELGESON JP, GT HABERLACH 1980 Disease resistance studies with tissue cultures. _In_ DS Ingram, JP Helgeson, eds, Tissue Culture Methods for Plant Pathologists. Blackwell Sci Publ, Oxford, pp 45-62
46. HIGHLEY L, TA TATTAR 1985 _Leptographium terebrantis_ and black turpentine beetles associated with blue stain and mortality of black and Scots pines on Cape Cod, Massachusetts. Plant Disease 69: 528-530
47. HINDAL DF, HS MCNABB, Jr 1973 Interactions of tissue cultures of _Ulmus_ and an isolate of _Cerocystis ulmi_. 2nd Int Cong Plant Pathol, Minneapolis, MN
48. HIRSCHBERG J, L MCINTOSH 1983 Molecular basis for herbicide resistance in _Ameranthus hybridus_. Science 222: 1346-1349
49. HOFF RF, GI MCDONALD 1980 Resistance to _Cronartium ribicola_ in _Pinus monticola_: reduced needle spot frequency. Can J Bot 58: 574-577
50. HOOYKAAS PJJ, RA SCHILPEROOT 1984 The molecular genetics of crown gall tumorigenesis. Adv Genet 22: 210-282

51. HOQUE E 1984 Growth inhibitors produced by the *in vitro* culture of *Heterobasidion annosum*. Eur J For Path 14: 43-51
52. HORGAN K, J AITKEN 1981 Reliable plantlet formation from embryos and seedling shoot tips of radiata pine. Physiol Plant 53:170-175
53. HORSCH RB, JE FRY, NL HOFFMAN, SG ROGERS, RT FRALEY 1985 A simple and general method for transferring genes into plants. Science 227:1229-1231
54. HORSCH RB, RT FRALEY, SG ROGERS, PR SANDERS, A LLOYD, N HOFFMAN 1984 Inheritance of functional foreign genes in plants. Science 223: 496-498
55. HOWELL SH 1982 Plant molecular vehicles: Potential vectors for introducing foreign DNA into plants. Plant Physiol 33: 609-650
56. INGRAM DS 1976 Growth of biotrophic parasites in tissue culture. *In* R Heitefuss, PH Williams, eds, Encyclopedia of Plant Pathology: Physiological Plant Pathology. Springer-Verlag, Berlin, pp 743-759
57. JACOBI WR, HVF AMERSON, RL MOTT 1982 Microscopy of cultured loblolly pine seedlings and callus inoculated with *Cronartium fusiforme*. Phytopathology 72: 138-143
58. JONES JB 1977 Commercial use of tissue culture for the production of disease-free plants. *In* WR Sharp, PO Larsen, EF Paddock and V Raghavan, eds, Plant Cell and Tissue Culture: Principles and Applications. Ohio State Univ Press, Columbus OH, pp 441-452
59. JUZWIK J, M HUBBES 1984 Association of bacteria with tarnished plant bug stem lesions on hybrid poplars in Ontario. Phytopathology 74: 803
60. KARNOSKY DF 1981 Potential for forest tree improvement via tissue culture. Bioscience 31: 114-120
61. KARNOSKY DF 1984 Regeneration of trees from tissue culture. Proc USDA For Ser Workshop Genet Eng For, Placerville CA, Dec 3-6, 1984 (In Press)
62 KARNOSKY DF, AM DINER 1984 A cotyledon culture system for cloning *Larix decidua* and *Pinus banksiana*. *In*: TAPPI Res Dev Conf TAPPI, Atlanta GA, pp 13-15
63. KENERLEY CM, K PAPKE, RI BRUCK 1984 Effect of flooding on development of phytophthora root rot in Fraser fir seedlings. Phytopathology 74: 401-404
64. KINLOCH BB, Jr, M COMSTOCK 1980 Cotyledon test for major gene resistance to white pine blister rust in sugar pine. Can J Bot 58: 1912-1914
65. KINLOCH BB, M COMSTOCK 1981 Race of *Cronartium ribicola* virulent to major gene resistance in sugar pine. Plant Dis 65: 604-605
66. KIRBY EG 1982 The use of *in vitro* techniques for genetic modification of forest trees. *In* JM Bonga, DJ Durzan, eds, Tissue Culture in Forestry. Martinus Nijhoff/Dr W Junk, Publ, The Hague, pp 369-386
67. KLEKOWSKI EJ, Jr, N KAZARINOVA-FUKSHANSKY 1984 Shoot apical meristems and mutation: Fixation of selectively neutral cell genotypes. Am J Bot 71: 22-27
68. KOLATTUKUDY PE, W KOLLER 1983 Fungal penetration of the first line defensive barriers of plants. *In* JA

Callow ed, Biochemical Plant Pathology. John Wiley and Sons, Ltd NY, pp 79-100

69. KNAUF VC, MF YANOFSKY, MP GORDON, EW NESTER 1983 Genetic analysis of host range expression by Agrobacterium. In A Puhler, ed, Molecular Genetics of the Bacteria-Plant Interaction. Springer-Verlag, Berlin, pp 240-247

70. KUHLMAN EG, H BHATTACHARYYA 1984 Vegetative compatibility and hypovirulence conversion among naturally occurring isolates of Cryphonectria parasitica. Phytopathology 74: 659-666

71. KUNIMI Y, H ARUGA 1974 Susceptibility to infection with nuclear- and cytoplasmic-polyhedrosis viruses of the fall webworm, Hyphantria cunea Drury reared on several artificial diets. Jap J Appl Entomol Zool 18: 1-4

72. LANE WD, M SHAW 1974 Isolation and axenic culture of poplar rust. Can J Bot 52: 2228-2229

73. LAWTON MA, M-D CHILTON 1984 Agrobacterium Ti plasmids as potential vectors for genetic engineering. HortScience 19: 40-42

74. LEONARD KJ, CC MUNDT 1984 Methods for estimating epidemiological effects of quantitative resistance to plant disease. Theor Appl Genet 67: 219-230

75. LESTER DT, JG BERBEE 1977 Within-clone variation among black poplar trees derived from callus culture. For Sci 23: 122-131

76. MARTIN NE 1980 Interactions of pectins and tissues of Cronartium ribicola and Pinus monticola in culture. US Dept Agric For Ser Res Paper INT-150

77. MCGRANAHAN GH, EB SMALLEY 1984 Influence of moisture, temperature, leaf maturity and host genotype on infection of elms by Stegophora ulmea. Phytopathology 74: 1296-1300

78. MCGRANAHAN GH, EB SMALLEY 1984 Colonial morphology, axenic growth, and sporulation of Stegophora ulmea. Phytopathology 74: 1300-1303

79. MCMORRAN A 1985 A synthetic diet for the spruce budworm (Choristoneura fumiferana (Clem.). (Lepidoptera: tortricidfae). Can Entomol 97: 58-62

80. MCPARTLAND JM, DF SCHOENWEISS 1984 Hyphal morphology of Botryosphaeria dothidea in vessels of unstressed and drought-stressed stems of Betula alba. Phytopathology 74: 358-362

81. MILLER JD, D Strongman, NJ Whitney 1985 Observations on fungi associated with spruce budworm infested balsam fir needles. Can J For Res 15: 896-901

82. MOLINE HE, KC GROSS 1984 Growth of five pathogenic fungi on cell wall-related monosaccharides. Plant Sci Lett 33: 15-21

83. MOREL G, C MARTIN 1952 Guerison de dahlias atteints d'une maladie a virus. CR Acad Sci 235: 1324-1325

84. MOTT RL, HV AMERSON 1981 Tissue culture plantlets produced from Pinus monticola embryonic materials. For Sci 27: 299-304

85. MOTT RL, HV AMERSON 1982 Phytopathology and tissue culture alliances. In JM Bonga, DJ Durzan eds, Tissue Culture in Forestry, Martinus Nijhoff/Dr W Junk, Publ, The Hague, pp 208-230

86. MORAWETZ W 1984 How stable are genomes of tropical woody plants? Heterozygosity in C-banded karyotypes of *Porcelia* as compared with *Annona* (Annonaceae) and *Drimys* (Winteraceae). Plant Syst Evol: 145: 29-39

87. MOTT RL, HV AMERSON 1982 A tissue culture process for the clonal production of loblolly pine plantlets. North Carolina Agric Res Ser Tech Bull No 271 pp 14

88. MOTT RL, HA THOMAS, G NAMKOONG 1978 *In vitro* rearing of southern pine beetle larvae on tissue-cultured loblolly pine callus. Ann Entomol Soc Am 71: 564-566

89. MULLICK DB 1977 The non-specific nature of defense in bark and wood during wounding, insect and pathogen attack. In FA Loewus, CC Runeckles, eds, The Structure, Biosynthesis and Degradation of Wood. Recent Advances in Phytochemistry, Vol 11, Plenum Press, New York, pp 395-441

90. MURAI N, DW SUTTON, MG MURRAY, JL SLIGHTOM, DJ MERLO, NA REICHERT, C SENGUPTA-GOPALAN, CA STOCK, RF BARKER, JD KEMP, TC HALL 1983 Phaseolin gene from bean is expressed after transfer to sunflower via tumor-inducing plasmid vectors. Science 222: 478-481

91. NEGRUTIU I, M JACOBS, A CATTIOR-REYNAERTS 1984 Progress in cellular engineering of plants: biochemical and genetic assessment of selectable markers from cultured cells. Plant Molec Biol 3: 289-302

92. NELSON ND, BE HAISSIG, DE RIEMENSCHNEIDER 1984 Applying the new somaclonal technology to forestry. Proc TAPPI 1984 Res Dev Conf TAPPI, Atlanta GA, pp 27-34

93. NORTON RA, GHN TOWERS 1984 Transmission of nopaline crown gall tumor markers through meiosis in regenerated whole plants of *Bidens alba*. Can J Bot 62: 408-413

94. ORCHARD LP, JE KUNTZ 1981 Disease reactions of walnut species to butternut canker. Univ Wisconsin For Res Notes No 246 pp 7

95 OTTEN, LH DEGREVE, JP HERNALSTEENS, M MONTAGU, O SCHIEDER, J SCHIEDER, J STRAUB, J SCHELL 1981 Mendelian transmission of genes introduced into plants by the Ti plasmid of *Agrobacterium tumefaciens*. Molec Gen Genet 183: 209-213

96. OWENS LD, DE CRESS 1985 Genotypic variability of soybean response to *Agrobacterium* strains harboring the Ti or Ri plasmids. Plant Physiol 77: 87-94.

97. PATTON RF 1972 Inoculation problems in testing eastern white pine for resistance in forest trees. In RT Bingham, RJ Hoff, GI McDonald, eds, Biology of Rust Resistance in Forest Trees, Proc NATO-IUFRO Advanced Study Institute. US Dept Agric For Serv Misc Publ 1221 pp 373-383

98. PARVLEVLEIT JE 1983 Can horizontal resistance be recognized in the presence of vertical resistance in

plants exposed to a mixture of pathogen races? Phytopathology 73: 379-380

99. PATEL KR, GP BERLYN 1982 Genetic instability of multiple buds of Pinus coulteri regenerated from tissue culture. Can J For Res 12: 93-101

100. PATEL KA, TA THORPE 1984 In vitro differentiation of plantlets of lodgepole pine (Pinus contorta Dougl ex Loud.). Plant Cell Tissue Organ Cult 3: 131-142

101. PETERSON GW 1984 Resistance to Dothistroma pini within geographic seed sources of Pinus ponderosa. Phytopathology 74: 956-960

102. PFEIFFER TW, ET BINGHAM 1984 Comparisons of alfalfa somaclonal and sexual derivatives from the same genetic source. Theor Appl Genet 67: 263-266

103. PING DR, MF CONDE, BG GENGENBACH 1981 Cytoplasmic genome variability in tissue culture-derived plantlets. Env Exp Bot 21: 369-377

104. QUAK F 1977 Meristem culture and virus-free plants. In J Reinert, YPS Bajaj, eds, Applied and Fundamental Aspects of Plant Cell Tissue and Organ Culture. Springer-Verlag, New York, pp 599-646

105. QUALSET CO 1982 Integrating conventional and molecular genetics. Calif Agric 36(8): 29-30

106. RADDI P, A PANCONESI 1984 Pathogenicity of some isolates of Seiridium (Coryneum) cardinale, agent of Cypress canker disease. Eur J For Pathol 14: 348-354

107. REAM LW, MP GORDON 1982 Crown gall disease and prospects for genetic manipulation of plants. Science 218: 854-859

108. RENFROE MH, GP BERLYN 1984 Stability of nuclear DNA content during adventitious shoot formation in Pinus taeda L. tissue culture. Am J Bot 71: 268-272

109. ROBB J, AE HARVEY, M SHAW 1975a Ultrastructure of tissue cultures of Pinus monticola infected by Cronartium ribicola. I Pre-penetration host changes. Physiol Plant Pathol 5: 1-8

110. ROBB J, AE HARVEY, M SHAW 1975b Ultrastructure of tissue cultures of Pinus monticola infected by Cronartium ribicola. II Penetration and post-penetration. Physiol Plant Pathol 5: 18-19

111. ROCA WM, JE BRYAN, MR ROCA 1979 Tissue culture for the international transfer of potato genetic resources. Am Potato J 56: 1-10

112. SCHAPER U, E SEEMULLER 1984 Recolonization of the stem of apple proliferation and pear decline-diseased trees by the causal organisms in spring. J Plant Dis Prot 91: 608-613

113. SCHOENWEISS DF 1981 The role of environmental stress in diseases of woody plants. Plant Dis 65: 308-314

114. SEDEROFF RR, AM STOMP, S CHILTON, L MOORE 1984 Two observations of crown gall infection in loblolly pine: Evidence for transfer and expression of foreign genes. Ann Rep Spec Project Tissue Cult. NC State Univ (Mimeo) pp 10

115. SEEM RC and JM RUSSO 1984 Simple decision aids for practical control of pests. Plant Dis 68: 656-660

116. SEQUEIRA L 1978 Lectins and their role in host-pathogen specificity. Annu Rev Phytopathol 16: 453-481
117. SHAIN L, U JARLFORS 1984 Ultrastructural histopathology of eastern cottonwood clones resistant or susceptible to leaf rust. Phytopathology 74: 803
118. SIMONS RA, MW NABORS, CW LEE 1984 A model of mutant selection in plant suspension cultures. J Plant Physiol 116: 95-102
119. SKELSEY AF 1984 Biotechnology in agriculture -- New tools for the oldest science. US Gov Print Off, 525-190 36 pp
120. SMITH CO 1935a Crown gall on conifers. Phytopathology 25: 894
121. SMITH CO 1935b Crown gall on sequoia. Phytopathology 25: 439-440
122. SMITH CO 1936 Crown gall on Araucaria bidwillii. Phytopathology 26: 400-401
123. SMITH CO 1937 Crown gall on incense cedar Libocedrus decurrens. Phytopathology 27: 844-849
124. SMITH CO 1938 Crown gall on Taxus baccata. Phytopathology 28: 153-155
125. SMITH CO 1939 Susceptibility of species of Cupressaceae to crown gall as determined by artificial inoculation. J Agric Res 59: 919-925
126. SMITH CO 1942 Crown gall on species of Taxaceae, Taxodiaceae, and Pinaceae, as determined by artificial inoculation. Phytopathology 32: 1005-1009
127. SPANO L, P COSTANTINO 1982 Regeneration of plants from callus cultures of roots induced to Agrobacterium rhizogenes on tobbacco. Z Pflanzenphysiol 106: 87-92
128. STERMER BA, RP SCHEFFER, JH HART 1984 Isolation of toxins of Hypoxylon mammatum and demonstration of some toxic effects on selected clones of Populus tremuloides. Phytopathology 74: 654-658
129. STRZELCZYK E, H ROZYCKI, M MICHNIEWICZ 1984 Effect of bacteria and actinomycetes on germination of conidia of Cylindrocarpon destructans isolates pathogenic to Abies alba and Pinus sylvestris. Eur J For Path 14: 265-270
130. SZOCS G, M TOTH 1982 Lepkek tenyesztese felszintetikus taptalajokon III Gamma bagoly (Autographa gamma L.). Novenyvedelem 18: 354-358
131. TOTH M, G SZOCS 1984 A simple semi-synthetic diet for rearing of the fall webworm, Hyphantria cunea Drury. Acta Phytopath Acad Scient Hungaricae 19: 3-7
132. VAN DER PLANK JE 1968 Disease resistance in plants. Academic Press, New York, 206 p
133. VANDERWAAL JA, HR HOLBO 1984 Needle-air temperature differences of Douglas-fir seedlings and relation to microclimate. For Sci 30: 635-644
134. VARDI A, P SPIEGEL-ROY, E GALUN 1975 Citrus cell culture: Isolation of protoplasts, plating densities, effect of mutagens and regeneration of embryos. Plant Sci Lett 4: 231-236
135. WANN SR, DW EINSPAHR 1985 In vitro isolation and

propagation of mammatoxin-resistant quaking aspen. For Sci 31:(In Press)

136. WATSON JD 1970 Molecular Biology of the Gene. 2nd Ed, WA Benjamin, Inc, New York

137. WHISTLECRAFT JW, CR HARRIS, AD TOMLIN, JH HOLMAN 1984 Mass rearing technique for a braconid parasite, *Aphaereta pallipes* (Say) (Hymenoptera: Braconidae). J Econ Entomol 77: 814-816

138. WHITE FF, G GHIDOSSI, MP GORDON, EW NESTER 1982 Tumor induction by *Agrobacterium rhizogenes* involves the transfer of plasmid DNA to the plant genome. Proc Natl Acad Sci USA 79: 3193-3197

139. WHITEHEAD HCM, KL GILES 1977 Rapid propagation of poplars by tissue culture methods. NZ J For Sci 7(1): 40-43

140. YAMAOKA Y, K KATSUYA 1984 Axenic culture of four *Melampsora* species on willows. Trans Mycol Soc Japan 25: 85-92

20. TUMORS

M WICKER

1. INTRODUCTION

This chapter deals with distinct outgrowths that are found occasionally on trunks, branches or roots of forest trees, without affecting the growth of the trees to any extent. Cankers showing necrotic or oozing zones are excluded from this review.

Literature on tumors of forest tree species published before 1950 does not deal with in vitro culture of tumor tissues, because tissue culture techniques of woody plants were still in their infancy. At that time, pathologists were primarily occupied with trying to find the causal agents of abnormal outgrowths. In general it was thought that the causal agents were pathogenic microorganisms.

Not much research has been carried out with tumors because they are not found in large numbers, except in some small areas, and are of little economic consequence. However, the difference in growth patterns between tumor and adjacent normal tissues poses interesting fundamental questions in the field of physiology, biochemistry and genetics. Tissue culture methods provide an excellent research tool to try to answer some of these questions. In vitro culture transforms tissues that are difficult to handle and slow growing in vivo into much faster growing ones, which renders these tissues more suitable for biochemical and genetic studies. Since it is important to have some appreciation of the natural causes and the natural behavior of tumors before initiating in vitro studies, I will discuss some general aspects of tumorogenesis first.

2. TUMORS OF FUNGAL ORIGIN

Brown has examined several tumors on oaks, *Carya* (12), elms and maples (13). With the exception of very young tumors they have a hard woody core and a rough bark. Amongst the elm tumors Brown found one that formed at the collar. She did not detect any *Agrobacterium tumefaciens* in any of the tumors investigated but instead, found them often to contain *Phomopsis*. Isolates reinoculated on trees from which the isolate was obtained, or inoculated on other trees of the same species, have resulted in tumor formation, except when maple was used. Isolates from such tumors have yielded spores of *Phoma* and *Phomopsis* in pure culture. Pycnids have never been found in tumors, and the perfect form of *Phomopsis* has never been isolated.

Ouellette, in Canada, has observed tumors on *Picea glauca* and *Picea abies*, that had often originated from necrotic zones (45). Six kinds of fungi were isolated from these tumors, which could have played a role in the formation of these tumors by modification of the rate of production of endogenous growth regulators in the host tissues. No reinoculation was attempted.

3. TUMORS OF BACTERIAL ORIGIN

Bacteria, either *Agrobacterium tumefaciens* or other identified or non-identified ones, are common in tumors. However, the tumorogenic capacity of these bacteria has not always been verified by inoculation of healthy plants with tumor isolates.

3.1. Tumors other than crown gall

In 1919, Dufrenoy found that tumors on Scots pine, Laricio pine, and Maritime pine are transmissible by pricking two-year-old branches with a needle that had previously punctured a tumor (22). In sections of the original tumors he found cocci that he could not identify. Later (23) he described tumors with bacteria on *Sequoia* and tumors on resinous trees in France and Italy (24), some of which contained a *Pseudomonas*. He also reviewed some of the older literature by Vuillemin (71), Prillieux (49), and Petri (48). More recently, Eskandari and Ride (28, 29, 53) isolated a *Corynebacterium* from tumors on Aleppo pine. Inoculation of long shoots (auxiblasts), epicotyls, and plantlets of Aleppo pine, with this bacterium resulted in the formation of tumors. Hansen and Smith (32, 33) have isolated a bacterium, which they named Bacterium pseudotsugae, from Californian Douglas fir, which, if inoculated during the growing season on fresh trees, induced new tumors.

Tumors caused by Pseudomonas savastonoi were noted on roots, branches, trunks and leaves of ash trees around the Mediteranean, Czechoslovakia, California, Mexico and Argentina (64).

3.2. Crown gall

Crown galls on specimens of cultivated herbaceous species have been studied in detail. This has provided a good model system for studies of the molecular biology of tumors. Furthermore, Agrobacterium tumefaciens, more precisely its plasmid, is already used as a vector to carry foreign DNA into isolated cells or protoplasts of some herbaceous plants (1, 14, 16, 34). Presently, an octopine synthetase promoter and a marker for kanamycin phosphotransferase have been introduced into micropropagated shoots of Douglas fir, and have been expressed (Dandekar, Durzan and Gupta, unpublished). Therefore, genetic engineering of trees appears to be possible as well.

Crown galls occur only infrequently on forest trees, and thus have little economic impact. Some tumors have been found on several gymnosperms, i.e., Cupressus arizonica (10, 11), Juniperus sabina (46), and Libocedrus decurrens (58, 60). Agrobacterium tumefaciens has been isolated from each of these and upon inoculation on fresh trees has formed new tumors, except in the case of Juniperus sabina. Martin (43) has found tumors on seedlings of Sequoia gigantea which could have been crown galls.

Some deciduous tree species also carry crown galls, although the causative agent was not always properly identified (25, 26, 27, 64, 67).

Smith has tested the sensitivity of about sixty species of gymnosperms to Agrobacterium tumefaciens (57, 59, 61, 62, 63). About 25% of these were immune, another 25% was slightly sensitive while others formed tumors, sometimes in great numbers. In Taxus-baccata for example, 70% of the inoculations were positive. De

Cleene and De Ley (15) have recently reviewed the results of later inoculations. Among the 112 forest species that were inoculated with *Agrobacterium tumefaciens*, 65 responded by tumor formation. *Fagus sylvatica* and all inoculated species of the genera *Picea*, *Pinus*, *Larix* and *Quercus* were immune. Personally, I have inoculated oak, beech and birch plantlets with the strains B6 and T37 of *Agrobacterium tumefaciens* and obtained one tumor per 200 inoculations on *Quercus robur* plantlets, 23 tumors per 136 inoculations on *Fagus sylvatica*, and 33 tumors per 144 inoculations on *Betula verrucosa* and *mandschurica* seedlings.

4. VIRAL TUMORS

Black (6, 7) has reported that a virus transmitted by hemiptera can induce tumors in clover by stimulating cell growth in the vascular system of the leaves. Artificial inoculations on about a hundred herbaceous species from 20 families have shown that fifty five of these are sensitive to virus infection (8). So far, viral tumors have not yet been surely reported for any tree species, nor have any viral inoculations been carried out with these. It is possible that under natural conditions, trees are not suitable hosts for the virus and insect vectors.

Recently Van Der Meer (68) noted small tumors under the bark of poplars. The disease was graft transmissible and, therefore, was possibly caused by a virus (69).

5. GENETIC TUMORS

On *Prunus* hybrids (*P. armeniaca* hybridized with *P. cerasifera* or *P. saliciana*), tumors will form in the absence of pathogens (56). Although the occurrence of genetic tumors has never been unequivocally established for forest trees, White (72) has suggested that tumors on spruce are of genetic origin.

6. TUMORS CAUSED BY INSECTS

In many areas of France some oaks have shown crown decay and black oozing tumors on the trunk up to 2 meters above the ground. These tumors were related to galleries bored between phloem and rhytidome by *Agrilus biguttatus* (coleoptera) larvae (36, 37). When the galleries touch the cambium, several annual growth rings of wood are deposited outside the phloem. Anatomically this wood is abnormal; its rings are wider than those of the trunk, it contains only a few vessels, the uniseriated rays are replaced by bi- and tri-seriated rays, and the large rays fissured by sheets of fibres (35, 37). From the frass of the larvae two substances have been extracted that are inactive in the avena coleoptile test, but that induce a strong tissue proliferation in tissues of the tuber of Jerusalem artichoke in vitro (37, 39). Injected into oak seedlings, one of these substances has produced tumor-like abnormalities, while the other has been comparable in its action to 5.10^{-4} M indoleacetic acid (IAA) (38, 39, 40).

Tumors known as "gout disease" are commonly introduced on branches of balsam fir in Canada by the aphid Adelges piceae (2). The insect injects its saliva into the cortex parenchyma of the branches, where it induces hypertrophy of the cortical and phloem tissues, and stimulates the cambium that produces a local enlargement of the xylem ring. The effects on the xylem have been duplicated by application of IAA in lanolin to the scarified bark surface (3). A growth inhibitor found in abnormal abundance in the bark of balsam firs attacked by Adelges piceae, has been suspected to be responsible for the hypertrophy observed in the phloem (9).

7. TUMORS OF UNKNOWN ORIGIN

7.1. Spruce tumors in North America

7.1.1. Description, development, origin. White and Millington (77, 78) described the often numerous and large tumors that occasionally occur on branches and trunks of Picea glauca. The affected trees are restricted in distribution to small areas, one of these being a narrow band along the coast of the State of Maine (77). The structure of the tumors shows that their development starts almost always when the tree is a first-year-seedling. Judging by the ease of peeling of the bark of the tumors, it was initially assumed that the wide growth rings in the tumor were caused by an extended period of cambial activity (72, 78). Later, White et al. (80) rejected this explanation, because they found mitosis only from May to June, both in the tumor and normal cambium, but two to four times more mitotic cells in the tumors. Therefore, the tumors were caused by more rapid growth rather than by prolonged growth of the cambium. The wood of the tumors has the same general organization as the normal wood. However, its tracheids are larger, more numerous and less regular in shape than those of normal wood, and the wood density of the tumor is lower than that of normal wood (66, 78). It is not likely that these tumors are caused either by a particular fungus, bacterium, virus or insect alone, but by a variety of different factors, including a combination of highly specific environmental conditions, common to all the sites where the tumors are found (78).

7.2. Tumors on other conifers in North America

Peterson (47) found a large number of tumors on the trunks and branches of several conifers east of the Rocky Mountains, in particular on *Pinus contorta*. In their macroscopic and microscopic characteristics these tumors resembled those described by White and his co-workers. No causal agent was identified.

7.3. Tumors in Europe

Sprengel (65), in Germany, has described small tumors on the trunks and branches of oaks, pines and spruces. He was unable to determine the cause of tumor formation, but suspected that hereditary factors were involved.

I myself have studied, mainly in the Forest of Fontainebleau near Paris, large tumors on the trunks of oak, beech, birch, hornbeam and Scots pine. Microscopic study of sectioned tumors of oak and beech showed that they contained all the cell types of normal wood, but that the cell size and orientation were different. The density of tumor wood is higher than that of normal wood. In particular in oak tumors one does not find the characteristic spring to summer wood alternations that one finds in cross sections of normal wood. The largest annual rings in tumor wood have a small number of narrow vessels and a large number of parenchyma ray cells. These rays are of a type intermediate between large rays and uniseriate rays (81), and are similar to those observed by Jacquiot (35) in tumors of oak caused by *Agrilus biguttatus*.

Densitometric variation of the radiographs of samples taken with an increment borer from several oak, beech and birch tumors, as well as from healthy parts of the trunks of these trees, corresponded with the earlier microscopic findings. The annual growth rings are much larger in the tumor than in the normal wood. Variations in optical density of the radiographs are small and unpredictable in the tumor wood, while they show a repetitive pattern in the normal wood. This shows that the rythm of cell division and cell differentiation are dissimilar in tumor and normal wood (82).

A study (85) has been carried out in the Forest of Fontainebleau to determine relationships between tumor distribution and species, soil type and position of the tumors on the trees. Soil type, mostly calcareous or sandy, did not affect tumor formation, but a relationship with soil moisture is a possibility. Often the trees with tumors appear in clusters.

In order to determine the physiological stimulus that initiates tumor formation, I have treated stems of oak seedlings with IAA (83). This has generated new cambia in the cortex parenchyma, some of which formed extra phloem and xylem in the treated areas. The annual ring formed during the treatment was larger in the treated than in the control stems, and showed abnormalities that were similar to those in the wood of tumors; i.e., only a few vessels were present, the tissue orientation was disturbed, and lignification was reduced.

The tumors have yielded several fungi and bacteria. However, none of these microorganisms induced tumors after their inoculation on oak, beech, and birch seedlings. There is no evidence yet which microorganism is responsible for the tumor formation.

8. IN VITRO CULTURE

8.1. Spruce tumors

To facilitate investigations of spruce tumors, in particular to help determine differences in physiological characteristics between tumor and normal tissues, White and his coworkers cultured these tissues *in vitro*. They encountered difficulties in developing an optimal culture medium. In their first efforts a highly complex liquid medium was used that contained numerous amino acids and vitamins. One of the main problems with these early cultures was the formation of large quantities of tannins in the tissues.

Tissue explants, containing normal cambial and adjacent tissues, excised from the trunk either near a tumor or from a nearby tree free of tumors, form callus with relatively constant growth rates and cellular characteristics. It was more difficult to establish callus from tumor explants, but once initiated the tumor callus grew faster than the normal callus. The response of the tumor callus to changes in the culture medium or culture conditions, was more unpredictable than that of callus obtained from normal tissues. The tumor callus specifically needed vitamin B_{12} for its growth (52). Callus cultures obtained from normal cambial explants became gelatinous in appearance on a medium lacking vitamin B_{12} and folic acid. Small groups of cells that became detached from these callus masses were used to establish individual tissue clones (50). Subsequent systematic efforts finally lead to a simplification of the original culture medium, and in more precisely determined optimal concentrations for the organic components of the medium (55). The physical environmental factors were optimized also, thus allowing better comparisons in growth behavior between normal and tumor callus *in vitro* (79). Sources of nitrogen other than glutamine, which is unstable when autoclaved, were investigated without success (75).

During these studies an important observation was made. Unlike tumors of herbaceous plants *in vitro*, those of *Picea glauca* cannot grow without auxin. In fact, the concentrations required for tumor callus are much higher than those for normal callus. The same was observed with regard to ascorbic acid (51). Normal callus tissue can proliferate without auxin, but it grows 50% faster with auxin in the culture medium. De Torok and Thimann simplified Reinert and White's medium considerably and found that the auxin concentrations required for tumor callus growth was twenty times lower in their culture medium than in the original one. They also found that tumor callus grows faster than normal callus when both are exposed to the same auxin concentration (20). When auxin antagonists were added the tumor callus growth rate decreased to that of normal callus. The normal callus on the other hand grew at the same rate in the presence of auxin alone, as in the presence of both auxin and auxin antagonist. The initial inhibition of tumor callus growth in the presence of antagonist was removed when the callus was returned to an antagonist free culture medium.

De Torok and White (21) determined chromosome numbers in squash preparations of the calluses. In the cells of the normal callus they found that 65% of the cells were diploid, 5-8% tetraploid, and a few cells aneuploid. Spindle abnormalities were not observed. In the tumor callus cells aneuploidy was common, with the chromosome numbers varying from 3 to 70. This cytological instability could explain the variability in nutrient requirements of the tumor callus _in vitro_. De Torok and Roderick (19) have compared, in the course of seven successive subcultures, callus derived from one explant each of normal and tumor cambium. In each of these cultures they found a correlation between the number of chromosomes and the rate of cell division and of weight increase. They suggested that there is a causal relationship between chromosome numbers and growth rate, through mitotic frequency. Subsequently De Torok (18) reconfirmed these observations in more reliable material, i.e., in clones each derived from a single normal and a single tumor cell. Again the tumor clones showed a much higher variation in chromosome number and rates of cell division and growth than the clones of normal cells. The clones with the highest number of chromosomes were the fastest growing ones. Therefore, the degree of tumorization of cells could be established with precision by determining mitotic rates and chromosome number changes. De Torok also grafted clones produced _in vitro_ on _Picea glauca_ seedlings. Only the plants that were grafted with clones with non-diploid cells produced tumors. Neither diploid tumor clones nor diploid normal clones produced tumors after being grafted on seedlings.

These experiments have shown that tumors of _Picea glauca_ are only partly composed of true, i.e., aneuploid tumor cells, the remainder of the cells being normal cells. Risser (54) on the other hand found that 90% of the cells of a tumor callus that had been maintained through 40 to 50 transfers, were diploid. De Torok (17) is of the opinion that these results indicated that the material in question was obtained from several rather than a single tumor cell.

Finally, cytological observations have been presented regarding cell wall formation and the capacity of the cells to accumulate glycosides (74). In tumor tissues in vitro there are fewer meristem cells than in normal callus. Cells rich in starch, which one can find only in the oldest regions of normal callus, are abundant over a wide area around the meristematic nodules in the tumor callus. The more active rate of growth of tumors is not based on their cells being auxin independent, but on a capacity of the cells to accumulate carbohydrates (76).

8.2. Birch tumors

To learn more about the physiology of tumor formation, I have simultaneously cultured cambial explants from non-tumor bearing trees and cambial explants from tumor and healthy parts of the same tree. I chose the birch Betula verrucosa for these experiments. Culturing was always started between May and August, when the cambium is succulent. The proceedings for cutting the samples was similar to Gautheret and Morel's (30). The nutrient medium was initially developed for normal cambium (84). The best results for subcultures were obtained with the mineral nutrients from Nitsch's medium (44), the vitamins inositol, aneurin, biotin, and para-aminobenzoic acid, NAA 5.10^{-6} M as an auxin, and 2% glucose. Cytokinin was not required. The cultures were maintained in tubes at 25°C under continuous fluorescent light (6.7 Wm^{-2}), and transferred to fresh medium every 40 to 50 days. The tumor cambial cultures were not totally auxin or cytokinin autotrophic. If immediately inoculated on a growth regulator free medium and subsequently subcultured through several passages on such medium, growth diminishes and eventually stops in the third subculture. If the tumor cambium culture is subcultured through seven to eight passages on a medium with 5.10^{-6} M NAA and 5.10^{-6} M kinetin, one can subsequently subculture the tissue through an additional ten to fifteen passages on hormone-free medium before the growth rate declines. Three strains of tumor cambium explants cultured, as well as another one obtained from a birch tumor that had been induced by inoculating a plantlet with T37 Agrobacterium tumefaciens, are being cultured. I have also cultured habituated normal cambium after first subculturing it once on a medium with 2.10^{-6} M or 4.10^{-6} M 2,4 dichlorophenoxyacetic acid. These latter two tissue sources, i.e., crown gall and habituated cambium have served as references in three studies in which I have attempted to determine the degree of tumorization in cambial explants from natural tumors.

In the first study I have grafted different fragments of cultured tissues on twigs of birch that were kept either in vivo or in vitro. The tumor cambium, as well as the habituated cambium, grew in one out of three cases after grafting, while the crown gall cambium grew in two out of the three. The normal cambium did not proliferate after grafting.

In the second study, I investigated the effect of an anti-tumor agent of plant origin, i.e., alstonine, on the growth of various tissues. Alstonine is an alkaloid extracted from one of the Apocynaceae and inhibits DNA replication in crown gall cells (5). This chemical imposes a degree of inhibition of DNA synthesis in habituated cells and has little or no effect on healthy tissue. Culture medium with alstonine was used for four explant types, i.e., tumor and healthy trunk cambium from one tree, crown gall cambium and habituated cambium.

Finally, I have tried to compare the stability of DNA extracted from the different tissues with the "oncotest" designed by Beljanski (4). This test is based on the fact that the stability of hydrogen bonds linking the two DNA strands can be evaluated by determining the degree of hyperchromicity, i.e., by measuring the extinction coefficient in UV light. The more destablized the DNA the stronger the hyperchromicity. A maximum of 40 to 50% extinction is obtained with 0.3 N KOH (4). This oncotest shows that the DNA of the tumor cells is less stable than that of healthy cells. Several carcinogenic substances, antimitotic substances, some antibiotics, auxins, guanidines extracted from crown gall tumors, increased the hyperchromicity of DNA of crown gall cells (41). We have studied the effect of three guanidines on the hyperchromicity of DNA extracted from several of our cultured tissues. This work was carried out in cooperation with Beljanski's team at the Faculty of Pharmacy of Chatenay - Malabry. In the three studies we found that the birch tumor tissues acted in much the same manner as the habituated tissues, but not like the crown gall tissues. These results will be published in the near future.

9. CONCLUSION

Tree tumors have not been studied often in vitro, partly, no doubt, because it is difficult to obtain aseptic cultures and because tree tumors do not pose a serious economic threat. However, such studies could be very helpful in fundamental research of tumor induction, of metabolic changes involved in tumor induction, and of autonomous growth. Furthermore, genetic modification during tumor induction could be studied in vitro. Tissue culture methods are of great value in tumor research because tissue culture allows separation and isolation of cells, and possibly the extraction of specific DNA or other important molecules.

10. REFERENCES

1. AN G, BD WATSON, S STACHEL, MP GORDON, EW NESTER 1985 New cloning vehicles for transformation of higher plants. EMBO J 4: 277-284
2. BALCH RE 1952 Studies of the balsam woolly aphid, Adelges piceae (Ratz), and its effects on balsam fir, Abies balsamea (L.) Mill. Pub 867 Can Dept Agric, Ottawa
3. BALCH RE, J CLARK, JM BONGA 1964 Hormonal action in production of tumors and compression wood by an aphid. Nature 202: 721-722
4. BELJANSKI M 1982 Oncotest: dépistage des potentiels cancérogènes et spécifiquement anticancéreux. Conceptions et perspectives nouvelles en cancérologie. Environ Nouv Méd 2: 18-23
5. BELJANSKI M, L LE GOFF, A FAIVRE-AMIOT 1981 Preventative and curative anticancer drug. Application to crown gall tumors. Acta Hortic 125: 239-248
6. BLACK LM 1944 Some viruses transmitted by agallian leafhoppers. Proc Am Phil Soc 88: 132-144
7. BLACK LM 1945 A virus tumor disease of plants. Am J Bot 32: 408-415

8. BLACK LM 1965 Physiology of virus-induced tumors in plants. Encycl Plant Physiol XV, 2: 236-266
9. BONGA JM, J CLARK 1965 The effect of β-inhibitor on histogenesis of balsam fir bark cultured in vitro. For Sci 11: 271-278
10. BROWN JG 1932 Crown gall-like hypertrophies on a conifer (Abstr). Phytopathology 22: 4
11. BROWN JG, MM EVANS 1933 Crown gall on a conifer. Phytopathology 23: 97-101
12. BROWN NA 1938 The tumor disease of oak and hickory trees. Phytopathology 28: 401-411
13. BROWN NA 1941 Tumors on elm and maple trees. Phytopathology 31: 541-548
14. DE BLOCK M, L HERRERA-ESTRELLA, M VAN MONTAGU, J SCHELL, P ZAMBRYSKI 1984 Expression of foreign genes in regenerated plants and in their progeny. EMBO J 3: 1681-1689
15. DE GLEENE M, J DE LEY 1976 The host range of crown gall. Bot Rev 42: 389-466
16. DE GREVE H, J LEEMANS, JP HERNALSTEENS, L THIA-TOONG, M DE BEUCKELEER, L WILLMITZER, L OTTEN, M VAN MONTAGU, J SCHELL 1982 Regeneration of normal and fertile plants that express octopine synthase from tobacco crown galls after deletion of tumor-controlling functions. Nature 300: 752-755
17. DE TOROK D 1967 Successful tissue transplantation of auxin-dependent tumors and some of its problems. Nature 216: 376-377
18. DE TOROK D 1968 The cytologic and growth characteristics of tumor and normal clones of Picea glauca. Cancer Res 28: 608-614
19. DE TOROK D, TH RODERICK 1962 Associations between growth rate, mitotic frequency, and chromosome number in a plant tissue culture. Cancer Res 22: 174-181
20. DE TOROK D, KV THIMANN 1961 The auxin requirement and the effect of an auxin-antagonist on tumerous and normal tissues of Picea glauca. Physiol Plant 14: 543-547
21. DE TOROK D, PR WHITE 1960 Cytological instability in tumors of Picea glauca. Science 131: 730-732
22. DUFRENOY J 1919 Sur les tumeurs résineuses expérimentales des pins. C R Acad Sci 169: 545-547
23. DUFRENOY J 1922 Tumeurs de Sequoia sempervirens. Bull Soc Pathol Vég Fr 9: 148-150
24. DUFRENOY J 1925 Les tumeurs des résineux. Ann Inst Nat Agron Paris 19: 33-201
25. EMERY M 1931 Origine bactérienne des tumeurs de l'orme. C R Acad Sci 193: 76-77
26. ERKAMO V 1944 Neue Fundorte der auf Salix caprea vorkommenden Pseudomonas tumefaciens-Galle. Ann Bot Soc "Vanamo" 19: 44-46
27. ERKAMO V 1974 Crown galls (Agrobacterium tumefaciens) on willow and their distribution in Finland. Mem Soc Fauna Flora Fennica 50: 31-43
28. ESKANDARI F 1967 Acquisitions nouvelles concernant les tumeurs du pin d'Alep. Thèse de Docteur Ingénieur, Paris
29. ESKANDARI F, M RIDE 1967 Étiologie des tumeurs du pin d'Alep (Pinus halepensis Mill.). C R Acad Sci 264: 2638-2641
30. GAUTHERET RJ 1959 La Culture des Tissus Végétaux. Masson Ed, Paris

31. GOUVERNEL-GUILLEMAIN J 1953 Sur l'identification de la substance de croissance provoquant les tumeurs des chênes attaqués par Agrilus biguttatus Fab. C R Acad Sci 236: 978-979
32. HANSEN HN, RE SMITH 1933 A bacterial gall disease of the Douglas fir. Science 77: 628
33. HANSEN HN, RE SMITH 1937 A bacterial gall disease of Douglas fir, Pseudotsuga taxifolia. Hilgardia 10: 569-577
34. HOOYKAAS PJJ, RA SCHILPEROORT 1985 The Ti-plasmoid of Agrobacterium tumefaciens: a natural genetic engineer. Trends Biochem Sci 10: 307-309
35. JACQUIOT C 1949 De quelques particularités anatomiques et cytologiques observées dans l'aubier des chênes dépérissants. C R Acad Sci 229: 241-243
36. JACQUIOT C 1950 Sur le déterminisme de la formation des tumeurs ligneuses observées chez les chênes attaqués par l'Agrilus biguttatus Fab. C R Acad Sci 231: 1552-1554
37: JACQUIOT C 1950 Des relations entre les attaques d'Agrilus biguttatus Fab. et certains cas de dépérissement des chênes. Rev Pathol Vég Entom Agric 29: 172-182
38. JACQUIOT C 1954 Action sur des tissus végétaux de facteurs naturels produits par des organismes parasites d'arbres. Ann Biol 30: 457-460
39. JACQUIOT C 1976 Tumors caused by Agrilus biguttatus Fab. attacks on the stems of oak trees. Marcellia 39: 61-67
40. JACQUIOT C, J GOUVERNEL-GUILLEMAIN 1954 Action sur les tissus des pousses de chênes de deux principes actifs extraits de la vermoulure des galeries d'Agrilus biguttatus Fab. C R Acad Sci 238: 924-926
41. LE GOFF L, M BELJANSKI 1981 Crown gall tumor stimulation or inhibition: correlation with DNA strand separation. Proc 5th Int Conf Plant Pathol Bact Cali (Columbia) 1981, pp 295-307
42. LE GOFF L, M BELJANSKI 1982 Agonist and/or antagonist effects of plant hormones and an anticancer alkaloid on plant DNA structure and activity. IRSC Med Sci 10: 689-690
43. MARTIN EI 1957 Neoplastisches Wachstum bei Sequoiadendron giganteum Bucholz. Phytopathol Z 30: 342-343
44. NITSCH JP 1969 Experimental androgenesis in Nicotiana. Phytomorphology 19: 389-404
45. OUELLETTE GB 1967 Sur l'origine et le développement des tumeurs chez l'épinette. Phytoprotection 48: 101-106
46. PERSONS TD 1934 New or unusual diseases reported or observed in Mississipi in recent years (Abstr). Phytopathology 24: 843
47. PETERSON RS 1961 Conifer tumors in the central Rocky Mountains. Plant Dis Rep 45: 472-474
48. PETRI I 1924 I tumori batterici del pino d'Aleppo. Ann R Istit Sup For Naz 9: 187-229
49. PRILLEUX 1889 Les tumeurs à bacilles des branches de l'olivier et du pin d'Alep. Rev Gén Bot 1: 293-303
50. Reinert J 1956 Dissociation of cultures from Picea glauca into small tissue fragments and single cells. Science 123: 457-458
51. REINERT J, H SCHRAUDOLF 1959 Uber die physiologischen Eigenschaften von Gewebekulturen aus Tumor und normalem Gewebe von Picea glauca. Planta 53: 18-24

52. REINERT J, PR WHITE 1956 The cultivation *in vitro* of tumor tissues and normal tissues of *Picea glauca*. Physiol Plant 9: 177-189
53. RIDE M, F ESKANDARI, R MEZILY 1968 Pathogenèse des tumeurs du pin d'Alep. 1st Int Congr Phytopathol, London
54. RISSER PG 1964 Somatic mitoses in cells of *Picea glauca* cultivated *in vitro*. Science 143: 591-592
55. RISSER PG, PR WHITE 1964 Nutritional requirements of spruce (*Picea glauca*) tumor cells *in vitro*. Physiol Plant 17: 620-635
56. SLUDSKAYA LA 1953 Tumor formation and other pathological phenomena after wide crossing in stone fruits (in Russian). Trudy Prikl Bot Genet Selek 30: 59-64
57. SMITH CO 1935 Crown gall on the *Sequoia*. Phytopathology 25: 439-440
58. SMITH CO 1935 Crown gall on conifers (Abstr). Phytopathology 25: 894
59. SMITH CO 1936 Crown gall on *Araucaria bidwillii*. Phytopathology 26: 400-401
60. SMITH CO 1937 Crown gall on incense cedar, *Libocedrus decurrens*. Phytopathology 27: 844-849
61. SMITH CO 1938 Crown gall on *Taxus baccata*. Phytopathology 28: 153-155
62. SMITH CO 1939 Susceptibility of species of Cupressaceae to crown gall as determined by artificial inoculation. J Agric Res 59: 919-925
63. SMITH CO 1942 Crown gall on species of Taxaceae, Taxodiaceae and Pinaceae as determined by artificial inoculations. Phytopathology 32: 1005-1009
64. SPAULDING P 1961 Foreign diseases of forest trees of the world. USDA Agric Handbook n° 197
65. SPRENGEL F 1936 Uber die Kropfkrankheit an Eiche, Kiefer und Fichte. Phytopathol Z 9: 583-635
66. TSOUMIS G 1965 Structural deformities in an epidemic tumor of white spruce, *Picea glauca*. Can J Bot 43: 176-181
67. UROSEVIC B 1977 Tumorous canker of some forest roots. Lesnictvi 23: 23-32
68. VAN DER MEER FA 1981 Mosaic, witches' broom and knobbiness of poplar and a virus-like growth reduction of willow. Populier 18: 51-59
69. VAN DER MEER FA 1982 Knobbiness of poplar, a disease caused by an unknown graft transmissible agent. Acta Bot Neerl 31: 248
70. VERVOITTE V 1984 Intérêt des cultures *in vitro* de tissus végétaux pour dépister de nouveaux composés anticancéreux. Application à l'alstonine. Bio-Sciences 3: 14-16
71. VUILLEMIN P 1888 Sur une bactériocécidie ou tumeur bacillaire du pin d'Alep. C R Acad Sci 107: 874-876; 1184-1186
72. WHITE PR 1958 A tree tumor of unknown origin. Proc Nat Acad Sci 44: 339-344
73. WHITE PR 1958 An epiphytotic tumor of white spruce, *Picea glauca*. *In* KV Thimann, ed, The Physiology of Forest Trees. Ronald Press Co, New York, pp 493-508
74. WHITE PR 1967 Some aspects of differentiation in cells of *Picea glauca* cultivated *in vitro*. Am J Bot 54: 334-353
75. WHITE PR, SN GILBEY 1966 Sources of nitrogen for spruce tissue culture. Physiol Plant 19: 177-186

76. WHITE PR, SN GILBEY 1968 Some differences between two strains of cells of the white spruce (Picea glauca), one from normal and one from tumor tissues, grown in tissue culture. Phytomorphology 18: 66-74
77. WHITE PR, WF MILLINGTON 1954 The distribution and possible importance of a woody tumor of the white spruce, Picea glauca. Cancer Res 14: 128-134
78. WHITE PR, WF MILLINGTON 1954 The structure and development of a woody tumor affecting Picea glauca. Am J Bot 41: 353-361
79. WHITE PR, PG RISSER 1964 Some basic parameters in the cultivation of spruce (Picea glauca) tissues. Physiol Plant 17: 600-619
80. WHITE PR, G TSOUMIS, F HYLAND 1967 Some seasonal aspects of the growth of tumors on the white spruce, Picea glauca. Can J Bot 45: 2229-2232
81. WICKER M 1970 Structure et densité de bois de tumeurs de chêne et de hêtre. Rev Gén Bot 77: 499-517
82. WICKER M 1972 Application de la radiographie à l'analyse de l'accroissement de tumeurs ligneuses. Rev Gén Bot 79: 187-196
83. WICKER M 1976 Structures pathologiques expérimentales chez le chêne en relation avec l'histologie de tumeurs naturelles. Rev Gén Bot 83: 53-71
84. WICKER M 1981 Utilisation de la culture de tissus pour l'étude des propriétés de tumeurs du tronc chez le bouleau. In Proc IUFRO Sect S2 01 5. Int Workshop "In Vitro Cultivation for Tree Species", Fontainebleu, France, pp 211-215
85. WICKER M 1982 Observation de tumeurs du tronc en forêt de Fontainebleu. Rev For Fr 34: 191-203

21. CELL SUSPENSION CULTURES FOR THE STUDY OF PLANT CELL SENESCENCE

R.J. ROMANI

1. INTRODUCTION

Plant senescence plays an important role in agriculture. Yields of cereal crops, stability and productivity of orchards, marketability of plant produce, all are affected by the senescence of plants or plant parts. Even successful somatic embryogenesis, an increasingly important methodology for agriculture and forestry and one far removed from senescence, must contend with cellular and tissue senescence (Durzan, this volume). The study of plant senescence, however readily justifiable, is difficult to pursue given its complexity with facets at the intracellular, cellular, organismal and population levels. Can plant cell cultures be manipulated in such a manner that the cells will undergo senescence and if so, will the cells be useful as prototypes for the study of this complex phenomenon? What are the potentials and the constraints of this approach? These are the questions addressed in this chapter.

2. TERMINOLOGY AND CONTEXT

2.1. Senescence and aging

The data to be discussed below are derived from studies with cultured fruit cells and pertain to senescence processes in determinant parent organs. It is therefore appropriate to utilize terminology for developmental stages in fruit as suggested recently by a committee of the American Society for Horticultural Science (43). In that context senescence is seen as the final phase of genetically controlled development and encompasses the physiological process that leads to death. Aging, on the other hand, connotes the mere passage of time which may or may not be accompanied by physiological change. The distinction between senescence and aging emphasized here is consonant with the views of Woolhouse (45) who makes his point with the following quote from Medawar (18), "Aging hereafter stands for mere aging, and has no other innuendo. I shall use the word senescence to mean aging accompanied by that decline of bodily facilities and sensitivities and energies which aging colloquially entails. Dorian Grey aged, but only his portrait disclosed the changes of senescence. I hope that makes it clear."

Before utilizing "senescence" and "aging" to describe changes in cultured cells it may be instructive to briefly review their use in a wider context.

2.2. Polycarpic and monocarpic senescence

While Medawar's distinction is clear its applicability to the senescence of complex organisms, be they human or plant, is much less so. Medawar also wrote, "No death is wholly 'natural'; no one dies merely of the burden of years". The quote had reference to humans; it applies equally to trees whose longevity is also highly variable. The general loss of vigor which occurs with aging of trees does not appear to be the result of pervasive and irreversible senescence. Witness, as put by Woolhouse (44), the vigor of "cuttings taken from old and decrepit trees" or clonal stability as evidenced by the maintenance of varietal characteristics such as those of the Pearman apple (600 years) or some varieties of figs (ca. 2000 years). The death of trees, as observed by Wareing and Seth (42), is most often the result of random causes or stochastic events such as disease, nutritional stress, draught, injury or some other insult.

Given these circumstances, there would appear to be little hope of discerning specific cellular events that control or contribute to the senescence of aging trees and other long-lived polycarpic plants. It is quite understandable that plant physiologists wishing to study senescence have turned to monocarpic annuals or biennials whose predictable and generally uniform senescence (ripening fields of wheat are the classic example) is clearly genetically programmed and may even be mediated by a yet illusive "senescence signal" (23). In such systems it is highly plausible that determinative events reside at the cellular - intracellular - molecular level.

The complex and ambiguous nature of polycarpic senescence notwithstanding, there remains an intuitive sense that the loss of vigor and increased susceptibility to stress in aging trees must also have its origins in altered cellular functions. Indeed, a thread connecting cellular events with polycarpic senescence derives from Leopold's (14) widely held view that senescence, death, and organismal turnover are the handmaidens of evolutionary change. [For a concise presentation of this and other views on the subject see Strehler's (38) chapter on the evolution of cellular aging]. Accepting Leopold's contention, it follows that the longevity of trees, among their other traits, must have been the result of selective pressure and hence become a part of the genetic information of each constituent cell. This would include, conceivably, a genetically controlled propensity for some cells and tissues to undergo age-related changes (i.e., loss of vigor) that in turn render the entire tree susceptible to a specific or to cumulative environmental stress. Such reasoning is in accord with Carr and Pate (5) who proposed that, "The causes of aging [senescence] at higher levels of organization, as Sorokin (37) points out, must eventually be sought at lower levels of organization, and in plants they are ultimately intracellular."

2.3. Cellular senescence

Mammalian cellular systems have been used for years both to study cellular senescence *per se* and as models for organismal senescence. The observation (10) that cultured human fibroblasts exhibit a loss in replicative ability after a number (ca. 50) of sub-cultures has provided a basis for much of mamallian cell-aging research. Recently Muhitch *et al.* (19) also observed an age-dependent loss in replicative activity of serially cultured rose cells and proposed that the cells may be useful for the study of plant senescence. However, the rather gradual loss of replicative ability shares many abiguities with "the loss of vigor" of an entire organism and includes, as noted above, the probable and confusing interplay of both stochastic and genetically controlled events.

Some of the "aging" ambiguities inherent in replicative systems are avoided by the use of non-replicating differentiated cells that, ostensibly, undergo senescence and death much as they would in the parent tissue. Numerous cell lines are being used for this purpose largely in mammalian aging research (21). Essential to this approach is the demonstration of phenomenological similarity between *in vitro* and organismal senescence (4). That is to say, to be experimently useful as a model system the cultured cells must exhibit senescence phenomena *in vitro* that bear a close resemblance to comparable phenomena in the senescent parent organism. This criterion is often difficult to achieve and has not, until recently been obtained for plant cells.

2.4. Fruit senescence in a phenomenological context

Because of its importance to successful commerce in fresh foods of plant origin, the postharvest physiology or postharvest senescence of fruits and vegetables has been the subject of extensive study for several decades. A brief perusal of recent compendia (9,15) will reveal that the ripening or senescence (the terms will be used interchangably) of many kinds of fruit and, especially climacteric fruits such as apple, avocado, banana, pear, and tomato, is accompanied by very well defined, well-described physiological transitions. Among these are changes in color, respiration rate, membrane integrity, enzymic components and activity, respiratory pathways, macro-molecular synthesis, messenger RNA's, ethylene susceptibility, and ethylene production. It is fair to say that fruits have been the most extensively characterized senescent plant organs, more so than senescent leaves whose dramatic losses of chlorophyll and protein may have led to an overely restrictive reliance on these two parameters as indicies of senescence. Accordingly, the information on fruit ripening, albeit largely descriptive, provides an excellent phenomenological background against which to view and assess the merits of model systems and experimental approaches.

3. SENESCENCE OF CULTURED PEAR FRUIT CELLS

In the 1970's scientists at Toulouse, France successfully cultured pear and apple fruit cells and proposed that they should prove useful in studying various metabolic events associated with ripening (24). As expected, auxin-deprivation of the fruit cells results in a cessation of division with subsequent cell death. In key experiments (6) it was observed that the addition of an osmotium to auxin-deprived cells resulted in an extended period of aging before the cells began to die. What is summarized below are the results of subsequent studies at Toulouse and at Davis, California, whose aim was to compare the phenomenology in aging (senescent?) cultured cells with that in senescent parent tissue.

3.1. Experimental

All experiments were conducted with a strain of cells established in 1972 by Pech *et al.*, (24) from 'Passe Crassane' pear (*Pyrus communis* L.) fruit. Growth media consisted of the mineral nutrients of Murashige and Skoog (20) and the organic components of Nitch *et al.*, (22), as slightly modified by Pech and Romani (26). The cells used for senescence studies were first deprived of 2,4-dichlorophenoxyacetic acid (2,4-D) by incubation for 9 to 11 days in growth medium lacking 2,4-D and then transferred to *aging* medium which consisted of 1/4 the concentration of nutrients in the growth medium, no 2,4-D, but supplemented with 0.4 M mannitol or 0.385 M mannitol and 0.015 M sucrose. The suspension-cultures were contained in large Erlenmeyer or Fernback flasks placed on a rotary shaker. The apparatus employed to facilitate measurements of respiration, media exchange and cell sampling is shown diagrammatically in Fig. 1.

To measure ethylene production 50 ml aliquots of cell suspension were placed in 125 ml Erlenmeyer flasks and an illicitor, either indoleacetic acid

(IAA), or $CuCl_2$, or precursor, 1-aminocyclopropane-1-carboxylic acid (ACC), added. After a specified time the flasks were flushed with an air stream, capped with a septum and 30 minutes later a head space sample was taken for the determination of ethylene by gas chromatography.

Assessment of cell vitality was based on the ability of live cells to exclude Evans blue dye. Cell counts were obtained from a calibration curve relating packed cell volume to cell number. Additional experimental details pertinent to each study are given in the figure legends or can be found in the indicated references.

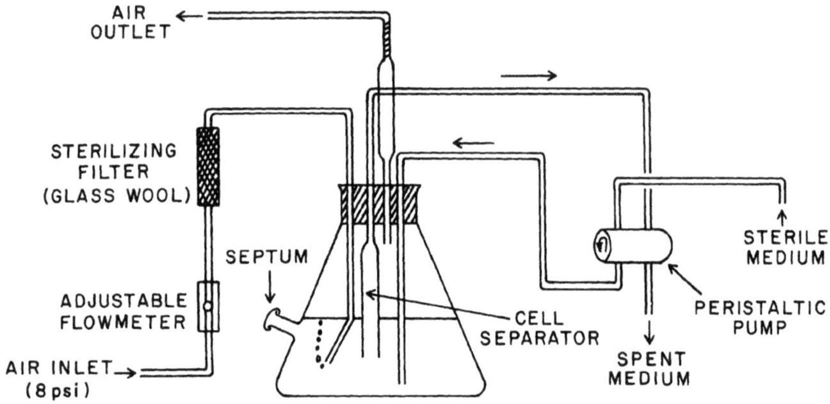

FIGURE 1. System for cell culture that allows for respiration measurement (airflow) and continuous renewal of medium without removal of cells (settlement rate of cells in the large diameter "cell separator" is faster than the outward flow of medium). The culture flask is placed on a rotary shaker whose speed is adjusted (100-120 rpm) to just maintain the cells in suspension. From Pech and Romani (26).

3.2. Synopsis of results

3.2.1. Pattern of cell growth and death. The general pattern of cell growth, quiescence, and death in culture is shown diagrammatically in Figure 2. Deprivation of 2,4-D and exposure to osmoticum delays the time of death (7,26). The resultant quiescent, pre-death period is variable in length although generally somewhat longer with continuous medium renewal (26). For the purpose at hand this quiescent period is the experimentally useful 'window' allowing one to search for the presence of physiological transitions which precede cell death. Evidence for several such transitions is summarized below.

3.2.2. Respiration. Cellular respiration remains high for a few days after transfer of the cells to aging medium and then declines to much lower levels where it remains relatively stable until cell death (26). No respiratory rise is observed that resembles the classic respiratory climacteric (12) characteristic for senescent pears and many other fruit. However, as discussed below (3.2.7.), respiration is affected by availability of carbohydrate substrate and is also responsive to ethylene. In these latter respects the cellular respiratory responses bear a resemblance to those of the senescent intact fruit.

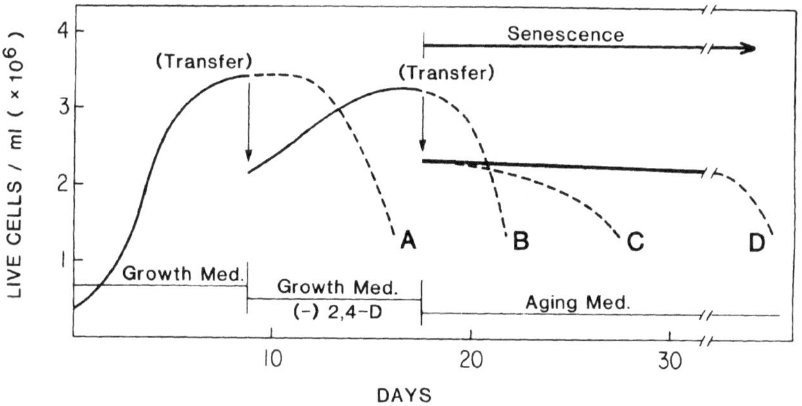

FIGURE 2. Diagramatic presentation of the periods of cell growth, auxin-depletion, and quiescence (or senescence) in mannitol containing aging medium. Cells remaining in growth or in auxin-depleted medium die rather soon (A,B), those transferred to mannitol containing aging medium enter a period of quiescence whose length will vary from experiment to experiment ranging from a few days (C) to over 3 weeks (D). See Puschmann et al. (29).

3.2.3. Macromolecular synthesis. Macromolecular (protein and RNA) synthesis declines after transfer of the cells to aging medium but then undergoes an interesting transitory rise prior to cell death (Fig. 3B). First reported by Pech and Romani (26), the age-dependent rise in protein synthesis has also been observed by Balagué et al., (1). Of interest is the existence of a chronological connection between the transient increase in synthesis and subsequent cell death (25,26). Moreover, both the increase in synthetic activity and cell death are averted by the addition of 2,4-D (1). That is to say, the rise in protein synthesis occurs as part of the death-oriented senescence processes and it does not occur when added 2,4-D stems senescence and reinstates cell division. Finally and importantly, an analgous rise in protein and RNA synthesis occurs in many fruits as they begin to ripen (8,13,30).

3.2.4. Cyanide resistant respiration (CN-R). The capacity of the alternative electron transport, cyanide resistant pathway has been found to increase appreciably during ripening and the climacteric respiratory flux in avocados (36) even though the pathway is apparently not in use i.e., it is not engaged unless the tissues are placed under additional stress (40). While using cycloheximide to inhibit protein synthesis in cultured cells we were surprised to find that the inhibitor also illicited an increase in CN-R along a temporal pattern shown in Figure 3C (32). The change in respiratory function was shown to be of mitochondrial origin and, as in ripening fruit, to be present but not normally engaged. Whether the magnitude of the cycloheximide-induced CN-R changes with cell age remains to be determined.

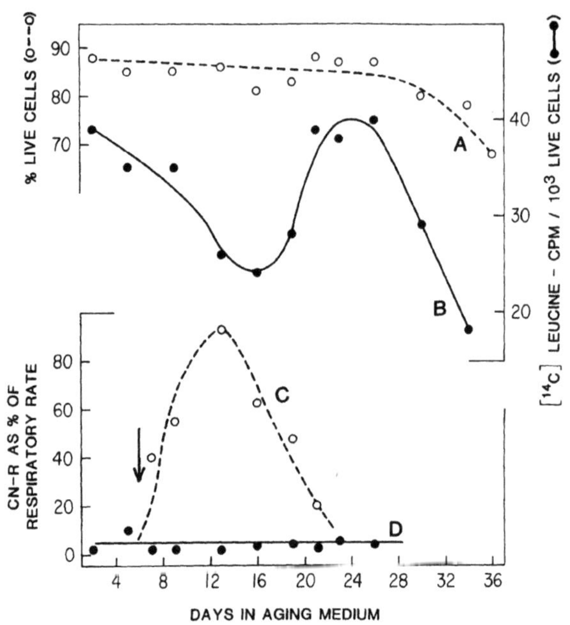

FIGURE 3. Senescent cells exhibit a transient rise in protein synthesis (B) and a transient but dramatic increase in capacity for cyanide resistant (CN-R) respiration (C), the latter in response to the addition (arrow) of cycloheximide. Without cycloheximide (D) the capacity for cyanide resistant respiration remains very low at 5 to 10% of the total respiratory activity. Cell suspensions (approx. 2 L in 3 L Fernback flasks) were maintained in aging medium with aeration (30 ml/min.) and slow (290 ml/day) medium renewal. Percent live cells (A), and (^{14}C)-leucine incorporation (B) were measured with aliquots of

cells taken from the culture flasks at the indicated intervals. For (C) cycloheximide (final conc. 1.75μm) was added to a culture flask on day 6. Control respiration rate (not shown) declined over the first 10 days to a basal rate (see Figure 5) but was not significantly affected by the addition of cycloheximide, i.e. engagement of the alternative electron transport pathway did not change total electron flow. From Pech and Romani (26), Romani et al. (32).

3.2.5. Ribosomal response to temperature stress. It has been shown that the polyribosomal population of pear cells, *in situ*, decreases dramatically upon exposure of the fruit to 40°C for several hours and recovers when the fruit are once again transferred to 20°C (31). The polyribosomal population of cultured pear cells demonstrates the same temperature dependent changes (unpublished data) although the transitions occur more rapidly in keeping with the faster temperature equilibration obtained in liquid culture as compared with intact fruit. Thus, with respect to the temperature senesitivity of protein synthesis and ostensibly the associated senescence transitions, cultured cells and intact fruit respond in a very similar fashion.

3.2.6. Ethylene biosynthesis. One of the distinguishing characteristics of climacteric fruit is their autocalytic synthesis of ethylene once exposed to threshold levels of the gas. Since it also stimulates respira- tory activity, ethylene is thought to be the prime inducer of the climacteric respiratory rise. Dividing pear cells in culture produce low levels of ethylene (Fig. 4A). As the pear cells are deprived of auxin (Fig. 4B) and then transferred to aging medium (Fig. 4C) they produce progressively less ethylene. However, during the depletion of auxin and especially after transfer to aging medium the cells acquire a greatly increased potential to produce ethylene when treated with either auxin, $CuCl_2$ (stress), or ACC (metabolic precursor) (29). This potentiation for ethylene production, which maximizes often several days after transfer to aging medium (27,29) is to be distinguished from the more modest ethylene production observed following the addition of IAA to auxin-deprived cells (3,17, or Fig. 4B).

The kinetics of ethylene production, its inhibition, and its requirements for macromolecular synthesis have been described in detail (28). Of importance here is the age-transition mediated by auxin depletion. The transition involves a significant physiological adjustment affecting ethylene production not only in response to added auxin but to chemical stress and also, to digests of cell wall (41). Moreover, the occurrence of the transition presages eventual cell death (27,29). It appears, therefore, that the increase in ethylene producing capability by the cultured pear cells is a senescence-related phenomenon as is the autocatalytic rise in ethylene production by ripening pear fruit.

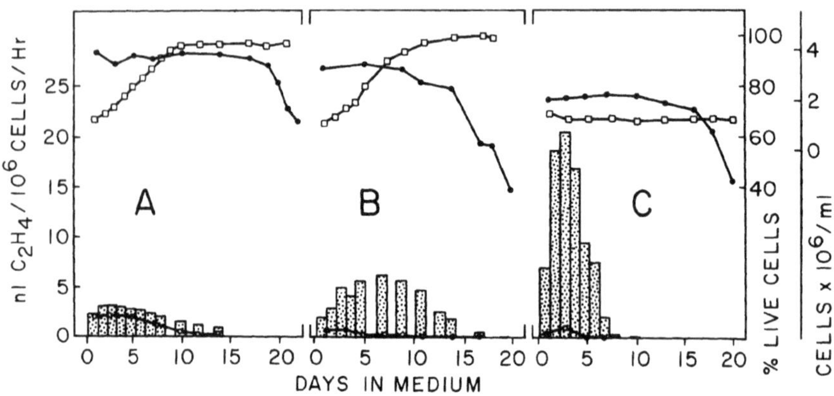

FIGURE 4. Maximum rates of ethylene production (bars) measured 12 hours after the addition of $50\mu M$ IAA to aliquots of cultured pear fruit cells being held in: (A) growth medium, (B) growth medium less 2,4-D and, (C) aging medium. Corresponding changes in number of cells ([]———[]), % live cells (o———o) and ethylene production by control, non-IAA-treated cells (o———o). From Puschmann <u>et al</u>. (29).

3.2.7. <u>Response to ethylene</u>. Exposure of most fruits and vegetables to 5-to-10 ppm ethylene results in increased respiration and increased rates of senescence. In mature, climacteric-type fruit sufficient exposure to ethylene, generally more than 6 hours, results in autocatalytic ethylene production, a climacteric respiratory rise, and an enhanced rate of senescence. Shown in Figure 5 is the response of suspension-cultured fruit cells in aging medium upon treatment with 5 ppm ethylene (C. Brady and R. Romani, unpublished results). Ethylene causes a modest but definite increase in respiration and accelerates senescence as measured by the incidence of cell death.

The addition of sucrose prolongs an elevated respiration that would otherwise decline rapidly to a basal rate. It is not clear at this point whether the respiratory response to ethylene is constrained by available metabolites or by overall respiratory capacity. Nonetheless, the response is real and coincident with accelerated senescence. The analogies to the effects of ethylene on intact fruit are unmistakable.

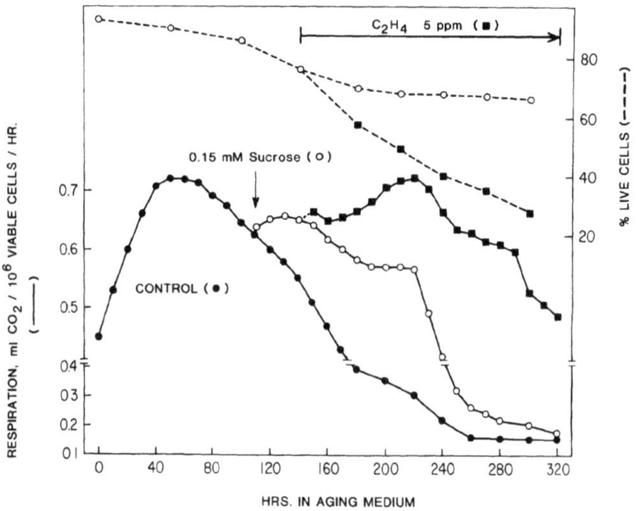

FIGURE 5. Responses of senescing pear fruit cells to sucrose and ethylene. At 110 hours after transfer to aging medium a portion of the cells received an additional 0.15 mM sucrose. Thirty hours later a portion of those cells were subjected to continuous exposure to 5 ppm ethylene. Respiration (——), % live cells (- - -), control (o), + 0.15 mM sucrose (o), + sucrose + exposure to 5 ppm ethylene ([]). (C Brady and R. Romani - unpublished).

3.2.8. Leakage and ultrastructural change. Balagué et al. (1,2) observed that beginning on or about the 5th day after transfer to aging medium cultured pear cells began to leak amino acids, notably serine. Leakage reached a maximum on the 12th day as the cells began to die. Coincident structural changes included swelling of the endoplasmic reticulum observed on day 5, vesicle formation and loss of ribosomes by day 13, and a large increase in vesicularization and distortions of the plasma membrane by day 16. Leakage and analogous membraneous changes are a recognized manifestation of senescence in fruit and other plant tissues (33).

3.2.9. Response to spermidine. In a recently completed study (11) 1mM spermidine was found to delay the senescence and death of cultured pear cells and to partially inhibit the production of ethylene. Although similar studies have not been conducted with pear fruit these effects of spermidine are very much like those observed with polyamine treatment of other senescent plant tissues (35, Bagni and Biondi volume 1).

4. DISCUSSION

4.1. Cultured cells as prototypes for the study of cellular senescence

It is clear from studies thus far that auxin-deprived cultured pear fruit cells cease to divide and thereafter undergo several physiological transitions terminating in cell death. Auxin-deprivation is a necessary pre-condition for the subsequent transitions to occur. In itself, a decline in endogenous auxin mirrors the status of cells in many fruit where IAA levels decrease before or coincident with the onset of senescence (16,42). What is in question, however, are the events following auxin-deprivation and whether or not they are analogous to those that occur in the senescent intact organ.

To be sure none of the transitions observed in the cultured cells matches exactly its counterpart *in situ* i.e., the phenomenology is not entirely similar. However, the parallels in each of the several instances described above are unmistakable and, I believe, sufficient to justify the use of auxin-deprived cultured pear cells as a model system for the study of certain specific senescence-related phenomena.

4.2. Constraints

Users of cell cultures must be wary of several sources of variability (34,39) which include endopolyploidy, mutations, cross-contamination with other cells, mycoplasma, and unintentional selection during repeated subculturing. These constraints notwithstanding, studies conducted over several years in France and in the United States obtained closely similar results with regard to the rise in protein synthesis and subsequent cell death. In our own studies, the kinetics of ethylene production have also remained unchanged over more than 4 years. Some variability has been manifested in the form of 2-to-3 fold differences in the amount of ethylene produced by successive cellular sub-cultures. A few sub-cultures, inexplicably, produced no ethylene at all in response to $CuCl_2$ or IAA. Variability has also been observed in the time lag between transfer to aging medium and the peak in inducible ethylene production. However, there is good indication (29) that this latter variability is closely associated to variable rates of senescence as early manifestations of inducible ethylene production are generally associated with early cell death.

In part, and perhaps in large part, these forms of variability arise from cellular sensitivity to the environment and our limited awareness and inadequate control of all pertinent variables. To the extent that that is true, much of the observed variability in cellular performance may, with time and further understanding, become the basis for a more discerning experimental use of cultured cells.

4.3. Potentials

Tractability is a clear experimental advantage of cultured cells. Radioisotopes, inducers, inhibitors, etc., can be readily administered or removed. Levels of nutrients or other components of the liquid environment can be monitored and maintained or adjusted at will. These attributes should prove especially useful in assessing the interplay of phytohormones and related 'metabolic networks' (Durzan, this volume) that control aspects of cellular development from somatic embryogenesis to cellular senescence.

With reference to the introductory remarks it may be well to emphasize the obvious, that is, the extent to which cultured cells will be useful in studies of organ or plant senescence will be directly dependent on the coincidence between the pertinent phenomenology *in situ* and in culture. In the context of postharvest senescence (ripening) the coincidence, as noted above, has been very encouraging. It is anticipated that cultured cell systems may also prove useful in the study of monocarpic plant senescence where the phenomena are largely programmed and often subject to hormonal regulation. However, suspension cultures as envisioned today can not simulate the complex interdependencies of cells, tissues, stochastic events and time, which affect the gradual loss of vigor and eventual death of polycarpic biennials and perennials. Nonetheless, for reasons noted above (section 2.2.) cultured cells may be useful in deliniating the cellular impact of environmental stresses or other identified, life-shortening events that affect the intact plant. In particular, studies of the cambium and shoot meristems may be of interest because senescence in these would affect such important characteristics as longevity of trees as well as tree form.

5. CONCLUSIONS

Auxin-deprived, suspension-cultured fruit cells that undergo a prolonged pre-death quiescent (senescent) phase are proving to be especially useful as prototypes for the study of certain aspects of determinant organ senescence. It is anticipated that these and comparable cell systems will prove useful in the study of senescence processes that are programmed or hormonally regulated or, in general, have a cellular basis. Seldom, if ever, will the senescence phenomenology *in vitro* and *in situ* be precisely the same, however, parallels are unmistakable and experimentally exploitable.

6. REFERENCES

1. BALAGUÉ C, A LATCHÉ, J FALLOT, J-C PECH 1982 Some physiological changes occurring during the senescence of auxin-deprived pear cells in culture. Plant Physiol 69:1139-1343
2. BALAGUÉ C, L SOSSOUNTZOV, M PETITPREZ 1983 Ultrastructural changes during senescence of pear fruit cells during auxin starvation. Cytobios 36:53-63
3. BALAGUÉ C, JC PECH 1985 Relationship between the activity of the ethylene forming enzyme and the level of intracellular 2,4-dichlorophenoxyacetic acid in pear cell cultures in vitro. J Plant Growth Regul 4:81-89
4. BLUMENTHAL GH, DK KASBIKAR 1979 Nonreplicating cultures of frog gastric tubular cells. In W Nichols and D Murphy, eds, Differentiated Cells in Aging Research, Academic Press, NY pp 191-212
5. CARR DJ, JS PATE 1967 Ageing in the whole plant. In Aspects of the Biology of Ageing, HW Woolhouse, ed, Soc Exp Biol Symp XXI, pp 559-599
6. CODRON H, LATCHÉ A, PECH J-C, FALLOT 1978 Mise au point d'un nouveau systeme d'etude de la sènescence des cellules vetegales. C R Acad Sci, Paris 287:21-24
7. CODRON H, A LATCHE, JC PECH, B NEBIE, J FALLOT 1979 Control of quiescence and viability in auxin-deprived pear cells in batch and continuous culture. Plant Sci Lett 17:29-35
8. FRENKEL C, I KLEIN, AND DR DILLEY 1968 Protein synthesis in relation to ripening pome fruit. Plant Physiol 43:1146-1153
9. FRIEND J, MJC RHODES (eds) 1981 Recent Advances in the Biochemistry of Fruit and Vegetables. Academic Press, London, 275 p
10. HAYFLICK L, PS MOREHEAD 1961 The serial cultivation of human diploid cell strains. Exp Cell Res 25:585-621
11. KE D, R ROMANI (0000) Effects of spermidine on the senescence and ethylene production of suspension-cultured pear fruit cells. Plant Physiol (submitted)
12. KIDD F, CA WEST 1930 Changes in the respiratory activity of apples during their senescence at different temperatures. Proc Royal Soc London Ser B, 106:93-109
13. KU LL, RJ ROMANI 1970 The ribosomes of pear fruit-their synthesis during the climacteric and the age-related compensatory response to ionizing radiation. Plant Physiol 45:401-407
14. LEOPOLD AC 1980 Aging and senescence in plant development. In KV Thimann, ed, Senescence in Plants. CRC Press, Boca Raton, FL, pp 1-12
15. LIEBERMAN M (ed) 1983 Post-harvest Physiology and Crop Preservation. Plenum Press, NY & London, 572 p
16. McGLASSON WB 1978 Role of hormones in ripening and senescence. In HO Hultin and M Milner, eds, Postharvest Biology and Biotechnology. Fruit & Nutrition Press, Inc, Westport, CN, pp 77-96
17. MACKENZIE IA, HE STREET 1970 Studies on the growth in culture of plant cells. I. Production of ethylene by suspension cultures of Acer Psuedoplatanus, L. J Exp Bot 21:824-834
18. MEDAWAR PB 1952 An unsolved problem in biology. Reprinted In Medawar PB 1957 The Uniqueness of the Individual. Methuen, London, pp 28-54
19. MUHITCH MJ, LA EDWARDS, JS FLETCHER 1983 Influence of diamines and polyamines on the senescence of plant suspension cultures. Plant Cell Rep 2:82-84
20. MURASHIGE T, F SKOOG 1962 A revised medium for rapid growth and bioassays with tobacco tissue culture. Physiol Plant 15:473-497
21. NICKOLS W, D MURPHY (eds) 1979 Differentiated Cells in Aging Research. Academic Press, New York, 215 p

22. NITSCH JP, T ASHIRA, LME ROSSINI, C NITSCH 1970 Bases physiologiques de la production de chair de pomme et de poire in vitro. Bull Soc Bot Fr 117:479-492
23. NOODEN LD 1980 Senescence in the whole plant. In KV Thimann, ed, Senescence in Plants. CRC Press, Boca Raton, FL, pp 220-258
24. PECH JC, C AMBID, A LATCHÉ, A DIARRA, J FALLOT 1975 Culture de tissus et de suspensions cellulaires de fruits. In Facteurs et Regulation de la Maturation des Fruits. Colloq CNRS Paris, pp 211-220
25. PECH, J-C, C BALAGUÉ, A LATCHÉ 1982 Senescence in plant cell cultures. In PF wareing, ed, Plant growth substances. Academic Press, NY, pp 549-558
26. PECH, J.C., R.J. ROMANI 1979 Senescence of pear fruit cells cultured in a continuously renewed auxin-deprived medium. Plant Physiol 64:814-817
27. PUSCHMANN R, RJ ROMANI 1981 Senescence of cultured pear fruit cells. Physiological transitions revealed by the response to auxin, copper or 1-aminocyclopropane-1-carboxylic acid. Plant Physiol (suppl) 67:116
28. PUSCHMANN R, RJ ROMANI 1983 Ethylene production by auxin-deprived suspension-cultured pear fruit cells in response to auxin, stress, or precursor. Plant Physiol 73:1013-1019
29. PUSCHMANN R, D KE, R ROMANI 1985 Ethylene production by suspension-cultured pear fruit cells as related to their senescence. Plant Physiol (in press)
30. RICHMOND A, JB BIALE 1966 Protein and nucleic acid metabolism in fruits: I Studies of amino acid incorporation during the climacteric rise in respiration of the avocado. Plant Physiol 41:1247-1253
31. ROMANI RJ, K FRENCH 1977 Temperature-dependent changes in the polysomal population of senescent (ripening) pear fruit. Plant Physiol 60:930-932
32. ROMANI RJ, TJ BOS, J-C PECH 1981 Cycloheximide stimulation of cyanide-resistant respiration in suspension cultures of senescent pear fruit cells. Plant Physiol 68:823-826
33. SACHER JA 1973 Senescence and postharvest physiology. Annu Rev Plant Physiol 24:197-224
34. SKIRVIN RM 1978 Natural and induced variation in tissue culture. Euphytica 27:241-266
35. SMITH TA 1985 Polamines. Annu Rev Plant Physiol 36:117-143
36. SOLOMOS T, GG LATIES 1974 Similarities between the actions of ethylene and cyanide in initiating the climacteric and ripening of avocados. Plant Physiol 54:506-511
37. SOROKIN C 1964 Aging at the cellular level. Experentia 20:353-416
38. STREHLER, BL 1962 Time, Cells and Aging. Academic Press, NY & London, 270 p
39. SUNDERLAND N 1977 Nuclear cytology. In HE Street, ed, Plant Tissue and Cell Culture. Univ Calif Press, Berkeley, pp 177-205
40. THEOLOGIS A, GG LATIES 1978 Respiratory contribution of the alternate path during various stages of ripening of avocado and banana fruits. Plant Physiol 62:249-255
41. TONG C, J LABAVITCH, S-F YANG 1985 The induction of ethylene by plant cell wall fragments. Plant Physiol (suppl) 77:63
42. WAREING PF, AK SETH 1967 Aging and senescence in the whole plant. In HW Woolhouse, ed, Aspects of the Biology of Ageing. Soc for Expt'l Biol Symp XXI pp 543-558
43. WATADA AE, RC HERNER, AA KADER, RJ ROMANI, GL STABY 1984 Terminology for the description of developmental stages of horticultural crops. HortScience 19:20-21

44. WOOLHOUSE HW 1967 The nature of senescence in plants. In HW Woolhouse, ed, Aspects of the Biology Ageing. Soc for Expt'l Biol Symp XXI, pp 179-213
45. WOOLHOUSE HW 1983 The general biology of plant senescence and the role of nucleic acids and protein turnover in the control of senescence processes which are genetically programmed. In M Lieberman, ed, Post-harvest Physiology and Crop Preservation. Plenum Press, NY & London, pp 1-43

22. PHYSIOLOGICAL STATES AND METABOLIC PHENOTYPES IN EMBRYONIC DEVELOPMENT

D.J. DURZAN

1. INTRODUCTION

A goal for the mass clonal propagation of trees is to control the developmental process accurately and to regenerate trees with elite genes that are expressed in a true-to-type fashion. Achievement of this goal will contribute significantly to the exploitation of genetic variability in existing germplasm. Since we now have several examples of somatic embryogenesis (SE) in forest trees (Tulecke, Durzan, this volume) and a basic understanding of zygotic embryogenesis (e.g., 28, 29, 84, 97) we can expect the emergence of models for large scale clonal propagation. This includes the need for improved diagnostic reasoning and quality control for <u>in vitro</u> procedures.

SE proceeds through variable sequences of growth and development depending on the origin of totipotent cells and how they are manipulated. The concept of totipotency has been linked to gene expression at the cell level and to developmental events throughout the life cycle (32, 38, 52, 99, 100). This review extends these concepts by proposing a new method for describing physiological states and metabolic phenotypes for evaluation and diagnostic purposes.

Factors, known to induce SE, now comprise the choice of explant, critical nutrients and growth regulators in the medium (53, 54, 55, 76). It is now becoming apparent that processional, regular development in SE depends on how cells respond to the sum total of all of a series of critical factors. This sensitivity and "pattern of responsiveness" is inherited as seen in the specificity, repetitiveness and repeatability of developmental processes. The predisposition of cells to the sum of factors at any one time describes their physiological state which has long been known to be a major determinant in the growth, development, and survival of clonal materials (31, 91, 114).

According to current dogma, the procession of physiological states in fully totipotent cells is based on the transfer of information with high fidelity from DNA to RNA and to protein (enzymes, structural proteins, reserves) for the formation of products of metabolism that contribute to the emergency of structure and function. For our purposes the physiological state may be defined as the past history of a plant system which affects its present and future behavior. Natural laws and behavior are represented as mathematical equations describing how the development of a system is uniquely related to its state at some earlier time. In our approach the equation of the state, at any one time, is normalized across networks of metabolites recovered from the soluble N pool. Specific states of cells that can be discerned in our laboratory by color-coded responses to diagnostic stains are: regenerative of SE; stationary, i.e., producing more callus; or degenerative, e.g., moribund. We have used a double-staining multi-color test to display physiological states in embryonic cell suspensions of Prunus cerasus (cited in 81) and four conifers (sugar pine, loblolly pine, Douglas-fir and Norway spruce).

The increasing frequency of soma- and gametoclonal variation in culture systems now suggests that propagation methods for some species may be aberrant more so than previously believed. Some of the new variation will be useful in cultivar development and in widening the genetic variation of trees such as red and Monterey pine (cf. Ahuja, volume one ; 16, 42, 94). However, these variations or aberrations in development must be better understood before their utility is accepted (32 and 89 for a glossary of terms).

In this review, a way of describing the metabolic phenotypes in SE and zygotic embryogenesis is also proposed. The description of activities of metabolites, including growth regulators, may someday be correlated with enzyme activities that produce discrete phenotypes and ultimately with the sequences of bases in DNA. It is also important to describe the functional interactions, fluctuations and spatio-temporal organization of such systems in dynamic terms. This is accomplished by graphical descriptions of physiological states and metabolic phenotypes that are combined to form a 3-dimensional map revealing the dynamic changes. Combined maps of this type have been designated as "adaptive landscapes," "signatures," "atlases," and even "repertoires" because of the wide range of activities and data that are portrayed.

Our concepts are based on earlier experiences with 3-dimensional maps for true-to-type zygotic embryo development in the pistachio (Pistacia vera cv. Kerman) (39). These maps were used to characterize embryo abortion in trees under field conditions. The diagnostic reasoning was based on changes in N metabolism and on the availability of zinc (section 3.4.). Someday a parallel model may be available for SE in cell suspensions of this species. Other uses of the model are discussed in relation to tree physiology and tree breeding and improvement programs (section 5.).

2. STATE-NETWORK MAPS

2.1. Gene expression in development

The control of enzymes coded by genes are usually demonstrated by induction, repression, activation, and inhibition (Fig. 1). Some aberrations among cells and tissues that are based on this control are listed in Table 1. At any time interval, each cell changes its physiological state according to the application of discrete laws and inputs, e.g., diffusion of substrates, pH optima, temperature thresholds, etc. (113). These changes comprise the gradients and positional information of development and may be important in the cell's final choice of morphogenetic programs.

We have developed a diagnostic model based on metabolic networks and physiological states during embryogenesis. This state-network model can be projected topologically using the mathematics of manifold or assemblage theory and/or by the software packages used by statisticians and engineers. Gene expression for sets of interacting genes (49) can be projected as a genetic net (62).

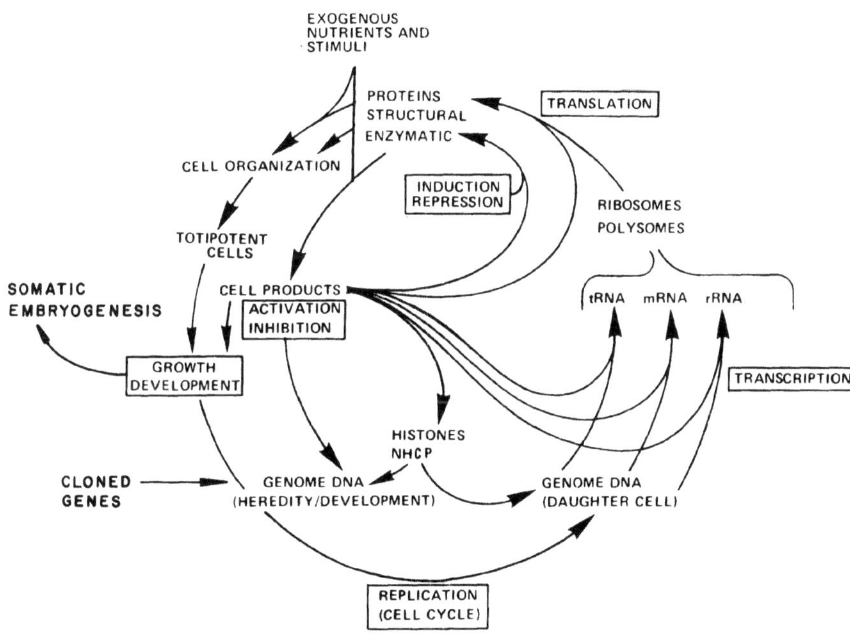

FIGURE 1. A molecular framework for characterizing gene expression in totipotent cells. Cells may be directed to produce somatic embryos or programmed for

other fates, e.g., callus and cell senescence. The chart describes how physiological states leading to growth and development (somatic embryogenesis), can be characterized by the composition of cells. Changes in composition arise by gene expression and in response to exogenous nutrients and stimuli. The embryo's responses to its environment do not involve, in general, the immediate products of gene action but can be the result of networks of processes. Products or metabolites typical of each physiological state and network may be projected geometrically to chart the behavior of the system (cf. Fig. 2). Cloned genes, e.g., for proline overproduction during the osmotic stress response or for production of bacterial toxins are now available for insertion into plant cells to perturb and study this model (e.g., 23). Expressions of these foreign genes may be followed by their metabolic phenotypes (e.g., Fig. 2A).

Table 1. Some sources of variation in cell and tissue culture systems.

A. Vegetative Propagation (63)	B. Soma- and gametoclonal variation (42, 94)
source material (genetic)	gross karyotypic changes
propagation system	cryptic chromosomal arrangements
planting site	somatic crossing-over and mutation
plant performance	sister chromated exchange
(gene expression)	transposable elements and stochastic variations
chimeras	gene amplification and diminuation
systemic infections	nuclear-cytoplasmic interactions
	multifunctional genes
	promiscuous genes among organelles
	selection-induced polymorphism

2.2. Map construction: Metabolic networks

Networks in metabolic phenotypes, based on gene expression and the principles of biochemical genetics, may be projected arbitrarily as a 3-dimensional net or map (Fig. 2A). Each raw variable in a data set is converted into a standardized variable with zero mean and unit variance. This preprocessing and normalization of data assists in establishing diagnostic criteria for pattern recognition. The approach and its alternatives have been described by C. Jurs (Science 232:1219-1224, 1986).

For example, the map format may comprise all developmental stages of the zygotic embryo (days) and the soluble N composition of embryos (e.g., distribution of N in each compound as a percentage of the total soluble N). The value of using parameters based on N metabolism underscores the importance of N assimilation in metabolism (98), growth (e.g., 59), and productivity (1). Alternatively, the carbon in metabolites can be employed instead of N. Comparison of carbon and nitrogen maps would permit the display of the C/N fluxes among metabolites over all stages of development. This would relate to historically important concepts of carbon-nitrogen balance in morphogenesis, e.g., stress and shade tolerance, flowering (114).

The time axis associated with the metabolic networks represents different physiological states of the developing zygotic embryo (section 2.3). This axis may be replaced to represent leaves, roots, and cell suspensions, or processes such as growth, photosynthesis, respiration, and critical environmental factors.

With metabolic phenotypes (Fig. 2A), we can extend the composition of the soluble N pool to include other compounds in physiological fluids, e.g., sugars, organic acids, phenolics, nucleotides, nutrient elements. Rather than portraying the distribution of N among metabolites, other variations include the flux of metabolites, N, carbon, sulphur, etc. This approach can be employed to reveal the more dynamic departures from equilibrium conditions (section 3.6.). Unidentified compounds may be included as a group or individually as arbitrary units on the metabolite axis to extend the map. The inclusion of equations describing the best-fit for data for the unidentified compounds improves the definition and characterization of gene expression especially where the metabolic basis of an elite trait is unknown (cf. section 5.).

Whenever possible, metabolites are arranged hierarchically and logically into families. Families of hierarchical networks are portrayed as best-fit equations on a 2-dimensional plane according to the genetically determined sequence of each metabolite in biosynthetic reactions (cf. 25, 38). In Figure 2A the free amino acids are arranged from left to right into families based on the compound that characterizes the "port of entry" of N into organic combination (38). Each numbered lane in a family represents the dynamic changes of one metabolite. The aspartate family contains 5 metabolites (lanes 1 to 5) derived from aspartic acid. Families are based on the number of carbons (1-5) arising from photosynthate, respiration and protein turnover that contribute the formation of the first amino acid in the biosynthetic sequence (cf. Fig. 3). Families of metabolites are separated arbitrarily by 5 blank intervening lanes. The same networking can be applied to any pivotal compound that contributes to the biosynthesis of subsequent products, e.g. methionine metabolism in Lemna (46).

It is possible to enhance the best fit lines by including additional lines (numbered lanes) that bridge the gap between the "dynamic" or equation for each logically related metabolite. For example, between citrulline and arginine, the equation that bridges the behavior of these two compounds could be useful to predict the behavior and levels of argininosuccinate. The latter is a very transient intermediate and remains enzyme-bound so that it is not normally extracted in significant amounts. The addition of unnumbered lines describing this spatio-temporal relationship between closely related metabolites is an arbitrary part of the model that enhances the "shape" of maps shown in Figure 2.

Other graphic aids, such as those based on polar coordinates, may be used (19). If metabolic networks are not arranged as a logical network using the principles of biochemical genetics, then irrelevant patterns and assumptions are introduced into the construction of metabolic phenotypes. For example, if we plot the sequence of amino acids as it elutes from chromatographic columns, the sequence becomes unrelated to metabolic networks. The elution pattern describes the capabilities of the analytical column and properties of compounds rather than the metabolic networking in the tissue. In developmental terms this pattern represents a distortion or misuse of abstraction.

Metabolite quantities in the vertical axis may be described in many ways, e.g., dry weight, fresh weight, molar, protein, N or carbon distribution, per cell or explant, etc. Quantities may even be related to genetic factors (dominance, recessiveness, fitness, etc.) or to dimensions (surface area or volume) of the developing tissues. These variables are inherent in Wright's adaptive landscapes and Richards' comprehensive growth equation (88).

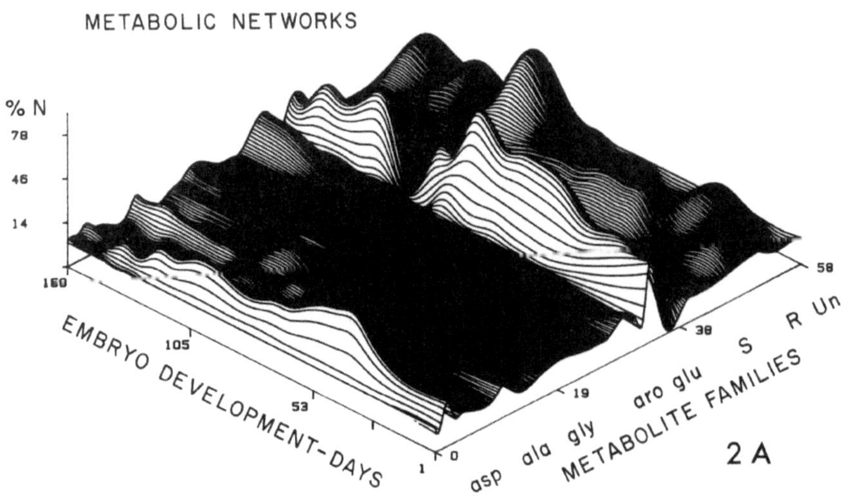

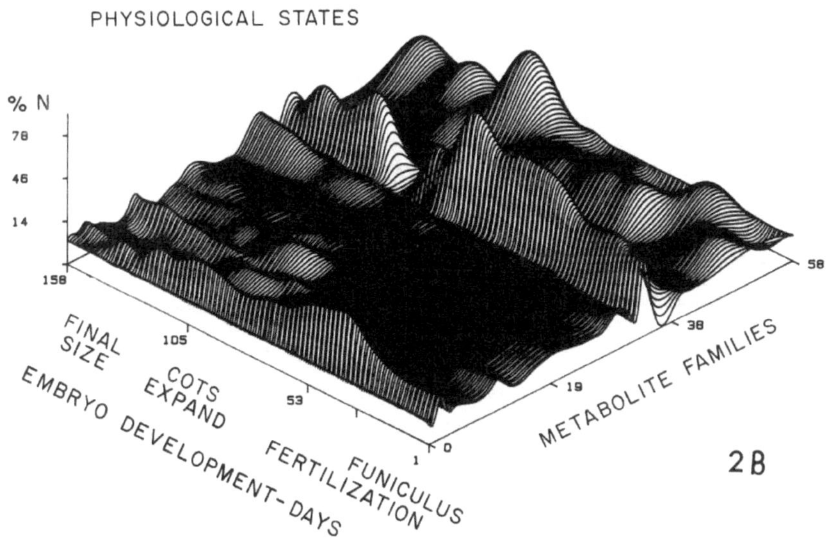

2B

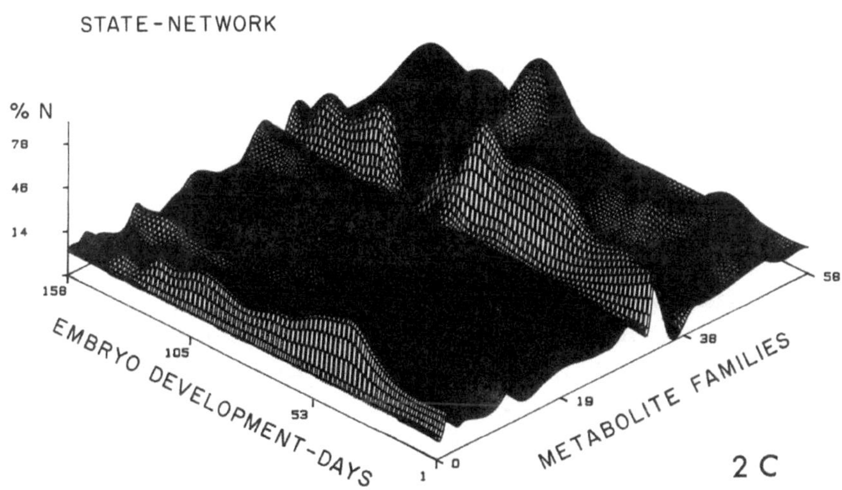

2 C

FIGURE 2. Construction of a combined state-network map based on the distribution of N compounds in the soluble pool of zygotic embryos (Pistacia vera) from fertilization to filling of the fruit at 158 days. A. Metabolites in physiological fluids are arranged arbitrarily into amino acid families, e.g., aspartate (lanes 1-5), alanine (10-12), glycine (17-19), aromatic amino acids

(24-27), glutamate (32-35), methionine and sulfur amino acids (40-43). Amino acids in families are arranged along the axis in the logical sequence of their biosynthesis. Where this is not known the remaining amino acids (R) are listed in order of elution from chromatograms (lanes 48-53). N in all unidentified compounds is summarized in lane 58. B. Each day of embryonic development is integrated over composition to describe the physiological state of the zygotic embryo. The maps start with the development of the funiculus and ovule (1-15 days). This is followed by fertilization (15-18 days) and the globular and heart-shaped stages (75 days). Final size is reached around 112 days. C. Integration of figures A and B into a single format to provide a state-network map viewed obliquely. Through computer-assisted technology, the map can be rotated and viewed at various degrees of magnification to reveal information hidden by peaks. The map can be magnified to enhance compounds that contribute to a very low percentage of the total soluble N. Features of these maps can be compared with mutant expressions or aberrations to provide by subtraction, quantitative differences.

2.2.1. Correlations with protein and nucleic acids

In Figure 2A, the percent distribution of N in free amino acids in the developing embryos is arranged into families of identified metabolites and a category (lane 58) having a total for the unidentified compounds (abscissa not shown). The β- and ω-amino acids have been assigned to the R (remainder) category. The distribution of identified amides and α-amino acids relates mainly to protein synthesis and turnover, whereas, the β-amino acid reflect the degradation of pyrimidine bases in nucleic acids (cf. 38). Levels of free amino acids were determined by ninhydrin on an automated amino acid analyzer designed for physiological fluids (56). Post-column modifications to this instrumentation have extended our analytical capabilities to include polyamines, monosubstituted guanidines, and to conjugates of phenols with nitrogenous compounds that exhibit natural autofluorescence. This means that the network-axis may be easily expanded to include over 100 compounds.

It is also now possible with large 2-dimensional gel electrophoresis to generate topological maps for proteins (2, 119). A complete description of the protein complement would include enzymes, regulatory elements, structural components, and storage proteins. Display of these proteins will relate to DNA structure, enzyme-forming systems, and state network maps. Maps showing the emergence of metabolic proteins particularly in organelles reveal new levels of "process control" by a variety of factors, e.g., calcium, calmodulin and protein phosphorylation (cf. section 3.5).

In another variation, the axis describing metabolic families could be replaced by DNA base sequences that lead to the sequential production of specific proteins over development. Mutation frequencies, Mendelian genes and quantitative genes could be mapped or arranged along one axis and their products

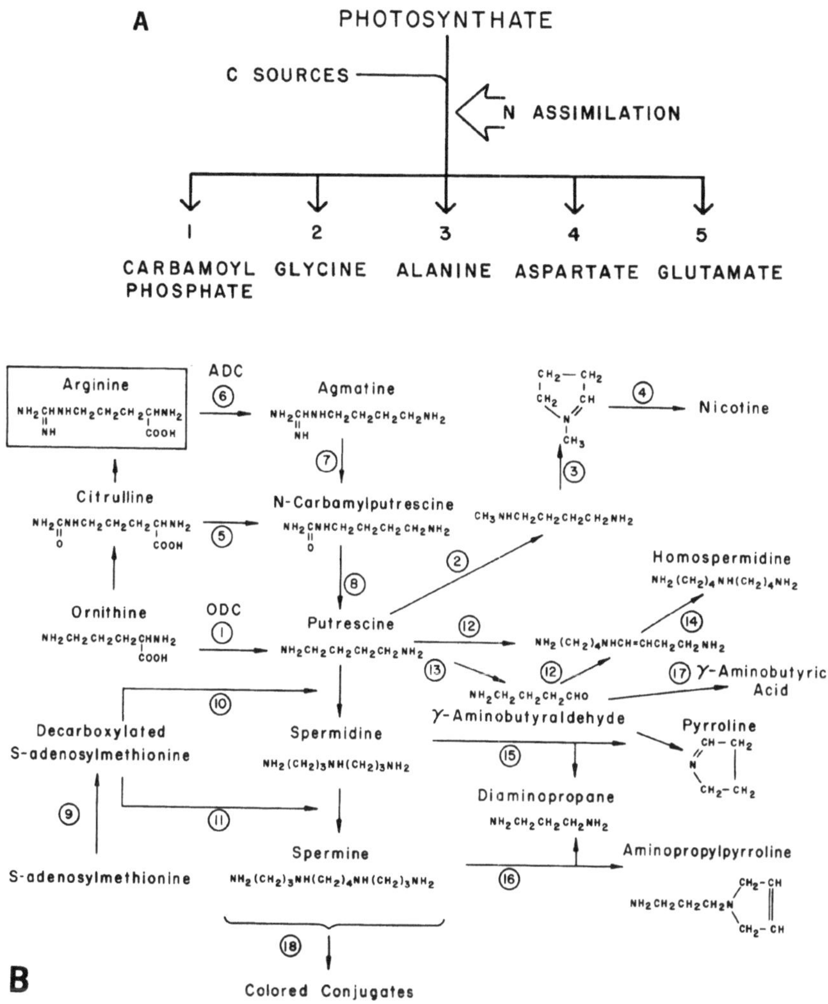

FIGURE 3. The arbitrary arrangement of amino acid families recovered from physiological fluids (the soluble N pool) into metabolic networks according to carbon sources (frameworks). A. Carbon from photosynthate combines with N to produce families of amino acids. Each family is represented by the number of carbons (1 to 5) in the acceptor for reduced N, e.g., pyruvate (3 carbons) yields the alanine family of amino acids (cf. 38). B. The number of metabolites in each family varies with the species and conditions. For example, arginine is a prominent N storage compound in woody species. While arginine has 6 carbons, it is derived from the glutamate family (5 carbons). Arginine contributes to the biosynthesis of monosubstituted guanidines, polyamines, and their colored conjugates (e.g., 43).

on the other axis. From maps of this type we can glean quantitative information regarding genes, *viz.* their elimination, amplification, rearrangement (directed or nondirected), base modifications, base-sequence polymorphism, nuclear-cytoplasmic interactions and their properties or map implications, e.g., stability, replication, transcription, processing, scheduling of expression, and transmission of information to daughter cells. By matching genetic maps with state-network maps for known and yet unidentified metabolites, we may even be able to detect genes whose expressions have eluded us to the present.

One specific example of the application of these concepts is as follows. A 3.7 kilobase mutant segment of DNA from *E. coli* that codes for the overproduction of proline has been found (23). This overproduction endows resistance to osmolyte stress. The proline genes have been cloned and sequenced with the aim of inserting it into a variety of *Agrobacterium* vectors. If this vector were introduced into forest trees it would be interesting to see if proline overproduction is expressed and if this phenomenon endows frost or drought resistance. Expression of proline overproduction in trees would indicate that less than a 4 kilobase DNA fragment accounts for the behavior of proline N (lane 34) on the map of the glutamate family of amino acids. By analogy, the display of 100 compounds on the map would represent a 4 x 100 or a 400 kilobase fragment of DNA.

For each metabolic phenotype, it is now possible to identify and probe the mRNA and proteins (e.g., 87, 89) generated by base sequences of DNA-gene fragments and vice-versa as follows:

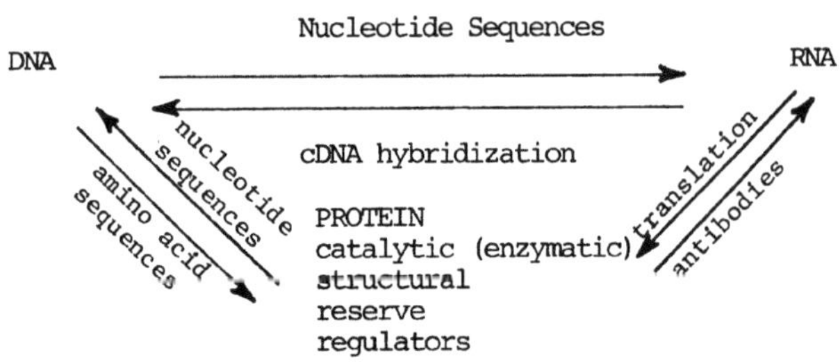

A common observation in embryonic tissue involves protein synthesis at the expense of the free amino acid pool. Protein degradation will lead to an increase in the free amino acid pool. Both aspects are observed during the development of the zygotic pistachio embryo (39). Integration of the information derived from interactions among macromolecules with diagnostic maps of metabolic phenotypes will not be a simple task. For woody perennials its total accomplishment may take years especially if we recognize how many yet unidentified metabolites and proteins exist and how little we know about the organization of genes and gene expression.

2.2.2. Networks as carriers of information

The information content of parallel arrays of metabolites in metabolic networks (Fig. 2A) can be defined formally in the strict sense of information theory (9, 85). The usefulness of this approach remains to be determined. In our example, the flow of N over carbon frameworks is directed through incompletely defined genetically determined sets of networks for a specific cultivar. Further interpretations are not easy. Data are incomplete and we will have to deal more effectively with the N source, quantity of N, the capacity of networks to carry N, and the output or messages conveyed as tissues reach a finite expectation in growth and development.

More formal mathematical definitions and models are available (26). These approaches involve the efficiency of biosynthesis (79) and the cost in ATP (adenosine triphosphate) for the biosynthesis of each amino acid in any given network or family (3). These parameters can be introduced into maps as coefficents that are to be multiplied into equations describing the best fit of data. More dynamic models can be based on metabolite flux and flux departure from equilibrium (107).

2.3. Map construction: Physiological states

Physiological states based on N metabolism can be portrayed by drawing lines (perpendicular to the lanes) and across the metabolic networks (Fig. 2B). From fertilization to the development of the ripe kernel we have a series of characteristic lines that normalize and integrate metabolite concentrations over all metabolic networks and unidentified components. The best-fit equations will now describe different thresholds of response (e.g., 0 to 1 as a probability or signal; 0 to 100% as a distribution). Each line in the horizontal plane provides a useful dynamic description of the physiological state that changes temporally over embryonic development.

Physiological states may be considered regenerative, degenerative, moribund, etc. (e.g., 17) and the overall state map referred to as a "developmental field." In reality, the equation describing the state in our model is not continuous. Each amino acid family is a discrete separate entity that may or may not be connected to another. Rearrangement of families on a map may be easily accomplished on the computer to observe interactions among families where these are known or predicted. Hence, the line represents a summation of the normalized behavior of amino acids within each family. The line is presented as a continuous function overall families for graphic simplicity. The temporal sequence of horizontal lines or state equations lead to a specific developmental expectation as described by its developmental field.

For cell and tissue culture, the inclusion along the metabolic axis of macro- and trace-mineral nutrients responsible for certain physiological states would be especially valuable. This would enable us to evaluate the effect of synthetic nutrient media on growth and development. We could compare patterns to see if the internal compositions of SE and true embryos are consistent or aberrant. Adjustments to the culture medium could be made according to the diagnosis. The applicability of this approach would be mainly to establish quality control largely because of the relatively high costs involved in producing such maps.

For the pistachio, stages of development before fertilization (Fig. 2B) represent the large funiculus and the attached small ovule. The funiculus is the structural link between the mother tree and the developing embryo (21, 67). Through it passes the soluble materials that nurture the embryo and signals for further development. Days 0-15 represent in large part the "flush" of soluble N moving through the funiculus to the ovule. Much of this soluble N is asparagine and arginine. After day 18 (fertilization) the complexity of the zygotic phenotype emerges.

2.4. State-network maps

When Figures 2A and 2B are combined and integrated into a state-network map (Fig. 2C), each area where coordinates cross will form a node (square or rectangle) depending on symmetry of the matrix. This node reflects gradients and some of the positional information that may be important in describing the local effects in spatio-temporal terms, i.e., in qualitative and diagnostic reasoning. Some computer programs will enable the investigator to subtract one pattern from another to give "difference phenotypes." Relative operating characteristics and "error signals" of developmental fields may be defined in this way (cf. section 3.7.).

Differences among state-network maps may help us to further define the concept of "adaptive landscapes" (118), i.e., mechanisms of gene expression based on the state of dominance of alleles. Errors in the system and gene x environment interactions could also be detected. However, it is yet difficult to obtain an accurate landscape of how dominant genes are expressed in this situation because there has been little careful study of degrees of dominance under varied genetic backgrounds.

There is a lack of generally applicable notions to describe the physico-chemical nature of developmental fields, adaptive landscapes and metabolic phenotypes. Hence, maps are no more than a convenient descriptive device (110). Table 2 contrasts attributes of maps with causal laws and morphogenesis. Table 3 describes useful information that may be drawn from such descriptions. Someday we may be able to compare and relate state-network maps to the actual subcellular surface topography of biological specimens at high resolution. A portrayal of surface topology could be achieved by scanning tunneling microscopy (4) and perhaps integrated to produce 3-dimensional models as has been done recently with membranes (41).

Table 2. Comparison among several useful attributes of A. metabolic phenotype (state-network map), (e.g., Figs. 2C, 4). B. causal factors in physiological systems. C. morphogenesis.

A. Metabolic phenotype	B. Causation	C. Morphogenesis
Statistical correlations	External factors, internal determinations	Template generated forms (copies and modifications)
Model-relevant and irrelevant components	Interaction law statements	Replication of basic pattern (redundancy)
Functional relations Composition & distribution	Dynamic law statements Trigger-action	Conditions generate pattern (spatial interactions)
Equilibrium configurations	Blocking-action	Instruction-generated pattern (strategic evolution of form)
Extremum law statements Kinematic laws	Global dynamic laws Dynamic statistics Dialectic principles Teleological law statements	Re-creation of potential (totipotency)

Table 3. Examples of the kinds of information and attributes of metabolic phenotypes that characterize physiological states.

Metabolic networks expressed in most of the tissue or only in a few cells
Wide area or local (restricted) metabolic networks for a given phenotype
Disproportionate synthesis and aging in cellular populations
Loss of information and malfunction, e.g., effect of gene deletion, failure of protein specificity
Epigenetic processes and homeostasis
Errors in assembly of cells or their components
Chemical deterioration in physiological states
Detrimental interactions among networks and developmental processes
Entrainment of networks and states (e.g., diffusive or nonlossy)
Extent of plastic-elastic (pleomorphic) states
Expression of dominant, recessive, and pleiotropic gene action
Coupling and patterning to evaluate coherence among states
Instabilities that break symmetry in metabolic flux
Periodic minimal surfaces
Synchronization of networks with states
Level of activity (macro vs. low molecular weight compounds, macro vs. micro elements)
Substitutions among reiterated patterns (e.g., homeotic control)

3. MAP FEATURES

3.1. Threshold types

This section deals mainly with the behavior of the vertical z axis and its many ways of representing and interpreting data. For example, under a given set of environmental conditions, specific combinations of characters to be cloned through the explant material have a "fitness" that may be greater than others. The result is a state-network map with one or more fitness optima. Optima are represented as thresholds by several possible map morphologies. Repetitively reoccurring metabolic phenotypes and developmental fields may represent evolutionary strategies (i.e., unbeatable under natural selection), developmental strategies (i.e., true-to-type expression), and cultural strategies (i.e., they change in responsiveness upon removal or introduction of constraints to growth and development).

Physiological thresholds can be characterized by signals based on maxima and minima, transport constants and proportionalities among metabolites that change according to some known or postulated "process control" in cells (Figs. 4 and 7). These thresholds represent the large-scale integration of physiological states over metabolic fluxes.

Namkoong (74) and Wright (118) have described some useful gene-action models. In these models, simple thresholds may be "point" or "threshold dicotomous" (118). A point dichotomy is characterized by sharply alternative states. By contrast, a threshold dichotomy is based on a complex or ratio of interacting factors at a critical time. It may involve a succession of thresholds. Thresholds may represent specific, temporal, sequential, hysteretic, mnemic, autopoietic, catastrophic and complex conditions (13, 14, 45, 48, 60, 86, 92, 102, 103, 111, 113, 115, 117, 120).

An initial inspection of most dynamic hierarchical control systems may require the identification of thresholds as being linear or nonlinear (57, 65, 102). By using information based on fixed and adaptive systems, some patterns can be recognized and even classified by machine (90). In complex interactions involving many traits, pleiotropic thresholds and their dynamic parameters may be interlocked. Interlocking ensures that peaks detected on maps of metabolic phenotypes would differ more or less in their selective values (118). Metabolic phenotypes in N metabolism (Fig. 2C) must also somehow relate to the biosynthesis and action of plant growth regulators involved in morphogenesis (cf. Figs. 3 and 4; 30, 58).

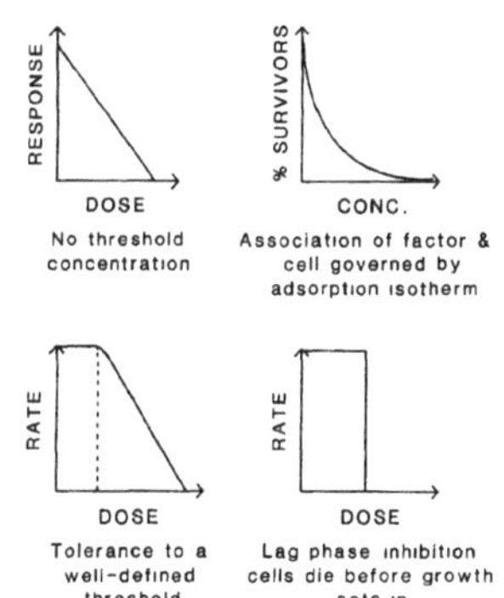

4

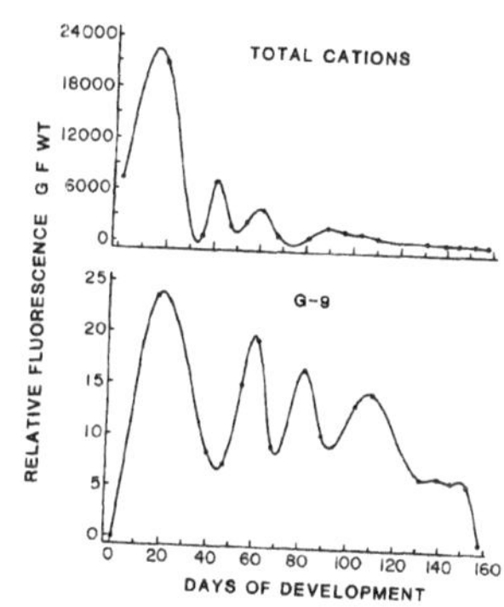

5

FIGURE 4. Simple threshold types derived from physiological variables identified on the surface of state-network maps. Conc. is concentration or dose of a governing factor.

FIGURE 5. Complex set of thresholds based on damped oscillations of levels of total monosubstituted guanidines (top) and an individual unidentified compound (6-9) in the embryo (bottom) during the development of the pistachio kernel (F. Ventimiglia and D.J. Durzan, unpublished). Maxima and minima of these functions could represent signals or thresholds inherent in zygotic embryogenesis.

Considerable evidence now exists that cellular metabolism involves various types of negative feedback control (110). Simple control is achieved by inhibition of enzyme activity, by the end-product repression of enzyme synthesis, mutual interdependence of regulator-operator systems of genes and so on (57). Such systems tend to depart from equilibrium. During departure they are liable to go into oscillation (Fig. 5; 64).

A damped oscillation is inherent in the mathematical function that describes divergence angles of phyllotaxis and in the number of leaf primordia at shoot apices (60), p. 140). In the pistachio, the behavior of the unidentified compound (Fig. 5) could be contiguous with developmental events that become damped with cellular development. The stability of threshold values in physical and biological systems have been described mathematically (24, 40, 47, 57, 69, 82, 102).

3.2. Stability

The concept of stability is widely used in considering dynamic systems of all sorts (22, 57, 62, 72, 73, 82). Stability relates to the development of forces that tend to restore the original condition or process. Stability provides useful information about the behavior of local or wide areas of the metabolic phenotype and physiological state.

Once the rules and set of differential equations describing the state-network map are known, the investigator may want to predict the operation of the developmental system by finding a mathematical solution for a set of equations. Where an equation is simple and linear, a solution can usually be found. However, in practice, such systems are complex and nonlinear with multiple factors varying over time (10, 24, 55, 116). Rarely can such systems be determined exactly and this is where the concept of stability becomes useful.

At the genetic level, gene over-amplification may lead to an unstable phenotype. Amplification of genes can be traced to different base sequences in DNA (93). Pea root-tip cells have double-stranded extrachromosomal DNA (exDNA) that remains associated with the nucleus as cells differentiate from the G-2 phase and occupy the elongation zone (66). The exDNA is capable of autonous replication via a strand displacement mechanism. This results in the release of single-stranded DNA (ssDNA) molecules. If the base sequence of the displaced ssDNA is identical to that of its template, DNA amplification occurs. If, however, the base sequences are different, any change in genetic information, could affect genetic stability and cell differentiation. Elongated suspensor cells of conifers may also have exDNA (J. Van't Hof, personal communication) but how the process relates to the phenotype of suspensors as opposed to embryonal initials remains unknown.

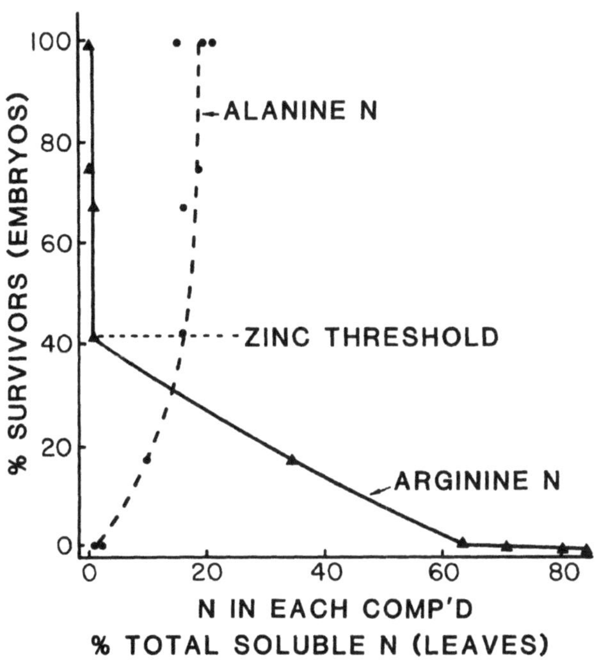

FIGURE 6. Effect of various degrees of zinc deficiency in 20-year-old pistachio trees on the composition of free arginine and alanine N in leaves in relation to the abortion of zygotic embryos (% survivors) (39). Low-zinc-induced syndromes have elevated arginine N levels (glutamate family) that are associated with the abortion of embryos. Completely developed embryos are characterized by lower levels of arginine and high alanine N levels (cf. Fig. 2B). Zinc is a cofactor for DNA polymerase-1 (77). Zinc is bound to an arginyl residue in the enzyme. Conceivably the buildup of free arginine could reflect the failure of the DNA polymerase in the zinc-deficient pistachio.

The stability represented in true-to-type state-network maps can be related to the effect of disturbances on the developing embryo. Comparisons among the physiological states at three different times for 12 trees over the range of zinc deficiency reveals the extent of the metabolic blocks (39). At the more advanced stage of embryonic development, it is the relative stability of the glutamate family of amino acids that is disrupted most by zinc deficiency as judged by an extraordinary buildup of arginine N in leaves. In extremely zinc-deficient trees the zygotic embryos abort.

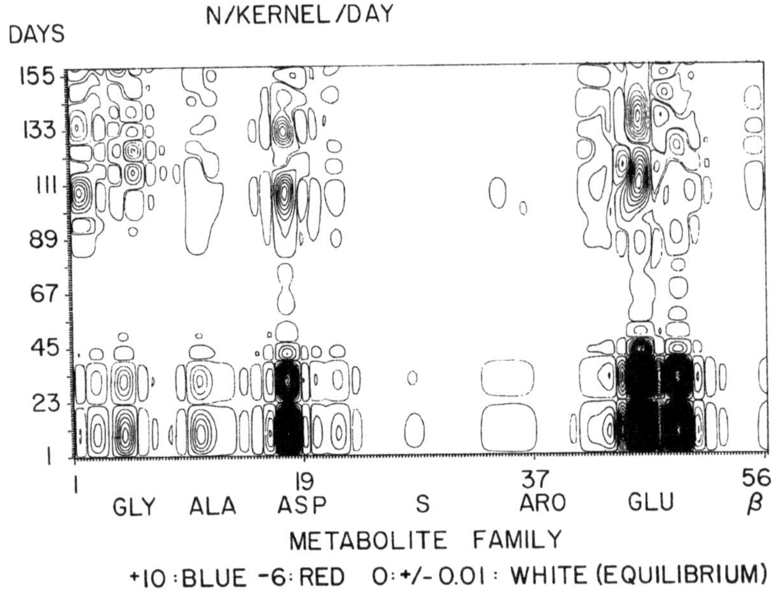

FIGURE 7. Positive and negative departures from equilibrium within networks of amino acid families during the development of the zygotic embryo (pistachio). The axes are the same as for state-network maps except that departures from equilibrium (mean value at fertilization) are indicated. Departures are based on the flux of N associated with each amino acid in the appropriate families (glycine, alanine, aspartate, sulfur amino acids, aromatic, glutamate, β-amino acids). Positive and negative departures from equilibrium are indicated by contours around maximum (blue) or minimum (red) values at or near the center of isobar contours. In practice, the positive and negative trends are always color-coded to facilitate inspection. The equilibrium value representing the steady state is indicated by a white background around contours. Metabolic networks are separated physically from one another by 5 intervening lanes.

Results show that as biosynthesis proceeds from left to right and over time, symmetry-breaking instabilities arise within each network.

If the availability zinc level differs, only very little from what is needed for the normal process, some branches will bear fully developed kernels but at a very reduced yield. Even so, kernels are quite similar in composition to the normal developing embryo.

Instability in morphogenesis has been described by Thom (103) in topological terms by catastrophe theory. Surfaces of state-network maps representing catastrophic events in metabolism could be coined as "Thom prints." Oscillations in metabolic networks (e.g., Figs. 6 and 7) may be viewed as energy or matter waves (e.g., 101). These activities can lead to the stabilization or collapse of biological systems (24, 86, 105).

3.3. Sequential signals

Canny (15) has called for new postulates to explain how plant hormones provide a variety of signals for growth and development. With metabolic fluxes (Figs. 4 and 7), it is possible that differences among thresholds (signal patterns based on maxima and minima) may be identified by subtracting one phenotype from another. Differences at any threshold level describes an "error signal" or "difference signal" in metabolic flux and in physiological states when a state-network map is used.

The "relative operating characteristics" of most systems of signals can be displayed as a signal space or lattice (12, 27) in 2- or 3-dimensions (e.g., Fig. 7). In many developmental problems we are dealing with a very long time horizon, complex equations and a very large numerical matrix to describe these sequential signals. When the flux or state operates in a "nonlinear" fashion, the mathematic models may require too many iterations, i.e., the computation time can be lengthy. In attempting to formulate a computational algorithm, our initial strategy has been to show the existence of some optimal pattern or solution based on a finite developmental goal, e.g., completion of somatic or zygotic embryogenesis. We eliminate useless information to reduce computations. This helps to avoid non-optimal or equivocal signals (e.g., Fig. 8).

Since it may be hopeless to look at all feasible solutions for the programming of metabolic signals in development, more practical solutions are often sought. The description of an "optimal signal pattern" without an exhaustive search involves some knowledge of concepts of dynamic programming. Notions in dynamic programming, particularly of sampled-data control processes, can help to keep the logic inherent in the system distinct from biochemical and biophysical plausibility although all must be compatible.

For example, for any metabolite in a family we can identify maxima and minima of the function describing departure from equilibrium as positive and negative integers (Fig. 8). As metabolism unravels, we can add these integers to estimate the "threshold costs" to complete a finite developmental process. By the same method we can estimate the costs across metabolic networks to maintain the cellular system in a stable physiological state. To proceed, we apply the notion that at any node on the map, the process can fail because of a number of constraints. So far, there are no ideas to tell us the best strategy on how to proceed. Computational experience and intuition are the only guides to the development of algorithms based on sequential metabolic signals. For the above reasons, we need more evidence for correlations among recurring metabolic phenotypes (seasonal, diurnal, circadian, etc.) with levels of growth regulators and other factors. Correlations should be supported by a kinetically compatible coincidence of signals. Kinetics and coincidence should be predictable by experimental designs.

Unfortunately good examples do not exist where morphogenesis in cell suspensions has been compared with the normal development of the population from which explants were taken. Where this has been attempted (18, 34, 35, 36, 37) a remarkable similarity in the sequence of accumulation of free amino acids is seen. In growing zygotic embryos and their derivative cell suspensions, the free amino acid N is initially dominated by the amide, glutamine. This pattern is followed by the elaboration of free amino acids for protein synthesis. Midway during the growth phase (near-exponential and autocatalytic), cells tend to set aside some of the soluble N as arginine. Toward the end of the decline of the relative growth rate, another amide, asparagine, accumulates. In my laboratory this sequence has been observed repeatedly for the growth of cell suspensions, germinating embryos, and expanding leaves. Differences among fluctuations in metabolites relate mainly to their amplitude (maxima and minima) and frequency rather than to their sequence.

Some compounds in metabolic networks at any given physiological state may be regulators of growth and development (promotors and inhibitors) (Fig. 5). Indeed, in cell suspension of jack pine, the monosubstituted guanidines derived from arginine are inhibitory to growth over many concentrations (36). Even so, we do not yet have a clear view of how morphogenesis relates to state-network maps. The consideration of unidentified compounds in metabolic phenotypes may help to sort out this dilemma. This approach assumes that all compounds responsible for the morphogenesis are included on the state-network maps and their significance in the development of embryos can be proven.

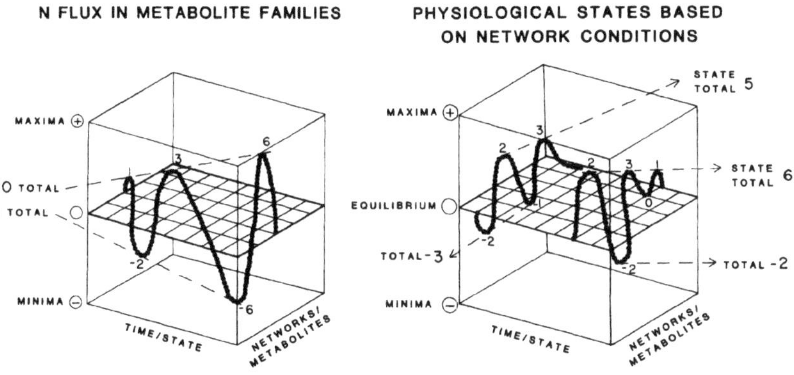

FIGURE 8. Costs of biosynthesis (left) and establishing physiological states (right) using maximal and minimal values (thresholds, signals) of the metabolic flux. Numerical values for maxima and minima are totalled over each function to simplify computation and to formulate "process control" algorithms. Costs can be reinterpreted to include the number of ATP's to generate patterns for each metabolite (3).

3.4. Bifurcations

Physiochemical models based on "vertical density currents" show patterns of bifurcation (8). For example, the vertical density currents created by solid, opaque or colored particles placed on top of viscous fluids produce inanimate suspensor-like structures and shapes as the particles sink under the influence of gravity. The elongated patterns generated by the currents are visibly analogous in shape to somatic polyembryos found in mucilaginous callus and in cell suspensions. The mushroom shape of particles at the base of currents resembles the densely nucleated embryos. Under some conditions these density currents of solid particles bifurcate and produce secondary density currents (8). The visual pattern produced is strikingly analogous to structures of suppressed secondary embryos in the developing conifer seeds. We are not yet sure how far such forces, currents, and bifurcations in the physiochemical model can be extended to viscous conifer cell suspensions having the capacity for SE.

As for the metabolic phenotypes in embryogenesis, the networking of carbon, nitrogen and other elements through interrelated metabolic pathways bifurcates as new compounds are produced (38). This may result in breaking the symmetry of contours around maxima and minima on maps (Fig. 7). Mathematical solutions to equations describing these phenomena also bifurcate (8, 64, 68). A solution can follow a path in a network or jump from one branch to another and form a cascade of bifurcations. Bifurcations would be more evident if we could trace the metabolic fate of a ^{14}C-labeled substrate. Recovery of the array of radioactive products reveals the fate of the substrate and the extent of metabolism. Patterns of recovery of products derived from radioactive substrates can be imposed graphically upon the map so as to provide the quantitative distribution of the tracer.

In biochemical genetics, phenotypic alternatives or adaptive landscapes arising from bifurcations in metabolic networks may relate to the dominance, recessiveness or pleiotropic action of genes responsible for steps in the pathway (118), i.e., reaction coordinate in the metabolic phenotype. Dominance is revealed through creating metabolic bottlenecks, partitioning of interacting substances and by competition between anabolic and catabolic reactions. Presumably the same can occur in physiological states.

3.5. Process control: global relatedness

In Figure 2B morphogenesis of the pistachio is represented by a temporal series of discrete, cooperating physiological states based on amino acid networks governing embryonic development. This alone does not do justice to the complexity of morphogenesis nor to the concept of totipotency (99). It may not distinguish clearly the significant and critical patterns due to growth and development or to specific physiological processes (e.g., 7, 45, 51, 80, 106). It is also difficult to see how morphogenesis achieves very large scale integration and global relatedness in development.

Once, however, we have sufficient detail and the proper correlations, the time axis of state-network maps may be more properly represented by "processes" or by algorithms of morphogenesis, e.g., gradients, positional information and bifurcations related to the action of growth regulators. More accurate geometric descriptions of global morphogenesis become spatially complicated. We now believe that an understanding of global morphogenetic process control and its algorithms will require so much more data that computerized displays will become essential (11).

A complete description of global relatedness among phenotypes, processes and components may not be obtained easily nor may it always be necessary (83). Through a sampling of critical features the truly effective metabolic phenotypes may be discerned. The "effective sampling index" should be identified by the number of features or attributes sampled over all of the known variables and components. Also some measure will be needed of the effective map configurations as opposed to the total possible configurations. To achieve an understanding of global relatedness, considerable knowledge, simulation and familiarity with the experimental material is required.

3.6. Mnemic theories: "Engrams"

The true-to-type and recurrent characteristics of embryonic development may be described by the continuity among physiological states and metabolic networks. The orderly succession of development and mechanisms that account for repetition of embryogenesis from generation to generation supports the notion that a memory is somehow encoded into the developmental process. Memory or mnemic theory must explain why in the first stages of development, specific structures are formed in advance of functioning, and why, in a general way, the embryo repeats through ontogeny, the ancestral history of the species. Algorithms based on this theory should also be useful in our understanding of senescence.

Normal embryonic development is viewed by many biologists as a repetition of an ancestral routine (92). When unusual stimuli are imposed on cells under in vitro conditions the responses to new conditions may be conditioned by past conditions (Bonga, volume one; 33, 91). This preconditioning does not explain the driving forces behind development nor the initiatives shown by the embryo in adapting itself to unusual conditions. It may account, however, for aberrations and bifurcations in development.

Figures 2 and 7 may be viewed as physicochemical records of true-to-type physiological events in the zygote. The topological record and its material vestiges can be postulated as containing evidence or traces of local external stimuli. In our model of the soluble N pool of the pistachio, the main stimuli are based on the influx of soluble N from the mother tree via the funiculus and the elaboration of amino acid networks in the developing embryo. Ancestral routines or traces arising from these networks may involve stimuli that are transmitted to somatic daughter cells as a "memory." In this narrow context, memory depends upon a material trace or impression made by soluble N upon the system depicted as local area networks on the map. The trace may be threshold costs, constraining events, configurations of factors or stimuli that are somehow remembered and locked-in (Table 4).

From a physiological viewpoint, any stimulus acting on the embryo not only produces an immediate physical effect but leaves behind it an enduring physical trace. This locked-in trace has been termed an "engram" by earlier biologists (cf. 92). Stimuli of a similar or associated kind are postulated to evoke an ecphoric response, i.e., induction of a latent engram or system of engrams to an active state. In this context levels of some metabolites in networks may be quantized as part of the memory "recall process."

Table 4. Some events, configurations and stimuli that may determine or lock-in physiological states reversibly or irreversibly and contribute to large scale integration in growth and development.

Factor	Example
Stable intermediate in a reaction sequence	RNA polymerase-DNA-protein complexes in nucleus. Stable protein configuration in epigenetic situations. Daughter cell adhesion. Reinforcement centers in cell walls.
External environment	Light, temperature, pH, and colligative effects on enzyme activity. Keto acid/amino acid conversions.
Ionic milieu	Distribution of nutrients, physical forces (osmotic). Trace metal chelates with amino acids. Surface tension, etc.
Hierarchical and sequential changes metabolic building blocks	Post-translational modification of in proteins. Methylation of DNA to inactive genes. Modification of DNA base and signal sequences in proteins to schedule and allocate signals. Informational macromolecules in cellular networks and compartments.
Residence-time of transient intermediates	Rates of hydrolysis, oxidation, reduction, assimilation, etc. Auto-self regulatory activities to keep genes "on." Cystolic metabolic arrays in genetic nets.
Accumulation of metabolic intermediates	Blocks in metabolism. Nutrient limitation. Stress responses in plants. Convolution of metabolic arrays and their connectivity.
Thresholds and metabolic control	Diffusion. Simple and complex bifurcating metabolic systems. Relationships among shape, volume, and respiration. Failure of error-prone repair systems.

Table 4 (continued)

Positional information and cell differentiation	Specialization of cells for metabolism, storage, structure, etc. Microtubular organization of cells. physiological gradients. Deposition of materials extracellular to the cell wall and in the cell wall.
Genetic control	Mutation. Somatic recombination. Promoter and enhancer DNA sequences. Boolean processes in enzyme-forming systems. Cell cycle events. Epigenetic factors.
Growth regulators and regulatory molecules	Messengers and signal molecules. Evolving catalytic activities of metabolic networks. Oligosaccharin activity. Optimization strategies.

The memory associated with developmental stages and physiological states may be characterized by the ability to retain and recall sets of revivable traces that can be characterized by metabolic phenotypes. Once these traces can be identified on our maps, we may be able to define a number of mnemic laws that are equally valid for a range of similar developmental fates.

Corollaries of mnemic laws are usually based on associations among single components of an engram complex. Association between and among engrams has been termed "engraphy." The smallest map unit able to contain the entire engram as a function of input signal needs to be defined. We can expect that such an imprint on our metabolic phenotypes will have a genetic and subcellular basis. The resultant patterns should help us to characterize potentially totipotent cells.

4. PROCESS CONTROL

4.1. Metabolic control

Early notions of metabolic control have related diffusion, osmosis, or other transport processes at surfaces to metabolism and respiration based on the shape of the organism (50, 75, 78, 88). Later concepts pertain more to process control and are based on a) nonreversibility of catabolic sequences, b) compartmentation of reactions and reactants, c) control by amounts of enzyme, d) control by pyridine nucleotide cofactors (NAD and NADP), e) control by ADP supply, f) circadian, wave-like and pacemaker reactions, g) allosteric regulation at metabolic branch points, e.g., feedback inhibition (e.g., 5, 44, 57, 71, 110). These controls are imposed on numerous physiological processes (photosynthesis, assimilation, differentiation, gene expression, etc.).

"Process controls" should initiate, program, and schedule sequences of events through recurrent mechanisms involving genetic elements (70). When challenged, the stability of these control systems, may lead to far-from-equilibrium processes, genome shock, unscheduled DNA synthesis, and recombination of one type or another (72, 73), developmental plasticity (104), and to senescence (17).

The recognition that cells are oscillatory metabolic systems opens a new approach to the metabolic control of development (48, 57; cf. Fig. 5). In aggregates of cells, the oscillations of one cell system can affect those of its neighbors. Some control may be achieved by constrained diffusion (50). Some overall pattern of oscillations may tend to build up (Fig. 6). According to Waddington (110) it is only necessary to suppose that the rate of fluctuation of cellular metabolism influences the type of synthetic process that will be favored.

4.2. Simple feedback model and local optimization

State-network maps describe hierarchies within a system. Optimization of processes among hierarchies must somehow involve local search by some master process to minimize trial and error. Notions based on feedforward and feedback mechanisms are required for process optimization to occur.

Metabolic optimization may arise by simple feedback and feedforward systems. Feedback results from a sequential flow of self- and cross-coupled signals along metabolic networks and physiological states (57). In a simple model based on enzyme-forming systems, new signals may be derived from some function such as the level of 2,4-D. For embryogenesis, a new physiological state seems to be "set" by 2,4-D and its removal to optimize for embryonic development. Cells engaged in this process can be identified by their strong reaction with acetocarmine. For the model to operate, the original state of cells must be compared with reactions describing the new physiological state. The difference between the new state commanding embryogenesis and the original physiological state can be viewed as a deviation or an error signal. This difference acts as the "forcing signal" which is somehow recognized by genetic elements in process control. These elements in turn generate controlling actions, which may depend on amplifying or differentiating the error signal (see

Durzan, this volume; Fig. 6 in Chapter on ammonia in somatic embryogenesis).

A characteristic feature of this type of model (61) is that the individual components in the system contribute significantly to the overall optimal behavior. This may assure that global relatedness is established in developing embryos. Undoubtedly other models will appear and improve on the above.

4.3. Metabolic pursuit

Relationships among internal cellular and external stimuli are critical for regulating developmental and mnemic processes (section 3.8.). Explants from mature trees are often attuned to conditions in the tree at the time of excision and as defined environmental and nutritional interactions. As a result, the responsiveness of the explant may include effects due to explant variation, e.g., topophysis, periphysis and cyclophysis (cf. 31). The response may be unconditional, i.e., not influenced by experiment, or conditional, i.e., depending on genetic and environmental influences. This dependency may be characterized by relationships that tend to follow, track or pursue one another (e.g., 24, 116).

One approach to a clearer definition of explant response involves concepts of "hereditary mechanics" and "metabolic pursuit" (cf. 24, for analogies in mathematics). These concepts are aimed at taking into account the non-genetic "hereditary" or "memory" influences which may persist in physiological states during growth and development for precise genetic adaptations.

Within a metabolic phenotype it is inevitable that one metabolite or variable will be linked somehow to another. The metabolite appears to follow the influence of some forcing factor possibly arising from the physiological state. The process may involve continued sampling by a "sensor" of one system upon another. The concept of pursuit derives its theoretical strength from sampled-data digital controller systems. This type of control is not unlike problems of growth in two populations conflicting with one another (24).

In a curve of pursuit, i.e., a curve by which a variable "tracks" or moves in pursuit of another, we are faced with several types of approximate solutions to the topographic problem. These include conditions of capture and ones where the motions of one variable tends asymptotically to the other. If a lag occurs in the relationship between two variables it is customary to say that they display "hysteresis." Notions of this type add significance to the spatio-temporal lines arbitrarily introduced into our maps (section 2.2).

5. SOME IMPLICATIONS AND USES

5.1. Tree physiology

Woody perennials are complex experimental organisms that can be described in many ways. We now know that the performance of explants taken at different stages of the life cycle often depends on the physiological states of cells when the experimental and cultural treatments begin. These initial states embody parameters that approximate a memory. This memory forms the basis for physiological preconditioning, i.e., these parameters determine the future responsiveness of cells in the system for precise adaptation (91, 92). Hence metabolic phenotypes composed of physiological states and metabolic networks are of value in understanding and diagnosing the overall irritability and process control of totipotent cells and tissues.

Variations in metabolic phenotypes that arise in vitro under the influence of specific cultural parameters can now be explored for their information content, pattern, thresholds, stability, signals, bifurcations, global relatedness, engrams, and process control. Valuable correlations can be established with biosynthetic capacities of embryos as in the case of zinc-deficient pistachio trees. In our approach, diagnostic evaluations should involve genetics, gene expression and evolutionary strategies (92) with the following interactive parameters:

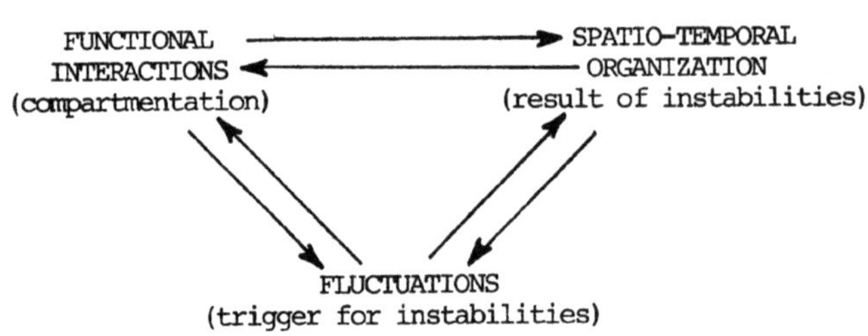

The essence of these interactions is captured by the dynamic representations of state-network maps and their diagnostic variations. As for the strategies inherent in these maps we can postulate that in evolutionary terms, and for any one cultivar or variety the net effect portrayed by the pattern is unbeatable under natural selection. In development, the patterns should represent true-to-type gene expression. As for cultural practices, we should look for the change in responsiveness upon the removal or introduction of constraints. This will enable us to evaluate performance and irritability through the behavior of the responsive parameters. This approach can be termed "diagnostic imagineering" because it involves the images that emerge from interactive parameters.

5.2. Tree breeding and improvement

In forestry, the capturing of genetic gains in trees having elite traits or derived through recombinant-DNA technologies are often described by population genetic terms (74, 121, Volume 1). Unfortunately in limited cases, none of the gains, even those involving simple traits, can be satisfactorily described at the molecular level although progress is being made (e.g., 6). Models based on metabolic phenotypes that include yet unidentified metabolites could facilitate this description especially when related to macromolecules (section 2.2.). This knowledge will assist the selection of elite but more complicated and valuable gene systems from a population.

The establishment of a "selection differential" for tree improvement is still not easy because the useful traits are indeed dynamic, complex and pleiotropic. We also have little molecular understanding for designing simple selection methods involving precise genetic adaptation (e.g., 74, 96). Steward (98) has suggested that environmental and nutritional factors make their metabolic impact by affecting enzymatic rates at sites where α-keto acids receive nitrogen. Products of this activity relate directly to the elaboration of characteristic families of amino acids and their secondary products (Fig. 1).

Since state-network maps display the known and yet unidentified metabolites, it is conceivable that through correlation of composition with the trait, e.g., disease resistance, we could identify phenotypic expressions with unique local or wide area networks or states of cells with elite or potentially pathological traits. These expressions could reaffirm direct (95) or indirect markers (108) for selection of trees in the breeding and improvement programs.

Where trees are cloned by *in vitro* methods, the stability, origin of somaclonal aberrations and points of bifurcation need to be evaluated (32, 63, 94). A state-network map for the development of propagules that are true-to-type would help to establish grower and industry standards. Over the long run, this approach could reduce the time for field testing of progeny and assists in establishing juvenile-mature correlations. Metabolic phenotypes could be useful especially where there is a need to determine which cells are destined for plant regeneration so that these could be selectively genetically modified. This technology should contribute to removal of constraints associated with the long, complex life cycle of tree when genetic gains are being captured.

6. CONCLUSIONS

Currently, the construction of state-network maps and their variations are costly and technically difficult. Applications are mainly theoretical and diagnostic. Their value relates to the geometries protraying the behavior of system components. We appear to be at least a decade or so away from application of useful models in cell and tissue culture systems. Nevertheless, comprehensive phenotypes describing the algorithms or metabolic command languages for the development of embryos are now being attempted. More comprehensive maps of gene expression based on current dogma and improved algorithms will be useful to further our understanding of growth, differentiation and morphogenesis. Utility is seen particularly in the "process control" of systems employing mass clonal propagation and in genetic approaches seeking to introduce new and useful genetic variation.

7. ACKNOWLEDGMENTS

Special thanks go to Frank Ventimiglia for computer simulations of the developing pistachio kernel.

8. REFERENCES

1. ÅGREN GI 1985 Theory for growth of plants derived from the nitrogen productivity concept. Physiol Plant 64: 17-28
2. ANDERSON NL, JP HOFMANN, A GEMMELL, J TAYLOR 1984 Global approaches to quantitative analysis of gene-expression patterns observed by use of two-dimensional gel electrophoresis. Clin Chem 30: 2031-2036
3. ATKINSON DE 1977 Cellular Energy Metabolism and its Regulation, Academic Press, New York
4. BARO AM, R MIRANDA, J ALAMAN, N GARCIA, G BINNING, H ROHRER, C GERBER, JL CARRASCOSA 1985 Determination of surface topography of biological specimens at high resolution by scanning tunnelling microscopy. Nature 315: 253-254
5. BEEVERS H 1974 Conceptual developments in metabolic control. Plant Physiol 54: 437-442
6. BERLYN GP, YC BATTEY 1985 Metabolism and synthetic function of cambial tissue. In T Higuchi, ed, Biosynthesis and Biodegradation of Wood Components, Academic Press, New York, pp 63-88
7. BLUM H 1973 Biological shape and visual science. J Theor Biol 38: 205-287
8. BRADLEY WH 1965 Vertical density currents. Science 150: 1423-1428
9. BRILLOUIN L 1956 Science and Information Theory, Academic Press, New York
10. BROWN FA Jr 1959 The rhythmic nature of animals and plants. Am Sci 48: 147-168
11. BROWN CM 1984 Computer vision and natural constraints. Science 224: 1299-1304
12. BUFFHAM BA 1985 Residence-time distributions in regions of steady-flow systems. Nature 314: 606-608
13. BUNGE M 1979 Treatise on Basic Philosphy. Vol. 4, Ontology II, A World of Systems. Reidel Publ Co, Dordrecht-Holland

14. BUNGE M 1960 Causality, chance, and laws. Am Sci 49: 432-448

15. CANNY MJ 1985 Ashby's law and the pursuit of plant hormones: a critique of accepted dogmas, using the concept of variety. Aust J Plant Physiol 12: 1-7

16. CARLSON PS, BF CONRAD, JD LUTZ 1984 Sorting through the variability. HortScience 19: 388-392

17. CERUTTI PA 1985 Prooxidant states and tumor production. Science 227: 375-381

18. CHALUPA V, DJ DURZAN, C VITHAYASAI 1976 Growth and metabolism of cells and tissue of jack pine (Pinus banksiana). 2. The quantitative analysis of the growth of callus from hypocotyls and radicles. Can J Bot 54: 446-455

19. CHENG DK 1983 Field and wave electromagnetics. Addison-Wesley Publ Co, Menlo Park, California

20. CHRISTIANSON ML 1985 An embryogenic culture of soybean: towards a general theory of somatic embryogenesis. In RH Henke, KW Hughes, MJ Constantin, A Hollaender, eds, Tissue Culture in Forestry and Agriculture. Plenum Press, New York, pp 83-103

21. CRANE JC, B IWAKIRI 1981 Morphology and reproduction of pistachio. Hort Rev 3: 376-393

22. CUNNINGHAM WJ 1963 The concept of stability. Am Sci 51: 425-436

23. DANDEKAR AM, D LERUDLIER, LT SMITH, MW JALCOWEC, LS GONG, RC VALENTINE 1983 Rhizobium as vectors for genetic engineering of salinity and drought tolerant legumes. In RB Goldberg, ed, Plant Molecular Biology. AR Liss, Inc, New York, 12: 277-289

24. DAVIS HT 1962 Introduction to nonlinear differential and integral equations. Dover Publ, New York

25. DEAN ACR, G SIR HINSHELWOOD 1966 Growth, Function and Regulation in Bacterial Cells. Oxford Univ Press

26. DEMONGEOT J, E GOLÈS, M TCHUENTE 1985 Dynamic Systems and Cellular Automata. Academic Press, New York

27. DENDRICK RF 1960 Random processes in control and communications. Science 132: 865-870

28. DOGRA PD 1978 Morphology, development and nomenclature of conifer embryo. Phytomorphology 28: 307-322

29. DOGRA PD 1967 Seed sterility and disturbances in embryogeny in conifers with special reference to seed testing and tree breeding in Pinaceae. Stud For Suec (Stockholm), Vol 45, 99 pp

30. DURZAN DJ 1982 Nitrogen metabolism and vegetative propagation of forest trees. In JM Bonga and DJ Durzan, eds, Tissue Culture in Forestry, Martinus Nijhoff/Dr W Junk, The Hague, pp 256-324

31. DURZAN DJ 1984 Special problems: Adult vs. juvenile explants. In WR Sharp, DA Evans, PV Ammirato and Y Yamada, eds, Handbook of Plant Cell Culture, Crop Species. MacMillan Publ Co, New York, Chap 17, 2: 471-503

32. DURZAN DJ 1984 Potential for genetic manipulation of forest trees: Totipotency, somaclonal aberration, and trueness to type. In Proc Int Symp Recent Adv For Biotechnol, July 10-13, 1984, Michigan Biotechnol Inst, Traverse City, Michigan, pp 104-125

33. DURZAN DJ, V CHALUPA 1968 Free sugars, amino acids, and soluble proteins in the embryo and female gametophyte of jack pine as related to climate at the seed source. Can J Bot 46: 419-428

34. DURZAN DJ, V CHALUPA 1976 Growth and metabolism of cells and tissue of jack pine (Pinus banksiana). 3. Growth of cells in liquid suspension cultures in light and darkness. Can J Bot 54: 456-467

35. DURZAN DJ, V CHALUPA 1976 Growth and metabolism of cells and tissue of jack pine (Pinus banksiana). 4. Changes in amino acids of callus and in seedlings of similar genetic origin. Can J Bot 54: 468-482
36. DURZAN DJ, V CHALUPA 1976c Growth and metabolism of cells and tissue of jack pine (Pinus banksiana). 5. Changes in free arginine and Sakaguchi-reactive compounds during callus growth and in germinating seedlings of similar genetic origin. Can J Bot 54: 483-495
37. DURZAN DJ, V CHALUPA, AJ MIA 1976 Growth and metabolism of cells and tissue of jack pine (Pinus banksiana). 1. The establishment and some characteristics of a proliferated callus from jack pine seedlings. Can J Bot 54: 437-445
38. DURZAN DJ and FC STEWARD 1983 Nitrogen metabolism. In FC Steward, ed, Plant Physiology, An Advanced Treatise, Academic Press, New York, VIII:55-265
39. DURZAN DJ, K URIU 1986 Metabolic networks in developing pistachio embryos (Pistacia vera cv. Kerman). In H.C. Slavkin, ed., Progress in developmental biology. Part A. Proc Int Soc Dev Biol. 10th Int Cong ISDB, Aug. 4-9, 1985, Los Angeles, CA, 199-202
40. ELSASSER WM 1958 The Physical Foundation of Biology. Pergamon Press, NY
41. ENGEL A, A MASSALZKI, H SCHINDLER, DL DORSET, JP ROSENBUSCH 1985 Porin channel triples merge into single outlets in Escherichia coli outer membranes. Nature 317: 643-645
42. EVANS DA, WR SHARP, HP MEDINA-FILHO 1984 Somaclonal and gametoclonal variation. Am J Bot 71: 759-774
43. FLORES HE, ND YOUNG, AW GALSTON 1984 Polyamine metabolism and plant stress. In JL Key, T Kosuge, eds. Cellular and Molecular Biology of Plant Stress. Proc Arco/UCLA Symp, Keystone, Co., April 15-21. A.R. Liss, Inc., pp 93-114
44. FRIEDEN C 1970 Kinetic aspects of regulation of metabolic processes. J Biol Chem 245: 5788-5799
45. GARNER WR 1970 Good patterns have few alternatives. Am Sci 58: 34-42
46. GIOVANELLI J, SH MUDD, AH DATKO 1985 Quantitative analysis of pathways of methionine metabolism and their regulation of Lemna. Plant Physiol 78: 555-560
47. GLANSDORFF P, I PRIGOGINE 1971 Thermodynamic Theory of Structure, Stability, and Fluctuations, J Wiley, New York
48. GOODWIN BC 1963 Temporal organization in cells. A Dynamic Theory of Cellular Control Processes. Academic Press, New York
49. GRANT V 1975 Genetics of Flowering Plants. Columbia Univ Press, New York
50. GORDON R 1966 On stochastic growth and form. Proc Nat Acad Sci USA 56: 1497-1504
51. GREEN PB 1980 Organogenesis. A biophysical view. Annu Rev Plant Physiol 31: 51-82
52. GRIESBACH RJ 1984 An introduction to somatic cell genetics. HortScience 19: 367-371
53. GUPTA PK, DJ DURZAN 1985 Shoot multiplication from mature trees of Douglas-fir (Pseudotsuga menziesii) and sugar pine (Pinus lambertiana). Plant Cell Rep 4: 177-179
54. HAKMAN I, LC FOWKE, S VON ARNOLD, T ERICKSSON 1985 The development of somatic embryos in tissue cultures initiated from immature embryos of Picea abies (Norway spruce). Plant Sci Lett 38: 53-63.
55. HAMNER KC, A TAKIMOTO 1964 Circadian rhythms and plant photoperiodism. Am Nat 48: 295-322
56. HARE PE, PA ST JOHN, MH ENGEL 1985 Ion-exchange separation of amino acids. In GC Barrett, ed, Chemistry and Biochemistry of Amino Acids, Chapman and Hall, London, pp 415-425

57. HIGGINS J 1967 The theory of oscillating reactions. Ind Eng Chem 59: 19-62
58. HORNBERG C, EW WEILER 1984 High affinity binding sites for abscisic acid on the plasmalemma of Vicia faba guard cells. Nature 370: 321-324
59. INGESTAD T 1977 Nitrogen and plant growth, Maximum efficiency of nitrogen fertilizers. Ambio 6: 146-151
60. JEAN RV 1984 Mathematical Approach to Pattern and form in Plant Growth. J Wiley and Sons Inc, New York
61. JOHNSON EF 1962 Automatic process control. Science 135: 403-408
62. KAUFFMAN SA 1969 Metabolic stability and epigenesis in randomly constructed genetic nets. J Theor Biol 22: 437-467
63. KESTER DE 1983 The clone in horticulture. HortScience 18:831-837
64. KOLATA GM 1975 Cascading bifurcations: The mathematics of chaos. Science 189-984-985
65. KOVACH LD 1960 Life can be so nonlinear. Am Sci 48: 218-225
66. KRIMER DB, J VAN'T HOF 1983 Extrachromosomal DNA of pea (Pisum sativum root-tip cells replicates by strand displacement. Proc Nat Acad Sci USA 80: 1933-1937
67. LIN TS, VS POLITO, JC CRANE 1984 Embryo development in 'Kerman' pistachio. HortScience 19: 105-106.
68. MACDONALD N 1983 Trees and Networks in Biological Models. J Wiley and Sons, New York
69. MACKAY AL 1985 Periodic minimal surfaces. Nature 314: 604-606
70. MAGASANIK B 1982 Genetic control of nitrogen assimilation in bacteria. Annu Rev Genetics 16: 135-168
71. MARKOWITZ D 1971 Collective modes and wave-like solutions to cellular control equations. J Theor Biol 31: 475-492
72. MCCLINTOCK B 1967 The role of the nucleus. Genetic systems regulating gene expression during development. Dev Biol (suppl) 1: 84-112
73. MCCLINTOCK B 1984 The significance of responses of the genome to challenge. Science 226: 792-801
74. NAMKOONG G 1979 Introduction to quantitative genetics in forestry. For Serv USDA, Washington DC, Tech Bull 1588
75. NETTER H 1969 Theoretical Biochemistry. Wiley-Interscience, New York
76. NORSTOG K 1982 Experimental embryology of gymnosperms. In BM Johri, ed, Experimental Embryology of Vascular Plants. Springer-Verlag, New York, pp 25-51
77. OLLIS DL, P BRICK, R AHMLIN, NG XUONG, TA STEITZ 1985 Structure of large fragment of Escherichia coli DNA polymerase-1 completed with dTMP. Nature 313: 762-766.
78. PALADE G 1975 Intracellular aspects of the process of protein synthesis. Science 189: 347-358
79. PENNING DE VRIES FWT, AHM BRUNSTING, HH VAN LAAR 1974 Products, requirements, and efficiency of biosynthesis: A quantitative approach. J Theor Biol 45: 339-377
80. PLATT JR 1961 Properties of large molecules that go beyond the properties of their chemical sub-groups. J Theor Biol 1: 342-358
81. POWLEDGE TM 1984 Biotechnology touches the forest. Bio/Technology 2: 763-772
82. PRIGOGINE I, R LEFEVER, A GOLDBETER, M HERSHKOWITZ-KAUFMAN 1969 Symmetry breaking instabilities in biological systems. Nature 223: 913-916
83. QUASTLER H 1959 Information theory of biological integration. Am Nat 93: 245-254
84. RAGHAVAN V 1983 Biochemistry of somatic embryogenesis. In DA Evans, WR Sharp, PV Ammirato, Y Yamada, eds, Handbook of Plant Cell Culture. McMillan, Vol 1, 655-671

85. RAISBECK G 1964 Information Theory. MIT Press, Cambridge, Mass
86. RASHEVSKY N 1960 Mathematical Biophysics. 3rd rev ed, Vols 1 and 2, Dover Publ, New York
87. REARDAN DT, CF MEARES, DA GOODWIN, M MCTIGUE, GS DAVID, MR STONE, JP LEUNG, RM BATHOLOMEW, JM FRINKE 1985 Antibodies against metal chelates. Nature 316: 265-268
88. RICHARDS FJ 1969 The quantitative analyses of growth. In FC Steward, ed, Plant Physiology: An Advanced Treatise. Academic Press, New York, VA, 3-76
89. RIEGER R, A MICHAELIS, M GREEN 1976 Glossary of Genetics and Cytogenetics. Springer-Verlag, New York
90. ROSEN CA 1967 Pattern classification by adaptive machines. Science 156: 38-44
91. ROWE JS 1964 Environmental preconditioning, with special reference to forestry. Ecology 45: 399-403
92. RUSSELL ES 1930 The Interpretation of Development and Heredity. Oxford Univ Press, Oxford
93. SCHIMKE RT, B HAMKALO, A HILL, B MARIANI, R JOHNSON, S SHERWOOD, T TLSTY 1986 Dihydrofolate reductase gene amplification in somatic cells. Proc 10th Int Cong Soc Dev Biol, Los Angeles, CA, August 4-9 (1985) (in press)
94. SCOWCROFT WR 1984 Genetic variability in tissue culture: Impaction germplasm conservation and utilization. Int Board Plant Genet Res, Rome, pp 1-41
95. SHIGO AL 1985 Wounded forests, starving trees. J For 83: 668-673
96. SILEN RR 1982 Nitrogen, corn and forest genetics. The agricultural yield strategy--implications for Douglas-fir management. USDA For Serv Gen Tech Rep PNW-137
97. SINGH H 1978 Embryology of Gymnosperms. Enc Plant Physiol, Gebrüder Borntraeger, Berlin
98. STEWARD FC 1963 Effect of environment on metabolic patterns. In L Evans, ed, Environmental Control of Plant Growth. Academic Press, New York, pp 195-214
99. STEWARD FC 1968 Growth and Organization in Plants. Addison-Wesley Publ. Co., Reading, Mass
100. STEWARD FC 1971 Plant physiology: The changing problems, the continuing quest. Annu Rev Plant Physiol 22: 1-22
101. SYNGE JL 1962 Water, waves and hydrons. Science 138: 13-16
102. TABOR M 1984 Modern dynamics and classical analysis. Nature 310: 277-282
103. THOM R 1983 Mathematical Models and Morphogenesis. Halsed Press, Chichester, England
104. TREWAVAS AJ, R SEXTON, P KELLY 1984 Polarity, calcium and abscission: molecular bases for developmental plasticity in plants. J Embryol Exp Morphol 83(Suppl): 179-195
105. TURCOTTE DL, RF SMALLEY Jr., SA SOLLA 1985 Collapse of loaded factal trees. Nature 333: 671-672
106. VEEN AH and A LINDENMAYER 1977 Diffusion mechanism for phyllotaxis. Theoretical physico-chemical and computer study. Plant Physiol 60: 127-139
107. VOLKENSHTEIN MV 1983 Biophysics. Transl A Beknazarov, Mir Publishers, Moscow
108. VON WEISSENBERG K 1976 Indirect selection for improvement of desired traits. In JP Miksche, ed, Modern Methods in Forest Genetics. Springer Verlag, New York, pp 217-228
109. WADDINGTON CH 1962 New Patterns in Genetics and Development. Columbia Univ Press, New York

110. WADDINGTON CH 1965 Autogenous cellular periodicities as (a) temporal templates and (b) the basis of morphogenetic fields. J Theor Biol 8: 367-369
111. WATSON MR 1972 Rate effectors and their role in metabolic control. J Theor Biol 36: 195-202
112. WEINBERG H and R COOPER 1972 The use of correlational analysis for pattern recognition. Nature 238: 292
113. WEISBUCH G 1985 Modelling natural systems with networks of automata: the search for generic behaviors. In J Demongeot E Golès M Tchuente, eds, Dynamic Systems and Cellular Automata. Academic Press, New York, pp 293-304
114. WHYTE RO 1949 Crop Production and Environment. Faber and Faber Ltd, London
115. WINFREE AT 1985 Organizing centres for chemical waves in two and three dimensins. In RJ Field, M Burger, ed, Oscillations and Travelling Waves in Chemical Systems. John Wiley, 441-472
116. WITHROW 1959 Photoperiodism and related phenomena in plants and animals. Am Assoc Adv Sci, Publ No 55, Washington, DC, pp 439-471
117. WOLFRAM S 1984 Cellular automata as models of complexity. Nature 311: 419-424
118. WRIGHT S 1968 Evolution and the Genetics of Populations. Univ Chicago Press
119. YOUNG DA 1984 Advantages of separations on giant two-dimensional gels for detection of physiologically relevant changes in the expression of protein gene products. Clin Chem 30: 2104-2108
120. ZELENY M 1977 Self-organization of living systems: A formal model of autopoiesis. Int J Genet Syst 4: 13-28
121. ZOBEL B, J TALBERT 1984 Applied Forest Tree Improvement. J Wiley, New York

TREE SPECIES INDEX
VOLUME 2

GENERAL INDEX
VOLUME 2

MIX
Papier aus verantwortungsvollen Quellen
Paper from responsible sources
FSC® C105338

If you have any concerns about our products,
you can contact us on
ProductSafety@springernature.com

In case Publisher is established outside the EU,
the EU authorized representative is:
Springer Nature Customer Service Center GmbH
Europaplatz 3, 69115 Heidelberg, Germany

Printed by Libri Plureos GmbH
in Hamburg, Germany